DEVELOPABILITY OF BIOTHERAPEUTICS

Computational Approaches

DEVELOPABILITY OF BIOTHERAPEUTICS

Computational Approaches

Edited by

Sandeep Kumar
Biotherapeutics Pharmaceutical Sciences
Pfizer, Inc.

Satish Kumar Singh
Biotherapeutics Pharmaceutical Sciences
Pfizer, Inc.

CRC Press
Taylor & Francis Group
Boca Raton London New York

CRC Press is an imprint of the
Taylor & Francis Group, an **informa** business

CRC Press
Taylor & Francis Group
6000 Broken Sound Parkway NW, Suite 300
Boca Raton, FL 33487-2742

First issued in paperback 2021

Version Date: 20151020

ISBN 13: 978-1-03-209820-3 (pbk)
ISBN 13: 978-1-4822-4613-1 (hbk)

Publisher's Note
The publisher has gone to great lengths to ensure the quality of this reprint but points out that some imperfections in the original copies may be apparent.

Visit the Taylor & Francis Web site at
http://www.taylorandfrancis.com

and the CRC Press Web site at
http://www.crcpress.com

Contents

SECTION I Principles of Biopharmaceutical Informatics

SECTION II   Developability Practices in the Biopharmaceutical Industry

Foreword

This book is about developability. But what a funny term *developability* is. A neologism, it can only be a transitional term, for, like its cousin, *manufacturability*, it implies that issues of biopharmaceutical drug development (and by analogy issues of manufacturing) should be taken into account during the discovery phase. It implies the need for achieving the "seamless interdependent whole" discussed in the first chapter, whereas at the same time acknowledging that we are still so far from it that we need to force ourselves to take into account parts of the process of creating new biopharmaceuticals while we pursue other parts without fully accounting for their interdependence.

Why are we in this situation? The answer was given by Francis Bacon more than 400 years ago. Before we give the answer, a little background is in order. Bacon is the thinker who transformed human aspirations from "abstract meditation" to "fruit and works." In doing so, he invented the experimental method with the major purpose of improving human health and extending life. Shortly thereafter, René Descartes built upon Bacon's method so "that we could be spared an infinity of diseases, of the body as well as of the mind." In order to accomplish this, he introduced the wholesale application of mathematics into science. It is only in the past few decades, however, that we have obtained the tools for applying the Baconian–Cartesian method to the creation of new pharmaceuticals. Even with these tools, we have been slow to apply them, to bring about holistic pharmaceutical creation.

Bacon tells us why. In discussing his famous idols of the mind, which run deep and hold us back from progress, he writes, "The human intellect, from its own character, easily supposes that there is more order and regularity in things than it finds." Do we really think that focusing on the properties that make a molecule bind to its target will absolve us from designing that molecule to do the other things that we wish it to do? Do we really think that our focusing on design of binding to a target will lead to the molecule interacting as we wish it to with other parts of the body, or having the stability to be made into a pharmaceutical product, or having suitable properties for manufacturing? Clearly, we do not think there is such regularity, but our minds lead us to believe that these issues will sort themselves out, as if there were such regularity.

This volume addresses the disparity between what we know is true, on the one hand, and how we act, on the other. It shows us how, in the field of biopharmaceutical creation, our mind's idols can be overcome. The idea is that if we learn about the wide-ranging and effective technical solutions available today, we will be motivated to apply them, and therein bring about the regularity that will not happen of its own accord. The editors and contributors of this volume discuss solutions that run the gamut, from chemical and physical stability to pharmacokinetics/pharmacodynamics (PK/PD) analysis and prediction to supply chain issues, immunogenicity, formulation, modality selection, polydispersity, effects of post-translational modifications, upstream and downstream production, and even epitope predictions. They show us that the Baconian–Cartesian tools that have formed the basis of the tremendously successful mathematical physics over the last several hundred years have now been

extended to the realm of complex biological systems and, as such, have the potential
to enhance human health in a hitherto unprecedented way.

Two key concepts form the twofold theme: informational and knowledge based. We
need accurate information for our models to be able to describe reality, and we have
reached the point where, while we can never have enough information, we have enough
to make much faster progress. Moreover, we need to base our models on knowledge,
not merely facts or data, but mechanistic understanding, as close to first principles as
possible. We may not be at the point of having the ultimate mechanistic understanding of these complex systems, but we are much further along than what our current
approach to creating biopharmaceuticals presumes. The editors go even deeper still in
identifying the key aspects that need to be addressed: the knowledge-based approach
is not just about mechanistic understanding but is "comprised of human intentions and
culture as much as it is of tools and technologies." Bacon also said: "The human intellect swells and cannot stay still or rest, but aspires to go further, in vain." If we would
only embrace the approach described here, not just the tools, but also the mindset, what
we could accomplish may indeed be limitless.

In the Preface, the editors discuss the aim of remaining in Well Country as
opposed to being forced to visit Sick City. Such metaphors emphasize the universal-
ism of what we are trying to accomplish, for after all, biopharmaceuticals that can
help anyone who needs them potentially help everyone. They also remind us of what
we are trying to accomplish by creating new biopharmaceuticals. The Greek word
for happy is *eudaimon*—possessing a "well" spirit. A necessary condition to being
eudaimon is possessing a well body, in order words, being healthy. Applying the
knowledge in this volume will make us healthier and therein happier.

Bernhardt L. Trout
Raymond F. Baddour, ScD (1949)
Professor of Chemical Engineering, MIT

Preface

Dear Reader,

If you consider that access to modern innovative medicines is a right of all patients and are interested in understanding how recent advances in computational sciences can help in this effort, then this is the book for you! A major portion of the developmental cost of new drugs is incurred during clinical trials, whose outcomes depend, in part, on the choices made during the discovery and selection (design) of the drug molecule, formulation, manufacturing, dosing, clinical trial design, and patient population selection. As you flip through the pages of this book, the use of computation in novel ways to improve the overall process of drug discovery and development will become evident. This book is focused mainly on developability of monoclonal antibody candidates and is organized in two sections. The first section describes applications of computational approaches toward discovery and development of biopharmaceutical drugs; the second section presents the best practices in developability assessments of early-stage drug candidates being followed by leading companies in this business.

Although we prefer to live in Well Country forever, forced visits to Sick City do happen every now and then. Advances in medicines have significantly improved the health and well-being of millions of people, particularly in the last century. The explosive growth of knowledge in biology and genetics has been driving this effort. However, the success rate for novel therapeutic entities is falling, raising their cost, especially for innovative biotherapeutics and vaccines. The consequent effects on the organizations involved in drug discovery and development; the medical systems, such as clinics, hospitals, insurance companies, and government-sponsored health-care; and ultimately the individual patients and society at large are enormous.

Each year the research labs in pharmaceutical companies discover several promising novel drug candidates, but only a few of these newly discovered compounds reach the clinic after several years. Projects that succeed must recover costs for their own development as well as of the failed ones, and make reasonable profits from sales over the remaining duration of their exclusivity and after the loss of exclusivity, so that the business of bringing new medicines to patients can be sustained. This already low success rate, however, continues to fall and points to the risks and inefficiencies inherent to drug discovery, development, testing, and approval processes. While biological activity, rightly, is a major focus during the discovery of biopharmaceutical candidates, the macromolecular sequence–structural properties of these candidates can also inform us about their cell line expression levels, potential degradation routes, interactions with extractables and leachables, behaviors of highly concentrated solutions, immunogenicity, pharmacokinetics/dynamics, and so forth. Such insights, when they come via computation at early stages of lead candidate design or selection, can help make discovery and development of biologic drugs more efficient by removing empiricism and reducing developmental costs and attrition rates.

We are thankful to Hilary LaFoe for inviting us to edit this book, to Kari Budyk for coordinating the effort, and to Prof. Bernhardt Trout of MIT for providing the

Foreword. The enthusiastic contributors to the book chapters represent the vanguards of biopharmaceutical informatics performing cutting-edge research on developability issues in biopharmaceuticals. Without the generous time and effort put in by these very busy and outstanding scientists, this book would not have been possible. Discussions with numerous colleagues spread over nearly a decade are also gratefully acknowledged. It goes without saying that we could not have undertaken this journey without unwavering support, affection, and encouragement from our families and friends.

Most pharmaceutical industry executives, professionals, postdocs, students, and enthusiasts can appreciate the potential of computational approaches toward biopharmaceutical discovery and development, but awareness of the many different contributions that computation can make is generally lacking. This book provides examples and focuses on filling this void. This is still a nascent field and you, dear reader, are encouraged to explore it on your own.

S.K. and S.K.S.

Contributors

Anson K. Abraham
Pharmacokinetics, Pharmacodynamics
& Drug Metabolism (PPDM)—Early
Stage QP2
Merck Research Laboratories
West Point, Pennsylvania

Neeraj J. Agrawal
Amgen, Inc.
Thousand Oaks, California

Juan C. Almagro
Centers for Therapeutic Innovation (CTI)
Pfizer, Inc.
Boston, Massachusetts

Nicolas Angell
Product Attribute Sciences
Amgen, Inc.
Thousand Oaks, California

Andreas Arnell
Applied Protein Services
Lonza Biologics plc.
Great Abington, Cambridge, United
Kingdom

Matthew P. Baker
Antitope Ltd. (an Abzena company)
Babraham Research Campus
Babraham, Cambridge, United
Kingdom

Patrick M. Buck
Pharmaceutical Research and
Development
Biotherapeutics Pharmaceutical
Sciences
Pfizer, Inc.
Chesterfield, Missouri

Anuj Chaudhri
Computational Research Division
Lawrence Berkeley National Laboratory
Berkeley, California

Naresh Chennamsetty
Bristol-Myers Squibb Company
New Brunswick, New Jersey

Kristine Daris
Drug Substance Development
Amgen, Inc.
Thousand Oaks, California

Surjit B. Dixit
Zymeworks, Inc.
Vancouver, British Columbia, Canada

Thomas R. A. Gallagher
Applied Protein Services
Lonza Biologics plc.
Great Abington, Cambridge, United
Kingdom

Ramón Gómez de la Cuesta
Research and Technology
Lonza Biologics plc.
Great Abington, Cambridge, United
Kingdom

M. Michael Gromiha
Department of Biotechnology
Indian Institute of Technology Madras
Chennai, India

Kannan Gunasekaran
Therapeutic Discovery
Amgen, Inc.
Thousand Oaks, California

Steffen Hartmann
Integrated Biologics Profiling Unit
Biologics Technical Development
  and Manufacturing
Novartis Pharma AG
Basel, Switzerland

Timothy P. Hickling
Pharmacokinetics, Dynamics and
  Metabolism—New Biological
  Entities
Immunogenicity Sciences
Worldwide Research and Development
Pfizer, Inc.
Andover, Massachusetts

Tim D. Jones
Antitope Ltd. (an Abzena company)
Babraham Research Campus
Babraham, Cambridge, United Kingdom

Anette C. Karle
Integrated Biologics Profiling Unit
Biologics Technical Development
  and Manufacturing
Novartis Pharma AG
Basel, Switzerland

Randal R. Ketchem
Therapeutic Discovery
Amgen, Inc.
Thousand Oaks, California

Hans P. Kocher
Integrated Biologics Profiling Unit
Biologics Technical Development
  and Manufacturing
Novartis Pharma AG
Basel, Switzerland

Sandeep Kumar
Pharmaceutical Research and
  Development
Biotherapeutics Pharmaceutical Sciences
Pfizer, Inc.
Chesterfield, Missouri

Li Li
Pharmaceutical Research and
  Development
Biotherapeutics Pharmaceutical
  Sciences
Pfizer, Inc.
Andover, Massachusetts

Ben Locwin
Lonza Biologics, Inc.
Portsmouth, New Hampshire

Alessandro Mascioni
Centers for Therapeutic Innovation (CTI)
Pfizer, Inc.
Boston, Massachusetts

Olga Obrezanova
Applied Protein Services
Lonza Biologics plc.
Great Abington, Cambridge, United
  Kingdom

Jason W. O'Neill
Therapeutic Discovery
Amgen, Inc.
Thousand Oaks, California

Russell H. Robins
Analytical Research and Development
Biotherapeutics Pharmaceutical
  Sciences
Pfizer, Inc.
Chesterfield, Missouri

Pratap Singh
Pharmacokinetics, Dynamics and
  Metabolism—New Biological
  Entities
Translational Modeling and Simulation
  (TMS)
Worldwide Research and Development
Pfizer, Inc.
Cambridge, Massachusetts

Satish K. Singh
Pharmaceutical Research and
    Development
Biotherapeutics Pharmaceutical Sciences
Pfizer, Inc.
Chesterfield, Missouri

Noel Smith
Applied Protein Services
Lonza Biologics plc.
Great Abington, Cambridge, United
    Kingdom

A. Mary Thangakani
Department of Crystallography and
    Biophysics
University of Madras
Chennai, India

Abhinav Tiwari
Pharmacokinetics, Dynamics and
    Metabolism—New Biological
    Entities
Translational Modeling and Simulation
    (TMS)
Worldwide Research and Development
Pfizer, Inc.
Cambridge, Massachusetts

Bernhardt L. Trout
Department of Chemical Engineering,
Massachusetts Institute of Technology
Cambridge, Massachusetts

Jesús Zurdo
Research and Technology
Lonza Biologics plc.
Great Abington, Cambridge, United
    Kingdom

Anup Zutshi
Pharmacokinetics, Dynamics and
    Metabolism—New Biological
    Entities
Translational Modeling and Simulation
    (TMS)
Worldwide Research and Development
Pfizer, Inc.
Cambridge, Massachusetts

Motivation for This Book

Biopharmaceuticals, particularly monoclonal antibodies and antibody-based biotherapeutics, have emerged as best-selling medicines in recent years, thereby delivering on promises from the early days of biotechnology. However, these innovative medicines are costly to develop and produce on a commercial scale. The biologics possess heterogeneous molecular structures and are vulnerable to physical and chemical degradation, such as aggregation, oxidation, deamidation, and fragmentation, because of stresses encountered by these macromolecules during manufacturing, shipping, and storage. At the same time, a drop in the number of novel small-molecule drugs being discovered and the failure of candidates during late-stage drug development (clinical trials) have led to unprecedented highs in the cost of bringing new medicines to use. These concomitant developments are among the major drivers for the rising costs of healthcare in the United States, Europe, emerging markets, and elsewhere in the world. At the same time, demand for innovative medicines is rapidly growing in both the developed and developing world. This is especially true for life-threatening diseases such as cancer, cardiac failure, and chronic diseases such as diabetes. Long considered to be the bane of the developed world, these diseases have now emerged as a major challenge to human health in the emerging markets and other developing countries as well. Therefore, several conflicting issues are being faced today by the pharmaceutical industry and by society at large in regard to continued access to new medicines. The most pressing challenges are sustainability of business via realization of costs associated with bringing new drugs to clinic and fulfillment of pharmaceutical companies' profit expectations, and the ability of payers, for example, governments, hospitals, insurance companies, individual patients, and their families, to afford these advanced medicines.

Innovative medicines do not have to be costly. Certainly, no patient should die just because the cost of modern life-saving medicines is beyond his or her reach. The high prices of innovative medicines, particularly anticancer biotherapeutics, have become a major issue in emerging markets and other developing as well as developed countries, and are proving to be a barrier to their widespread use, notwithstanding huge demand. Yet, much can be done to reduce the cost of biopharmaceutical drugs. *This is a winnable war.* Attention should be paid to details of protein sequence and structure during the development of biopharmaceuticals, and considerations regarding manufacturability, formulation development, flexible delivery options, and immunogenicity should be included, alongside potency and efficacy, at the early stages of lead candidate discovery and design. The development and use of appropriate computational biophysics techniques such as multiscale molecular modeling, dynamic simulations, and prediction can help de-risk the drug development pipelines at preclinical stages. Similarly, statistically robust planning and execution of clinical trials, including the design of appropriate bioassays and careful selection of patient groups that are most likely to positively respond to a candidate medicine, can go a long way toward preventing costly late-stage drug failures.

Over the past couple of years, an increasing number of pharmaceutical companies have begun to embrace the umbrella concept of developability. At its core, developability is a risk assessment and mitigation exercise that seeks to improve the likelihood that a biotherapeutic drug candidate discovered to be efficacious against a target will be successfully developed into a medicine available in clinics. The major regulatory agencies, such as the U.S. Food and Drug Administration (FDA) and European Medicines Evaluation Agency (EMEA), are also emphasizing quality by design (QbD) approaches to improve biopharmaceuticals. Several computational methodologies and biophysical techniques are being adapted, and examples of successful designs are beginning to emerge. Simultaneously, there is a growing awareness of improving clinical trial designs using statistical and mathematical modeling to prevent unnecessary drug candidate failures. However, this is still a very nascent field, and few books dealing with this subject are currently available in the market. As far as we are aware, there is currently no book that details the applications of computational and molecular modeling techniques toward biopharmaceutical drug development at preclinical stages. Yet, the importance of this subject and the need to increase awareness of industry leaders, regulators, clinicians, and the general scientifically interested public of matters related to biopharmaceutical drug development cannot be overstated. Beside these, there is also a need to train next-generation pharmaceutical, biophysical, and medical scientists in emergent issues facing the drug industry and on how innovative uses of computation can lead to highly efficacious, affordable, and safer medicines that are also convenient to use.

PURPOSE OF THE BOOK

This book serves as a primary reference and textbook for computational applications addressing the issues in biopharmaceutical development. The targeted audience for this book is pharmacy, medicine, and life science students and educators at the tertiary level, industrial research and development mid- and senior-level management, regulatory agencies, and scientists concerned with public health issues. Biopharmaceutical drug discovery and formulation professionals, as well as scientists interested in bioinformatics and computational biophysics, may also find this book of interest.

It is expected that this book will be bought by libraries supporting the schools of pharmacy, medicine, and life sciences in the United States and international universities and by companies invested in biopharmaceutical research and development. It is also hoped that this book raises awareness about the promise of computational research among pharmaceutical scientists and becomes a catalyst for innovative applications of computational design to biopharmaceutical drug development and delivery.

DESCRIPTION OF CONTENTS

Section I: Principles of Biopharmaceutical Informatics

Chapter 1: Biopharmaceutical Informatics: Applications of Computation in Biologic Drug Development

Chapter 1 defines the term *biopharmaceutical informatics* and describes the applications of computational biophysics toward understanding the challenges encountered during biopharmaceutical drug development.

Chapter 2: Computational Methods in the Optimization of Biologic Modalities

Once a target has been validated as druggable, there are a number of small-molecule and biologic-based modalities that can potentially be used. The process by which a molecular modality is optimized has a bearing on the overall success of the program. Therefore, this chapter reviews the considerations involved in the optimization of biologic modalities at the early stages of drug discovery.

Chapter 3: Understanding, Predicting, and Mitigating the Impact of Post-Translational Physicochemical Modifications, including Aggregation, on the Stability of Biopharmaceutical Drug Products

Several physicochemical modifications, such as aggregation, oxidation, deamidation, glycation, glycosylation, and disulfide scrambling, can adversely impact the molecular integrity of the active ingredient in biopharmaceutical drug products. These instabilities can arise from several sources, including extractables and leachables from drug delivery components, such as glass/silica from vials, silicon oil on prefilled syringes, and metal ions from injection needles. This chapter describes the consequences of physicochemical degradation on the stability, efficacy, and pharmacokinetics/pharmacodynamics (PK/PD) of biopharmaceuticals and attempts to identify potential degradation sites in the sequence and structure of the biotherapeutic candidates. An important part of this chapter is the issue of aggregation encountered during commercial manufacturing, storage, and shipping of biotherapeutics and how it can be mitigated via rational protein design. This chapter describes the computational efforts to understand the aggregation mechanism and predict aggregation-prone regions in proteins. A distinction is made between aggregation due to colloidal properties of liquid biopharmaceutical formulations and the one due to inherent conformational (in)stability of the protein needed to withstand the insults faced by the protein molecule during manufacturing, shipping, and storage.

Chapter 4: Preclinical Immunogenicity Risk Assessment of Biotherapeutics

A great advantage of biotherapeutics over small-molecule drugs is highly specific target binding and nearly complete absence of non-mechanism toxicity. However, administration of biotherapeutics, including recombinant and plasma-derived human proteins, often leads to undesirable immune responses among patients. These responses can vary from transient non-significant injection site inflammations to life-threatening events in rare instances. Another common immune response is the development of anti-drug antibodies. This chapter describes the computational ability to predict B- and T-cell immune epitopes in biotherapeutics and how such tools can help in preclinical immunogenicity risk assessments and the design of deimmunized biologics.

Chapter 5: Application of Mechanistic Pharmacokinetic–Pharmacodynamic Modeling toward the Development of Biologics

The PK/PD and distribution of monoclonal antibodies and other biopharmaceuticals in human tissues depend on their sequence and structural properties, as well as route of administration. This chapter describes the computational efforts aimed at mathematical modeling of PK/PD profiles of biotherapeutics.

Chapter 6: Challenges in High-Concentration Biopharmaceutical Drug Delivery: A Modeling Perspective

Subcutaneous delivery of high-concentration biopharmaceutical drugs is desirable from the perspective of patient compliance and convenience. However, successful development of such products requires overcoming several challenges related to colloidal behavior of the drug substance, such as viscosity and syringeability. This chapter focuses on the computational efforts to understand viscosity issues in biotherapeutics.

SECTION II: DEVELOPABILITY PRACTICES IN THE BIOPHARMACEUTICAL INDUSTRY

Chapters 7–9: Best Practices

These three chapters describe the best practices for developability assessment being followed by three of the major biopharmaceutical companies: Novartis, Amgen, and Pfizer.

Chapter 10: Developability Assessment Workflows to De-Risk Biopharmaceutical Development

Several pharmaceutical companies and contact research organizations (CROs) are beginning to develop *in silico* methods tailored toward improving bioprocess yields, biophysical stability, and safety profiles of biopharmaceuticals. One of these companies was contacted to describe its methods and technologies.

Section I

Principles of Biopharmaceutical Informatics

1 Biopharmaceutical Informatics

Applications of Computation in Biologic Drug Development

Sandeep Kumar,[*] Russell H. Robins,[†]
Patrick M. Buck,[*] Timothy P. Hickling,[‡]
A. Mary Thangakani,[§] Li Li,[¶] Satish K. Singh,[*]
and M. Michael Gromiha[**]

CONTENTS

[*] Pharmaceutical Research and Development, Biotherapeutics Pharmaceutical Sciences, Pfizer, Inc., Chesterfield, Missouri
[†] Analytical Research and Development, Biotherapeutics Pharmaceutical Sciences, Pfizer, Inc., Chesterfield, Missouri
[‡] Pharmacokinetics, Dynamics and Metabolism—New Biological Entities, Immunogenicity Sciences, Worldwide Research and Development, Pfizer, Inc., Andover, Massachusetts
[§] Department of Crystallography and Biophysics, University of Madras, Chennai, India
[¶] Pharmaceutical Research and Development, Biotherapeutics Pharmaceutical Sciences, Pfizer, Inc., Andover, Massachusetts
[**] Department of Biotechnology, Indian Institute of Technology Madras, Chennai, India

1.1 INTRODUCTION

Biologics, particularly monoclonal antibodies and antibody-based therapeutics (fragments of antigen binding [Fabs], antibody–drug conjugates [ADCs], and fragment crystallizable [Fc] fusion proteins), have emerged as an important class of therapeutics in the last couple of decades. Based on information available from the journal *mAbs*, more than 30 antibody-based therapeutics have been approved in the United States (US) and European Union (EU) as of February 2014 for the prevention or cure of diverse human diseases, including several cancers, respiratory syncytial virus (RSV) infection, rheumatoid arthritis, psoriasis, asthma, macular degeneration, multiple sclerosis, bone loss, and systemic lupus erythematosus. Besides this, another 10 antibody-based therapeutics are currently in review by US, EU, or Japanese regulatory agencies. According to a recent forecast,[1] four or five therapeutic monoclonal antibody (mAb) candidates currently in phase 3 clinical trials are expected to transition into regulatory review, and another three or four molecules are expected to be approved for marketing in the US or EU during 2014. Therefore, a robust growth in antibody-based therapeutics available to treat human diseases is anticipated in the near future. The success of antibody-based therapeutics in clinics is being enabled by several technological leaps in biology, antibody design, manufacturing, analytical characterization, formulation development, and delivery devices for such candidates. These biotechnological advances are occurring at an increasingly rapid pace in recent years and are facilitating development of both innovator and follow-on biologics. The success of biologics in the clinic is also fueling demand for these medicines throughout the world. However, these biotech successes have also been accompanied by sharp increases in drug development costs, regulatory hurdles, and attrition rates of therapeutic candidates at preclinical and clinical stages.[2] In addition, the rate of translation of therapeutic candidates discovered in laboratories into viable drugs available in clinics is declining.[3] Overall, high drug development costs and profit expectations of drug manufacturers are resulting in high pricing regimes that payers (governments, insurance companies, and individual patients) find increasingly difficult to afford.[2] Therefore, the current practices in biologic drug discovery and development need innovation for reducing costs of drug development and improving safety and flexible, patient-friendly delivery options. Availability of affordable, safe, and easily deliverable biologics can significantly expand the reach of biologics to all parts of the world and improve health for all humans.

The purpose of this chapter and also of this book is to highlight how computational tools and analyses can aid in biologic drug discovery and development, and describe a new discipline called biopharmaceutical informatics. Unlike small-molecule drugs, the discovery and development of biologics has been mainly a compartmentalized enterprise dominated by experimental processes of trial and error. However, currently available computational tools can be utilized at every stage of the drug discovery and development cycle, from target validation to design/selection of lead compounds to bioprocess optimization and formulation development to safety, efficacy, and pharmacokinetics to clinical trial design. For example, molecular design/selection and bioprocess development can significantly impact the biological activity and safety of a biopharmaceutical drug product.[2] Therefore, rigorous molecular-level assessments of biologic candidates for physicochemical degradations, thermodynamic stability, immunogenicity, and other drug safety attributes at early stages of candidate design and selection can be very helpful in forecasting resources required to develop them and prevent late-stage drug failures. At the early stages, it may also be feasible to optimize the amino acid sequence of the lead candidate(s) for easier, cost-effective drug development and safety profiles than the candidates optimized only for target binding affinity.[4–8]

Biopharmaceutical informatics endeavors to use information technology, sequence- and structure-based bioinformatics analyses, molecular modeling and simulations, and statistical data analyses toward biologic drug development. In this chapter, we shall examine several applications of biopharmaceutical informatics toward biologic drug development. We focus on understanding potential molecular origins of physicochemical degradation of antibody-based biologics and how these may be related to product quality and safety. Development of databases containing experimental data on biophysical stability, safety, and preclinical/clinical immune observations, along with molecular sequence–structural analyses for several biologic candidates, can enable comparisons among molecular-level properties of well-behaved candidates with those of poorly behaved ones. Such comparisons will improve our understanding of how molecular-level properties of the biologic candidates impact their development as drug products. Here, we present case studies from our own work. Several aspects of this chapter shall be described in greater detail in subsequent chapters of this book. Therefore, this chapter does not review the literature comprehensively. Instead, our goal is to spark the reader's interest in different aspects of computational applications to biopharmaceutical drug development by highlighting interesting scientific advances and case studies. It is pertinent to mention here that computation is also being increasingly applied for structure-based design of biologic candidates during drug discovery to improve binding affinity and selectivity of these molecules toward their cognate targets. This is an important aspect of biopharmaceutical informatics. However, it is beyond the scope for this chapter. In yet another aspect of biopharmaceutical informatics, computational tools are also applied in drug development through systems pharmacology[9] and pharmacokinetic/pharmacodynamic modeling.[10] These areas are also not covered in this chapter.

1.2 APPLICATIONS OF INFORMATION TECHNOLOGY: INFORMATION SUPPLY CHAIN AND KNOWLEDGE-BASED DECISION MAKING

In essence, the development of biologic drugs is an information business. With the exception of physical clinical supplies, all development products are in fact informational, supplying the demands of primary (intended) and secondary (unintended) customers toward the development and licensure of high-value products for all stakeholders. Whether it is validating analytical methods, robust manufacturing process understanding, fit-for-purpose formulation design, quality assessments, clinical supply management, regulatory interactions, predictive simulations, or the like, it is imperative to develop and maintain an effective information supply chain to meet the needs of internal and external customers within research and development. Therefore, there must be constant focus on enabling a knowledge-based approach in which high-quality decision making, development, and execution are facilitated and underwritten by facile access to high-quality data throughout the development value stream. To achieve this, it is essential to cost-effectively capture (produce), manage (curate), and make available (distribute) all pertinent information in a manner that sustains and leverages our collective intellectual legacy. Essentially, an effective information supply chain manifests as a knowledge-based development engine in which requisite information/knowledge is accessible to the right people, at the right time, and in the right format to enable value creation throughout the development cycle.

The desire to take a knowledge-based approach is easy to state but remarkably difficult to describe in operational details, as it is comprised of human intentions and culture as much as it is of tools and technologies. The willingness to understand and exploit technology toward common goals is equally important, making this both a technical and a human enterprise. It is also important to note that knowledge management, as it is often called, is not something an organization builds or buys. Instead, it is merely the consequence of developing a fit-for-purpose information supply chain that embeds appropriate technologies within effective work processes operating within a culture of knowledge appreciation and awareness. All three are essential for success, though the tendency is to focus solely on technology.

Before dissecting the primary elements of an effective information supply chain (production, distribution, consumption), we must recognize two fundamental information markets, as they uniquely impact both design and culture. The primary data market is the most familiar one, in which the customer requesting that information be produced is also the customer in a functionally closed request–delivery loop exemplified by process development requesting sample analysis from analytical development using specific criteria and workflows. Such transactions are well described, repeatable, and readily embedded within both tools and processes. Data standards, quality, formats, and reporting are agreed upon up front. The hypotheses under testing, for example, product quality versus specifications, are also well defined. The secondary data market, on the other hand, is one in which the information consumer did not request that the information be created initially, but seeks to extract additional value from existing information created for other reasons, for example, correlating clinical event data with lot disposition/characterization data to establish key quality attributes. These information queries are unique in that the data were never collected or curated originally to test such hypotheses per se, and as a result, the requirements for data capture, curation, and reporting were not defined for such purpose when the information was created. Historically, the focus has been on the primary data market, with little cultural regard for the secondary data market. This cultural perspective is understandable, but it hinders our ability to extract full value from biologic drug development data. A broader learning perspective that enables hypothesis generation, not just hypothesis testing, is required for the growth of biopharmaceutical informatics.

In the simplest sense, the information supply chain consists of data capture (production), data curation (inventory management), and reporting (distribution channels). In most primary data markets, all three elements are contained within a platform tool such as Laboratory Information Systems (LIMs), electronic Labortatory Notebook (eLN), and database structures. These include data authoring, data archiving, retrieval, and reporting all within a tool common to both the producer and consumer under well-defined criteria. In secondary data markets, producer and consumer often do not share expertise or access to a common tool or share criteria on data standards or definitions, making the exposure and retrieval of pertinent information difficult, if not impossible.

Therefore, an effective information supply chain that faithfully serves both our primary and secondary data markets across myriad biologic drug development interests and partners must have the follow capabilities:

- A culture that appreciates that all data have legacy as well as specific value and treats them as such
- A culture that is enabled with the capability to capture, curate, and use the information it produces
- Sufficiently common data standards and quality criteria to facilitate data retrieval and reutilization
- Sufficient tool interoperability to enable data discovery and distribution
- Work process alignment promoting the distribution of data in appropriate formats across end users
- Information consolidation layer capability enabling disparate systems to work as one when necessary or more practical
- Appropriate data analytics to reduce and extract meaning from our information

To acquire these capabilities, the following investments are usually required:

- Tools for the exposure and retrieval of desired information without reliance on social networks, local experts, or deep knowledge of multiple curation points
- Capture of appropriate metadata across work streams and partners with sufficient context to facilitate meaningful interpretation
- Development of intuitive end-user interfaces that facilitate compliance with business rules with minimal training or specific tool expertise
- Portfolio management systems that enable rapid and informed decision making, and preserve legacy learning across projects
- Implementation of single sourcing of key information to reduce redundancy, inefficiencies, and cross-verification burden
- Information utilization that is fully leveraged beyond the written word—audio, video, and imaging
- Tools to access information on demand agnostic of technology

The above discussion has presented an aspirational road map for utilizing information technology toward enabling goals of biopharmaceutical informatics. Besides this, there is also a need to develop scientific understandings, tools, and techniques to fully realize the potential of this field. The subsequent sections in this chapter describe our initial attempts.

1.3 DEVELOPABILITY ASSESSMENTS OF BIOLOGIC CANDIDATES: PREDICTING POTENTIAL PHYSICOCHEMICAL DEGRADATION SITES

Traditionally, biologic drug discovery and development has been compartmentalized into discovery and development organizations. During drug discovery, the main objective is to identify highly potent candidate molecules that are most likely to achieve a desired therapeutic effect. At this stage, high binding affinity and selectivity toward a particular receptor are the major drivers for design/selection of molecular candidates. Once selected, the biologic candidate proceeds into development for various stages of animal and human testing for safety and efficacy. Drug development scientists endeavor to stabilize the biologic molecule for commercially viable production, long shelf life, and delivery in a user-friendly format. This traditional approach implies that the amino acid sequence of the selected molecule is fixed and cannot be changed once it enters the development stages. Therefore, drug product development mainly involves use of external processes like formulation buffer, pH and excipient screening, lyophilization, and drug delivery devices to minimize physicochemical degradation of the drug molecule during storage, shipping, and administration. A limitation of this paradigm is that the drug product development fails or stalls for problematic molecules with poor stability or solubility if finding optimum formulation and delivery combinations proves difficult. Even if the optimum biopharmaceutical drug product is developed from a molecule with poor biophysical attributes, its delivery options may be limited. For example, it may be feasible to develop a biologic drug product for parenteral, but not for subcutaneous administration due to viscosity and syringeability related issues at high concentrations. Another limitation of this paradigm is that opportunities to optimize the selected candidate for desirable attributes, such as high cell line expression yields, safety (low immunogenicity), improved pharmacokinetics/pharmacodynamics (PK/PD), less frequent dosing, and flexible delivery options, are lost. Furthermore, lack of consideration of the discovery and development efforts as a seamless interdependent whole leads to expensive development, if not outright failures. All of these limitations contribute to the currently high costs of developing and manufacturing biologics. To overcome the above-mentioned limitations and capitalize on all opportunities available to a biologic drug product development program, it is essential to modify the above-described compartmentalized paradigm. This can be done by understanding sequence and structural features of biologic drug candidates and optimizing them, not only for potency, but also for developability, manufacturing costs, and safety. Below, we describe how computation can be used to assess biologic candidates for developability by taking into consideration potential chemical degradation sites, aggregation, immunogenicity, and high-concentration-solution behavior.

Biologics comprise a variety of products, such as oligonucleotides, growth factors, cytokines, hormones, receptors, enzymes and clotting factors, prophylactic/therapeutic vaccines, monoclonal antibodies (mAbs), antibody components (Fabs), Fc fusion proteins, and antibody–drug conjugates (ADCs). These macromolecules possess highly complex heterogeneous three-dimensional (3D) molecular structures and are produced using recombinant DNA technologies in a variety of hosts or may

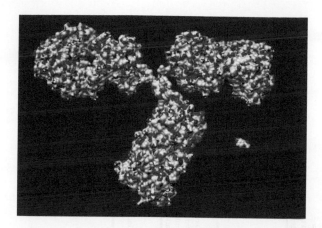

FIGURE 1.1 Molecular structure of a mAb versus that of a small-molecule drug is shown. The mAb shown in this case is a murine IgG2a mAb[18] whose crystal structure is available in the Protein Data Bank[19,20] entry 1IGT. The small-molecule drug shown here is acetaminophen.

be plasma derived. Figure 1.1 illustrates the complexity of biologic drug molecules by comparing the molecular structure of a mAb with a small-molecule drug. As stated earlier, mAbs are emerging as the most successful class of biopharmaceuticals, and it can be seen that their molecular structures are far bigger and much more complex than those of the small-molecule drugs. The mAbs and antibody-based drug candidates and products are the major focus of this chapter.

As a consequence of their size and structural complexities, biologics are vulnerable to numerous physicochemical stresses during manufacturing, shipping, storage, and administration. Degradation caused by these stresses can potentially compromise the potency and safety of these drug products. A number of physicochemical stresses potentially encountered by a biologic are shown in Figure 1.2. This figure illustrates that multiple stresses experienced by biologic molecules at various stages can result in common physical degradations, such as aggregation. Naturally occurring proteins found in organisms adapted to extreme environmental conditions, such as high and low temperatures and high acidity and salinity, also face similar stresses.[11] Therefore, strategies used by nature can potentially be applied to the molecular design and formulation of biologics. In particular, organisms adapted to high and low temperatures (thermophiles and psychrophiles) can teach us important lessons[12–17] toward improving protein stability and solubility without sacrificing potency or causing large-scale structural rearrangements. Striking a balance among protein activity, stability, solubility, and viscosity is consistent with the goal of biologic product development, which is to maintain molecular (physicochemical as well as structural) integrity of the protein coping with environmental stresses.

In a living cell, proteins can age due to several non-enzymatic covalent modifications that accrue gradually because of their oxygen-rich aqueous environment. Moreover, almost all amino acid residues found in natural proteins are vulnerable to one or another chemical degradation.[21] Likewise, it can be imagined that biologics can also age during storage. Therefore, biologic formulations must contain components

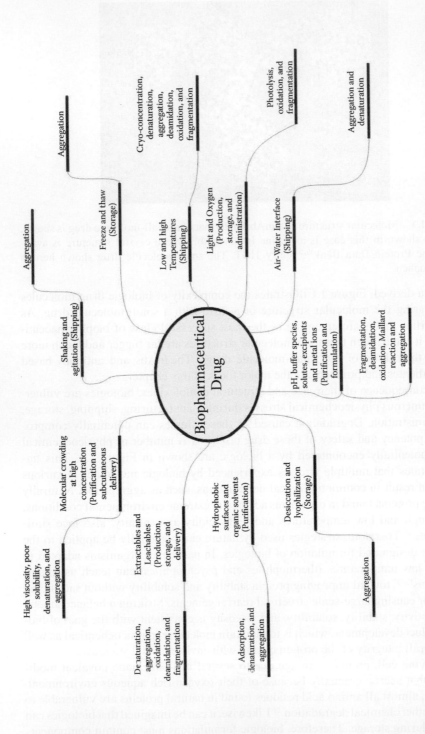

FIGURE 1.2 Physicochemical stresses (inner circle of bars) faced by a biopharmaceutical drug during various stages of production, shipping, storage, and administration and their consequences (outer circle of bars). Note that different stages of drug manufacturing, storage, and shipping may encounter common stresses.

that retard physicochemical degradation of the drug substances and maintain integrity of the biologic drug products over the duration of their shelf life. Several potential physicochemical degradations and their mechanisms have been described for biopharmaceuticals. Among these, deamidation, oxidation, isomerization, fragmentation, and aggregation are the common ones.[22] Most of the chemical degradation routes (i.e., deamidation, oxidation, etc.) arise from specific residues or residue pairs found in protein sequences.[23] Such degradation sites can be easily identified in the amino acid sequences of the biologic candidates. Aggregation-prone regions (APRs), susceptible to β-strand-mediated aggregation, are typically 5–10 residues long and can also be predicted using the amino acid sequence. However, APR prediction is complex and currently available methods are not 100% accurate.[6, 24, 25] Furthermore, β-strand-mediated aggregation can lead to several morphologies, ranging from amorphous β-aggregates to amyloid fibrils.[26, 27] Using experimental data available on hexapeptides that form either amorphous β-aggregates or amyloid fibrils, Thangakani et al. have developed an algorithm, called Generalized Aggregation Proneness (GAP).[28, 29] GAP scans a given amino acid sequence for amyloid fibril and amorphous β-aggregating hexapeptide segments based on propensities of amino acid residue pairs to occur together at the same or alternate faces of a β-strand that participates in aggregation.[29] Benchmarking studies using the available experimental data indicate that GAP performs at a significantly superior level than other APR prediction algorithms.[29] For example, Tsolis et al.[30] have recently compiled a set of 48 amyloid fibril forming peptide sequences found in 33 amyloidogenic proteins. These sequences were used here to benchmark performances of several freely available APR prediction tools. A peptide sequence was considered to be amyloidogenic if it contained at least one APR (six or more consecutive residues identified as aggregating). The results are shown in Table 1.1. GAP is considerably more accurate than the other methods. In summary, the above discussion indicates that physicochemical degradation sites can be predicted from amino acid sequences of biologic drug candidates. Such predictions can be further refined using structural models of the biologic candidates.

To elaborate on the use of amino acid sequences and structural models toward predicting potential physicochemical degradation sites in biologic candidates, the human b12 monoclonal antibody,[35] whose full-length crystal structure is available in the Protein Data Bank (PDB)[19, 20] entry 1HZH, is utilized. Consider that this is a biologic drug candidate at the initial stages of formulation development that needs to be assessed for potential physicochemical degradation routes. Figure 1.3 shows potential physicochemical degradation sites in the amino acid sequence of human b12 mAb, and Table 1.2 counts the number of such sites in variable and constant regions of the mAb. Figure 1.3 and Table 1.2 present the potential physicochemical degradation sites for a pair of heavy and light-chain sequences in the human b12 mAb. A full-length mAb contains two such pairs.

Figure 1.3 and Table 1.2 illustrate the complexity of macromolecules such as mAbs by pointing out that large portions of their amino acid sequences are inherently prone to one or another of the several physicochemical degradations. Chemical degradations often require that the involved residues be present on the protein surface so that they can interact with solvent, metal ions, redox agents, and so forth. Similarly, the APRs

TABLE 1.1

Performance of Different Aggregation Prediction Algorithms on 48 Sequences from 33 Amyloidogenic Proteins

Aggregation Prediction Algorithm	Number of Sequences Predicted to Contain at Least One APR	Number of Sequences Predicted to Contain No APRs	Total Number of Amyloidogenic Sequences	Accuracy[a] (%)
GAP[29]	40	8	48	83.3
Amylpred2[30]	30	18	48	60.4
AGGRESCAN[31]	32	16	48	66.7
TANGO[32]	14	34	48	29.2
WALTZ[33]	21	27	48	43.7
PASTA2[34]	14	34	48	29.2

[a] For each algorithm, accuracy was judged based on the number of correctly predicted sequences with at least one APR of six or more consecutive aggregation-prone residues. These benchmarks are for 48 amyloid fibril-forming sequences from 33 amyloidogenic proteins.[30] For a more comprehensive comparison, refer to Thangakani et al.[29]

need to lie at or near protein surfaces to be able to promote aggregation in response to a physical stress such as temperature. Therefore, building 3D structural models is essential to pinpoint which of the above marked sites are at greater risk of physicochemical degradations. For the human b12 antibody, a crystal structure of the full-length antibody is publicly available. 3D structures are often unavailable for most biopharmaceutical drug candidates. Moreover, crystallizing every candidate and its variants is costly and time-consuming. Computational techniques of protein structure prediction are commonly used during research and development of biopharmaceutical products. Homology-based models of biologic candidates can be rapidly derived if suitable templates are available. Homology-based protein structure prediction is a vast field with several applications to drug discovery and design,[38,39] and a review of this field is out of the scope for this chapter. Briefly, homology modeling relies on the premise that proteins with similar amino acid sequences have similar 3D structures. Therefore, it utilizes the similarity of a target protein's amino acid sequence with that of a template protein, whose experimental structure is available, to model the 3D structure of the target protein. Procedures for computational modeling of antibody-variable domains (Fvs and Fabs) have been developed in recent years, and these are proving helpful in structure-based design of antibody-based therapeutics.[40] Homology-based models of variable regions of antibodies and also of the full-length antibodies can prove useful in understanding physicochemical attributes of the candidates and for pinpointing potential physicochemical degradation sites.

Let us continue with our example of human b12 antibody and refine predictions made using the mAb sequence (Figure 1.3). As stated earlier, a 2.7 Å resolution crystal structure for this mAb is publicly available in the PDB entry 1HZH. This structure

>1HZH.H

QVQLVQSGAEVKKPGASVKVSCQASGYRFSNFVIHWVRQAPGQRFEWMGWINPYNGNKEFSA
KFQDRVTFTADTSANTAYMELRSLRSADTAVYYCARVGPYSWDDSPQDNYYMDVWGKGTTVIVS
SASTKGPSVFPLAPSSKSTSGGTAALGCLVKDYFPEPVTVSWNSGALTSGVHTFPAVLQSSGLYSLSSVVT
VPSSSLGTQTYICNVNHKPSNTKVDKKAEPKSCDKTHTCPPCPAPELLGGPSVFLFPPKPKDTLMISRT
PEVTCVVVDVSHEDPEVKFNWYVDGVEVHNAKTKPREEQYNSTYRVVSVLTVLHQDWLNGKEY
KCKVSNKALPAPIEKTISKAKGQPREPQVYTLPPSRDELTKNQVSLTCLVKGFYPSDIAVEWESNGQP
ENNYKTTPPVLDSDGSFFLYSKLTVDKSRWQQGNVFSCSVMHEALHNHYTQKSLSLSPGK

>1HZH.L

EIVLTQSPGTLSLSPGERATFSCRSSHSIRSRRVAWYQHKPGQAPRLVIHGVSNRASGISDRFSGSGSGT
DFTLTITRVEPEDFALYYCQVYGASSYTFGQGTKLERKRTVAAPSVFIFPPSDEQLKSGTASVVCLLNNF
YPREAKVQWKVDNALQSGNSQESVTEQDSKDSTYSLSSTLTLSKADYEKHKVYACEVTHQGLRSPV
TKSFNRGEC

FIGURE 1.3 Potential physical and chemical liabilities are mapped on the sequence of b12 human mAb. Keys to understanding the map are as follows: Big bold red font: Potential chemical degradation sites (see Table 1.1 for detailed counts for each liability type). Underlined: Automatically predicted CDRs. Yellow background: Aggregation-prone regions (APRs) predicted using a combination[5] of TANGO[32] and PAGE.[36] Cyan background: Additional APRs predicted by WALTZ.[33] In several instances, the TANGO/PAGE-predicted APRs overlap with those predicted by WALTZ. In such instances, only the additional residues are shown in cyan background. Gray background: Additional hexapeptide aggregation-prone regions detected by using experimentally derived patterns based on variants of an amyloid fibril-forming peptide, STVIIE.[37] Some of these hexapeptide patterns overlap with the APRs predicted from TANGO/PAGE and WALTZ. In such cases, only the additional residues are shown in gray background. Green background: Glycan attachment site.

has a few missing residues and a broken inter-heavy-chain disulfide bond. These were completed via molecular modeling. A full-length mAb contains two pairs of heavy and light chains that fold into 12 structural domains that are, in turn, organized into two Fab regions and an Fc region. Each Fab contains the variable (V_L and V_H organized into the Fv region) and constant (C_H1 and C_L) domains of a heavy-chain–light-chain pair. The Fc region contains the constant domains, C_H2 and C_H3, from each of the two heavy chains. The C_H2 domains in the Fc region are also glycosylated. Figure 1.4 shows a schematic diagram of the 3D structure of human b12 mAb.

Let us focus on the physicochemical degradation sites that are adjacent or overlap with the complementarity determining regions (CDRs) in this mAb. Degradations involving these regions of an antibody-based drug product are of immediate concern because these can potentially impair activity of the product. Figure 1.3 indicates that potential sites for β-strand-mediated aggregation, Asp isomerization, Asn deamidation, Met oxidation, metal-catalyzed fragmentation, and glycation (if formulation buffer contains sucrose) fall within or overlap with the CDRs. Availability of a structural model for human b12 mAb facilitated calculation of the solvent-accessible surface area

TABLE 1.2
Count of Potential Chemical Degradation Sites in a Heavy Chain:
Light-Chain Pair of Human b12 mAb

Liability Residue/Motif	Potential Degradation Risk/ Known Sites	V_H	C_H	V_L	C_L	Total
Unpaired Cys	Oxidation	0	0	0	0	0
Asp	Isomerization/cyclization	7	13	3	5	28
Phe	Oxidation	6	10	5	4	25
His	Oxidation	1	9	3	2	15
Lys	Glycation, if formulation contains sucrose	6	28	3	8	45
Met	Oxidation	3	2	0	0	5
Asn	Deamidation, isomerization/cyclization	6	15	1	5	27
Gln	Deamidation	8	11	5	6	30
Trp	Oxidation	5	5	1	1	12
Tyr	Oxidation	8	12	5	4	29
Asp-Pro	Fragmentation	0	1	0	0	1
Asp-Tyr	Fragmentation	0	1	0	1	2
His-Ser	Metal-catalyzed fragmentation	0	0	1	0	1
Ser-His	Metal-catalyzed fragmentation	0	1	1	0	2
Lys-Thr	Metal-catalyzed fragmentation	0	4	0	0	4
Asn-Gly	Deamidation, isomerization/cyclization	1	2	0	0	3
Asn-Ala	Deamidation, isomerization/cyclization	0	1	0	1	2
Asn-Ser	Deamidation, isomerization/cyclization	0	2	0	1	3
Asn-His	Deamidation	0	2	0	0	2
Asp-Gly	Isomerization/cyclization	0	2	0	0	2
Asp-Ser	Isomerization/cyclization	1	1	0	2	4
His-X-Ser	Metal-catalyzed fragmentation	0	0	0	0	0
Ser-X-His	Metal-catalyzed fragmentation	0	0	1	0	1
Asn-X-Ser	Glycosylation attachment motif	0	0	0	0	0
Asn-X-Thr	Glycosylation attachment motif	0	1	0	0	1

of the individual amino acid residues and molecular properties for the mAb under typical formulation conditions. By examining solvent exposure of the degradation-prone sites, one can refine the predictions made using the amino acid sequences alone:

Risk of Met oxidation: CDR3 in the heavy chain (H3 loop) contains Met 114 (contiguous sequence numbering). H3 loops contribute to antigen binding in a major way. Therefore, the sequence information suggests that oxidation of Met residues present in these loops may potentially impair activity of this mAb. However, accessible surface area (ASA) calculations show that Met 114 is solvent inaccessible in both heavy chains (ASA of Met 114 = 0.0%). Therefore, Met oxidation may not be a risk factor for this mAb as long as its 3D structure is preserved.

Risk of Asn deamidation: Four of the six CDR loops in this mAb contain Asn residues. In particular, CDR2 in heavy chain (H2 loop) contains a 55-NG-56

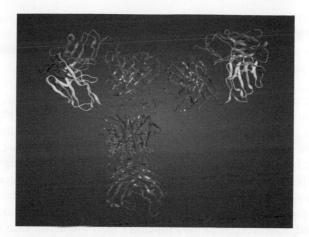

FIGURE 1.4   Schematic diagram of full-length human b12 mAb (PDB entry, 1HZH) is presented. The variable regions are shown in different colored ribbons. In these regions, the heavy and light chains are shown in green- and cyan-colored ribbons. Heavy-chain CDRs 1 and 2 are shown in brown, and the CDR3 is shown in red. Light-chain CDRs 1, 2, and 3 are shown in magenta. The constant regions in the mAb are shown in gray ribbons. The glycans are shown in stick representation.

motif. ASA calculations indicate that this motif is solvent exposed and is therefore susceptible to deamidation. In addition to these, Asn residues in H1 loop (N51) and in L2 loop (N54) are also solvent exposed in one of the two copies of heavy and light chains. Therefore, Asn deamidation may be a risk factor for this mAb, if Asn → Asp/IsoAsp substitution disturbs antigen binding.

Risk of fragmentation: CDR1 in the light chain (L1 loop) contains a metal catalyzed fragmentation motif 26-SHS-28. ASA calculations indicate that this motif is solvent exposed in both the light chains. Therefore, metal-catalyzed fragmentation may be a risk factor for this mAb.

Risk of Asp isomerization: The CDR3 in the heavy chain (H3 loop) contains an Asp isomerization motif, 105-DS-106. The ASA calculations indicate that this motif is also solvent exposed in both the heavy chains. Therefore, Asp isomerization may be a risk factor for this mAb, if Asp → IsoAsp substitution disturbs antigen binding.

Risk of glycation: Degradation of sucrose present in several biologic formulations into the reducing sugars, glucose and fructose, can cause glycation of Lys residues via Maillard reaction.[41] CDR2 in the heavy chain (H2 loop) of this antibody contains two Lys residues, K58 and K63. ASA calculations indicate that both of these residues are solvent exposed, charged, and have pKa in the range seen for Lys glycation.[41] Therefore, glycation of Lys residues may be a risk factor for this mAb if the formulation contains sucrose and the pH is such as to lead to its inversion.

Risk of aggregation: All three CDRs in the heavy chain (H1, H2, and H3) and the CDR3 in the light chain (L3 loop) either overlap with or contain APRs. The ASA calculations indicate that some residues in the APR

28-RFSNFVIHWV-37, which overlaps with the H1 loop, are solvent exposed. The issue of β-mediated aggregation was further probed for the b12 mAb via molecular dynamics simulations. These are described in the next section.

Overall, the above exercise demonstrates how using the structural information allows us to rule out several sequence-based predicted sites for risk of physicochemical degradation because those sites may not be solvent accessible. At the same time, this information also helps us to highlight the residues and regions that may be particularly at risk because of their location in the 3D structure of a biologic candidate. Additionally, the availability of structural models helps identify potential surface features that may not be apparent from the amino acid sequence alone. For example, solvent-exposed hydrophobic patches can act as aggregation-prone motifs. An algorithm called Spatial Aggregation Propensity (SAP) to detect such aggregation-prone motifs was developed and validated by Chennamsetty et al.[42,43] Commercially available molecular modeling and simulation software, such as MOE from Chemical Computing Group, are also able to detect solvent-exposed hydrophobic and charged patches. For example, Figure 1.5 shows solvent-exposed positively charged (blue), negatively charged (red), and hydrophobic (green) patches mapped on to the molecular surface of human b12 mAb.

In addition to answering questions about the specific physicochemical degradation sites, the availability of 3D structures also enables calculation of several useful solution properties, such as pI, charge, volume, diffusion coefficient, and dipole moment, for the biologic candidates. Table 1.3 presents such properties for the b12 mAb and its components, namely, Fv (heavy chain, Gln1-Ser127; light chain, Glu1-Lys108), Fab (heavy chain, Gln1-Cys230; light chain, Glu1-Cys215), and Fc (two heavy chains, Glu243-Lys457 each, and glycans) regions. These properties were calculated at pH 5.8 and 1 mM salt concentration. Such properties can be used to estimate solution behavior of the biologic candidates under dilute concentrations.

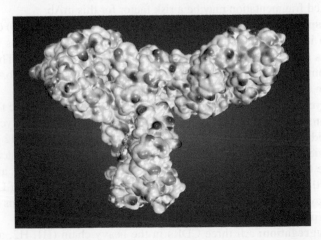

FIGURE 1.5 Solvent-exposed hydrophobic (green), positively charged (blue), and negatively charged (red) patches on human b12 mAb, detected using Patch Analyzer in MOE.

TABLE 1.3

Average Protein Properties for b12 mAb and Its Components at pH 5.8 and 1 mM NaCl

Property	Component			
	Fv	Fab	Fc Dimer	Full mAb
Protein mass (kDa)	26.24	48.58	48.66	148.14
Non-polar surface area ($Å^2$)	5,570.8	9,513.9	10,659.5	27,932.2
Polar surface area ($Å^2$)	4,437.9	7,617.4	7,951.1	21,293.2
VdW volume ($Å^3$)	24,102.6	44,654.0	44,984.4	13,6175.8
pI	9.74	9.50	7.97	9.17
Diffusion coefficient in water (cm^2/s)	1.3×10^{-6}	1×10^{-6}	9.9×10^{-7}	5.6×10^{-7}
Average net charge (Z)	8.50	16.27	10.39	44.34
Apparent charge	2.91	7.43	4.89	38.79
Zeta potential (mV)	28.62	55.41	35.44	159.40
Dipole moment (D)	215.39	52.46	584.84	2,424.87
K-D hydrophobicity moment	448.6	748.3	1,439.5	3,495.0
Number of positively charged patches with ASA > $100 Å^2$	9	14	9	29
Total number of positively charged patches	19	30	26	83
Number of negatively charged patches with ASA > $100 Å^2$	4	7	10	26
Total number of negatively charged patches	10	23	28	63
Number of hydrophobic patches with ASA > $100 Å^2$	3	1	7	11
Total number of hydrophobic patches	8	15	17	45

For example, all the regions in this mAb have a significant positive net charge and ξ-potential values. Their pIs are also above the formulation pH. In such conditions, colloidal interactions among different copies of this molecule are predominantly repulsive, and highly concentrated formulations of this mAb will likely demonstrate low viscosities.[44]

The availability of such detailed molecular-level information at the early stages of drug discovery and formulation can be used in several ways. For example, such information can be compared across several candidates with similar potencies and factored in during lead candidate selection. This information is also helpful in molecular redesign of highly potent lead candidates with relatively poor physico-chemical stability profiles. In particular, eliminating potential sites for aggregation could improve solubility of a biologic candidate. In the formulation stages where the molecular sequence is fixed and cannot be changed, this information can be used to define formulation strategies. For example, if there are solvent-exposed Met residues in the CDRs, then it may be useful to add methionine as an antioxidant in the formulation buffer to mitigate loss of potency arising from oxidation. Similarly, if there are potential metal-catalyzed oxidation and fragmentation sites that could impact potency, then addition of chelating agents may help. The bottom line is that the availability of this information beforehand can potentially save time and materials required for formulation development studies.

1.4 USING COMPUTATION TO ADVANCE OUR UNDERSTANDING OF PHYSICOCHEMICAL DEGRADATION OF BIOPHARMACEUTICALS

Biopharmaceutical candidates often show issues related to low thermodynamic stability, poor solubility, propensity to aggregate, and high viscosity in concentrated solutions. Among all physicochemical degradations shown by biologics, aggregation is the most common one, and it is encountered multiple times during purification, formulation, shipping, and storage (Figure 1.2). In addition to loss of material during purification steps, aggregation can potentially reduce potency and enhance immunogenicity risks associated with a biologic drug. Similar to folding, protein aggregation has emerged as a major unsolved problem in biochemistry. It is commonly encountered in laboratories during protein manipulations, and it has been implicated in several human neurodegenerative diseases. Comprehensive overviews of aggregation in biologics are available elsewhere.[6, 24, 25, 45–47] Here, we summarize our efforts to understand aggregation in mAbs using data analyses and molecular simulations.

Protein molecules can associate with one another in several ways, including the tendency to exclude solvent (water) due to surface-exposed hydrophobic hotspots, shape as well as charge complementarities among the interacting protein surfaces, and β-strand-mediated aggregates. Among these, β-strand-mediated protein aggregation may require destabilization of the native protein structure and result in irreversible amorphous β-aggregates or amyloid-like fibrils with well-defined morphological features.[48] The β-strand-mediated protein aggregation has been shown to underpin several neurodegenerative diseases in humans and involves formation of the cross-β steric zipper motif readily detectable by fluorescent staining dyes Congo red and thioflavin T (ThT).[26, 27, 48, 49] In a pioneering study, Maas et al. demonstrated that biopharmaceuticals past their expiry dates can also form β-strand-mediated aggregates, bind ThT, and elicit immune responses.[50] This observation motivated Wang and coworkers to apply APR prediction programs developed from the study of disease causing amyloidogenic proteins onto the sequences of commercially available therapeutic antibodies.[4] Several recurring APRs, capable of initiating β-strand-mediated aggregation, were found in the commercially available therapeutic mAb sequences. It has since been shown experimentally that mAbs, when exposed to stress conditions such as acid denaturation or elevated temperatures, do form β-strand-mediated aggregates detectable using ThT binding assays.[45, 51, 52]

An interesting observation from the study of Wang et al.[4] was that APRs often overlap with CDRs of therapeutic mAbs. This observation raised the possibility that residues found in APRs could also be involved in antigen binding. From the analyses of 29 non-redundant Fab–antigen complexes whose high-resolution crystal structures are available in the PDB,[19, 20] it was found that APRs contribute approximately one-fifth of the Fab surface that contacts the antigen.[5] The reason behind this coupling between aggregation and antigen recognition is propensity of the aromatic

residues, Tyr and Trp, to be found both in the APRs and the CDRs. This work suggested several rational structure-based design or selection strategies for improving
developability of antibody-based biologics. The goal for these rational strategies is
to reduce aggregation and improve the solubility and high-concentration-solution
behavior of the antibody-based therapeutics via disruption of strategically selected
APRs.[5,8,24,25,53,54]

It is essential to understand the contribution of APRs toward the native state of
the protein in order to select the APRs that may be suitable for disruption. In addition
to promoting aggregation, APRs also play several other roles in protein structure–
function. These roles were recently studied using multiple data sets that contained
randomly generated amino acid sequences, protein families, intrinsically disordered
proteins, protein monomers, and catalytic residues. It was found that amino acid
sequences in monomeric proteins have lower aggregation propensities than random
sequences generated using the same amino acid compositions, implying an evolu-
tionary bias against aggregation in the natural protein sequences. APRs contribute
toward the protein stability more than the average for similar sized segments, and
are found in the structural vicinity of catalytic residues more often than expected by
random chance. These observations indicate that APRs promote structural order and
are therefore found less frequently in intrinsically disordered regions. Interestingly,
proteins have evolved to optimize their risk of aggregation in the cell. The optimization strategy includes two opposing tendencies gleaned from families of homologous
protein sequences. On one hand, the incidence of APRs is sought to be minimized
among closely related protein sequences, and on the other hand, APRs that are
important for protein folding and function are conserved among more divergent
sequences.[8] Results from these studies led to formulation of rational strategies for
improvement in protein solubility via APR disruption.

An additional aspect of selecting an APR for disruption is to know if the APR
is located at or near the protein surface and would actively promote aggregation
in case the protein structure is destabilized in response to physical stresses. Such
studies often involve dynamical aspects of protein structure, and molecular dynamics (MD) simulations are a useful tool to examine these. By performing MD simulations for biologic candidates at elevated temperatures, we can attempt to mimic
accelerated stability experiments performed during protein formulation development
studies.[53-55] Overall, antibody conformational dynamics and the impact of glycan
truncation and de-glycosylation on the stability of antibody-based therapeutics can
also be studied using MD simulations. Figure 1.6 shows a fully solvated molecular
system (approximately 400,000 atoms) for a full-length murine IgG2a antibody[18]
whose crystal structure is available from the PDB entry, 1IGT. Below, we summarize
insights gained from molecular simulations of antibodies.

Molecular dynamics (MD) simulations were used to study conformational destabilization of the full-length murine antibody (1IGT) upon de-glycosylation and thermal stress.[54] C^α-atom root mean square deviation (RMSD) and backbone root mean
square fluctuation (RMSF) calculations were used to study changes in the mAb
structure over the course of simulations. Destabilization of mAb due to the removal
of glycan residues was found to be local in scope. It was restricted to the perturba-
tions in quaternary and tertiary structures of the C_H2 domains. Thermal stress, on

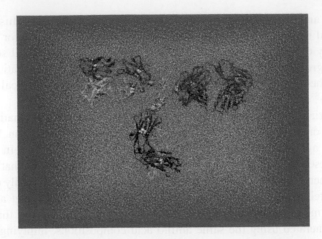

FIGURE 1.6 Fully solvated IgG2a murine antibody molecular system that was simulated by Wang et al.[54] The mAb is shown in ribbons, except for the disulfide bridges and glycans, which are shown in CPK representation.

the other hand, caused widespread destabilization of both the Fab and Fc regions. Our intention here was to destabilize, but not unfold, the antibody. Therefore, no large-scale structural melting was observed within the duration of the simulations. Interestingly, thermal stress and de-glycosylation appear to identify different APRs that could become active when the antibody's native structure is destabilized. An APR predicted to be present in strand A of the C_H2 domains becomes solvent exposed upon de-glycosylation. In the case of the thermal stress, residues present in two CDR overlapping APRs showed increased RMSFs and changes in average solvent exposure.[54]

The MD simulations of the full-length murine mAb were followed by analogous simulations on the Fab portion of human b12 mAb,[53] whose sequence- and structure-based potential physicochemical degradation sites have been described above (see Section 1.2). A Fab is a considerably smaller molecular system than a full-length mAb. This enabled us to study Fab dynamical behavior in greater detail. Along with APRs, deformation of individual structural domains and their interfaces were studied in response to thermal stress. Both variable (V_H:V_L) and constant domain (C_H1:C_L) interfaces deformed before the melting of individual structural domains. This was consistently observed in different simulation trajectories. Interestingly, β-strands present at domain interfaces resist melting for longer durations than all other β-strands and loops in the Fab. Therefore, APRs present in the domain interface regions are unlikely to be involved in the aggregation of this mAb. This study indicated that CDRs, framework region β-strands adjacent to CDRs, and the edge β-strands can deform substantially upon thermal stress, leading to solvent exposure of APRs present in these regions. Further analyses allowed rank ordering different sequence-predicted APRs for their ability to promote irreversible β-strand-mediated aggregation based on solvent exposure, structural location, mobility, and incidence of gatekeeping residues. Two predicted APRs, 28-RFSNFVIHWV-37 (overlapping with the CDR1 in the heavy chain) and 122-TTVIVS-127 (sequence region transitioning from the V_H to C_H1 domain), were found to be particularly at risk. Taking together the observations from Section 1.3 on the sequence and structure of the b12 antibody and the MD simulations described above, it can be concluded that the use of structural and dynamical information can help refine prediction of physicochemical degradation sites in biologic drug candidates. Insights obtained from the studies described above can be very useful in devising structure-based strategies for molecular design or selection of antibody-based biologics with improved developability attributes.

1.5 COMPUTATIONAL ASSESSMENTS OF ADVERSE IMMUNOGENICITY FOR BIOLOGIC CANDIDATES

Adverse immunological responses generated in patients receiving biologic drugs lead to safety and efficacy concerns for biologic drugs. Over the past three decades, immunological safety concerns have guided evolution of therapeutic mAb products from being of murine origin to chimeric to being fully human.[56] However, biopharmaceutical drug products containing fully human antibodies or recombinant/plasma-derived human proteins can still elicit adverse immune reactions in patients.[57,58] The complex nature of the immune system, born out of the need to maintain homeostasis in the presence of both pathogenic and commensal microbes, has inevitably led to many potential contributory factors to therapeutic protein immunogenicity. While the overall assessment of immunogenicity risk for a product must consider several individual risk factors, it is commonly accepted that T-cell epitopes represent a significant contribution to these unwanted immune responses, as they are required for class switching and significant expansion of B-cells.[59] Also because of the central role of T-cell responses in immunity, several attempts have been made to predict the T-cell epitopes present in the amino acid sequence of a protein via algorithms capable of identifying peptides that can potentially bind major histocompatibility complex (MHC) class II (human leukocyte antigen [HLA]). These algorithms were initially applied to microbial proteins as part of a vaccine design strategy, but are now frequently being applied to analyze the amino acid sequences of biological candidates.[60,61] When applying these tools to assess immunogenicity risk, care must be taken to ensure either that the sequence is foreign (e.g., CDRs of engineered mAbs or an endogenous protein that is missing in the recipient) or that a significant likelihood of a reactive T-cell would be present. This can be done, for example, via confirmatory T-cell assays in healthy donors or patient samples.[62] Further effort should be made with mAbs to determine whether germ line sequences are truly present in the population, as it is likely that patients receiving the therapy will possess different germ line sequences and therefore have different profiles for what constitutes foreignness. Here, these considerations were applied to the amino acid sequence of human b12 antibody (PDB entry, 1HZH) as an example (Figure 1.7). It can be seen that potential HLA binding peptides exist throughout the antibody heavy and light chains. However, potential T-cell epitopes derived from these sequences based upon a notion of foreignness number only three. It is commonplace to confirm predictions with *in vitro* assays. These have typically been HLA binding assays, but a new

>1HZH.H

QVQLVQSGAEVKKPGASVKVSCQAS<u>GYRFSNFVI</u>HWVRQAPGQRFEWMG<u>WINPYNGNKEFSA</u>KFQDRV

TFTADTSANTA<u>YMELRSLRSADT</u>AVYYCAR<u>VGPYSWDDSPQDNYYMDV</u>WGKGTTVIVSSASTKGPSVFPLA

PSSKSTSGGTAALGCLVKDYFPEPVTVSWNSGALTSGVHTFPA<u>VLQSSGLYSLSSVVTVPS</u>SSLGTQTYICNVN

HKPSNTKVDKKAEPKSCDKTHTCPPCPAPELLGGPSVFLFPPKPKDTLMISRTPEVTCVVVDVSHEDPEVKFN

WYVDGVEVHNAKTKPREEQYNST<u>YRVVSVLTVLH</u>QDWLNGKEYKCKVSNKALPAPIEKTISKAKGQPREPQ

VYTLPPSRDELTKNQVSLTCLVKGFYPSDIAVEWESNGQPENNYKTTPPVLDSDGSFFLYSKLTVDKSRWQQ

GNVFSCSVMHEALHNHYTQKSLSLSPGK

>1HZH.L

E<u>IVLTQSPGTLS</u>LSPGERATFSCRS<u>SHSIRSRRVA</u>WYQHKPGQAPRL<u>VIHGVSNRAS</u>GISDRFSGSGSGTDFTL

TITRVEPEDFALYYC<u>QVYGASSYT</u>FGQGTKLERKRTVAAPSVFIFPPSDEQLKSGTASVVCLLNNFYPREAKVQ

<u>WKVDNALQS</u>GNSQESVTEQDSKDSTYSLSSTLTLSKADYEKHKVYACEVTHQGLRSPVTKSFNRGEC

FIGURE 1.7 Predicted T-cell epitopes in the amino acid sequence of human b12 antibody (PDB entry, 1HZH). This example indicates prediction made by the Epivax EpiMatrix algorithm.[60,64] The sequence regions highlighted in yellow are potential HLA binding peptides, sequences highlighted in red are potential HLA binding peptides containing foreign (non-germ line) residues, and sequences highlighted in green are potential HLA binding peptides of germ line origin and are thought to potentially upregulate protective T regulatory cells. Residues in bold within the V_H and V_L regions differ from parental germ line sequences (V_H, IGHV1-3*01; V_L, IGKV3-20*01).

generation of assay involving measuring peptides processed and presented by DCs is emerging.[63]

Another potential risk for immunogenicity is the presence of B-cell epitopes. Predictive algorithms for identifying B-cell epitopes have lagged behind those for T-cell epitopes, perhaps most notably due to the T-cell epitope being linear. In contrast, many B-cell epitopes arise only in the 3D structure of the protein (conformational epitopes). However, some efforts to assess linear and conformational B-cell epitopes do exist, although their predictive capability ranges between 30% and 50%[65] when judged against known epitopes. As for validation of these methods, there are very few B-cell epitopes actually defined for therapeutic proteins, making the training set for any algorithm in this space rather limited. One example is the data on adalimumab suggesting that the B-cell epitopes reside in the CDRs.[66] Although the exact epitope has not been defined, the detection of anti-drug antibody (ADA) reactivity almost entirely to CDRs should be a useful guide as to how to develop these algorithms in the future. Figure 1.8 shows potential linear B-cell epitopes in the human b12 antibody. Note that the foreign linear B-cell immune epitopes are predicted to lie in the CDR2 of the heavy chain and in framework region 2 of the light chain of human b12 antibody. An algorithm designed to predict discontinuous B-cell epitopes, Discotope,[68] also identified potential B-cell epitopes in the same regions as the foreign linear B-cell epitopes. These discotopes are shown in Figure 1.9. Consistent predictions of B-cell epitopes by two different algorithms in the same sequence–structural regions (CDR2 in the heavy chain, FR2 in light chains) of the human b12 mAb give us greater confidence that these regions may be involved in adverse immune responses, if this antibody were to be developed as a

>1HZH.H

QVQLVQSGAEVKKPGASVKVSCQAS<u>GYRFSNFVIH</u>WVRQAPGQRFEWMG<u>WINP</u>▮YNGNKEF▮SAKFQDRV
TFTADTSANTAYMELRSLRSADTAVYYCAR<u>VGPYSWDDSPQDNYYMDV</u>WGKGTTVIVSSASTKGPSVFPLA
PSSKSTSGGTAALGCLVKDYFPEPVTVSWNSGALTSGVHTFPAVLQSSGLYSLSSVVTVPSSSLGTQTYICNVN
HKPSNTKVDKKAEPKSCDKTHTCPPCPAPELLGGPSVFLFPPKPKDTLMISRTPEVTCVVVDVSHEDPEVKFN
WYVDGVEVHNAKTKPREEQYNSTYRVVSVLTVLHQDWLNGKEYKCKVSNKALPAPIEKTISKAKGQPREPQ
VYTLPPSRDELTKNQVSLTCLVKGFYPSDIAVEWESNGQPENNYKTTPPVLDSDGSFFLYSKLTVDKSRWQQ
GNVFSCSVMHEALHNHYTQKSLSLSPGK

>1HZH.L

EIVLTQSPGTLSLSPGERATFSC<u>RSSHSIRSRRVA</u>W▮YQHKPGQ▮APRLVIH<u>GVSNRAS</u>GISDRFSGSGSGTDFTL
TITRVEPEDFALYYC<u>QVYGASSYT</u>FGQGTKLERKRTVAAPSVFIFPPSDEQLKSGTASVVCLLNNFYPREAKVQ
WKVDNALQSGNSQESVTEQDSKDSTYSLSSTLTLSKADYEKHKVYACEVTHQGLRSPVTKSFNRGEC

FIGURE 1.8 Prediction of linear B-cell epitopes by Emini surface accessibility algorithm[67] in the sequence of human b12 antibody (PDB entry, 1HZH). The sequence regions highlighted in yellow are self-epitopes, and those in the red are foreign B-cell epitopes. Residues in bold within the V_H and V_L regions differ from parental germ line sequences (V_H, IGHV1-3*01; V_L, IGKV3-20*01).

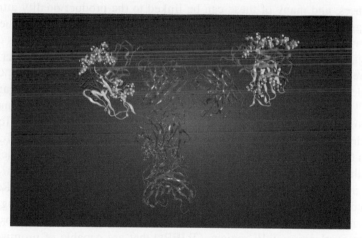

FIGURE 1.9 Prediction of B-cell epitopes by Discotope[68] in the 3D structure of human b12 antibody (PDB entry, 1HZH). The residues involved in the discotopes are represented in CPK. Only those discotopes that coincide with the foreign linear B-cell epitopes are shown here.

drug product. However, further characterization of anti-drug antibodies (ADAs) is needed to determine the accuracy of current B-cell epitope predictions. This will in turn lead to improvements in the predictive algorithms.

Multifaceted algorithms that integrate immunogenicity risk assessments with the principles of immunobiology can facilitate generation of predictive system biology models of the human immune system. Several such models exist (e.g., see Lee et al.[69]), though few have been applied to the problem of immunogenicity and predicting outcomes. A recent attempt to do this with adalimumab exemplifies what may be possible.[70] It remains to be seen whether we currently understand enough about the key players in the immune system to generate a truly predictive model, though it is likely that these models will provide a framework for generating hypotheses and guiding immunogenicity assessments.

1.6   LINKING PHYSICOCHEMICAL DEGRADATION WITH SAFETY OF BIOLOGICS: AGGREGATION–IMMUNOGENICITY COUPLING

Biopharmaceuticals bind their targets with high selectivity. This is a major advantage because non-mechanism toxicity (side effects) is substantially reduced in this case. However, immunological consequences of physicochemical degradation of biologics also lead to safety and efficacy concerns for biologics, in addition to the foreignness of the short-sequence regions discussed above. It must also be emphasized that product quality is only one of the many factors that lead to undesired immune responses among the patients receiving these therapeutics. Other factors include genetic constitution and medical treatment history of the patients.[58] Administration of several biologic drug products has been shown to lead to immune consequences, and many of these can be linked to the product quality attributes, such as aggregation.[58] A clinical manifestation of immunological consequences is the generation of anti-drug antibodies (ADAs) against the biologic. ADAs can target biologics and reduce their efficacy or even render them ineffective over time.[58] How aggregates present in the biologic drug products trigger immune responses is not completely understood. But, T-cell-mediated adaptive immunity appears to be involved.[71] As we have shown above, it is feasible to predict potential β-mediated aggregation-prone regions (APRs) as well as T-cell immune epitopes (TcIEs) in biologic drug candidates using their amino acid sequences. Like APRs, TcIEs are also often found in the CDRs and adjoining regions of therapeutic antibodies.[72] The question then becomes, do APRs and TcIEs overlap in protein sequences? If they do, what does this imply for biologic drug product quality and safety? Recently, we studied the potential coupling between APRs and HLA-DR binding TcIEs[7] among commercially available therapeutic mAbs.[73] It was found that most APRs fall within overlapping TcIE regions (TcIERs) that are capable of promiscuously binding several MHC class II alleles. These observations imply that common sequence–structural motifs contribute toward both physicochemical degradation and immunogenicity of these biologics. Therefore, it is feasible to simultaneously mitigate aggregation and immunogenicity risk for a biologic drug candidate. Such simultaneous optimizations of the drug product quality and safety attributes shall pave the way for next-generation biologic drug products.

Several human proteins known to be involved in autoimmune and neurodegenerative diseases also display coupling between aggregation and immunogenicity.[7, 73] Such evidence of overlap between the APRs and TcIEs is available at the crystallographic level for human insulin. A peptide from the insulin B chain has been crystallized in complex with MHC class II allele HLA-DQ (PDB entry, 1JK8).[74] The peptide in the co-crystal contains an amyloid fibril–forming portion

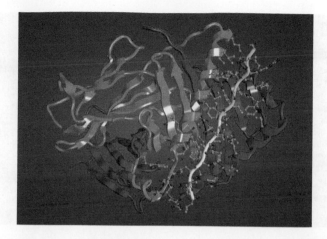

FIGURE 1.10 An example of aggregation and immunogenicity coupling is shown. The structure of an MHC class II allele HLA-DQ8 in complex with an insulin B chain is shown (PDB entry, 1JK8). All the residues in the peptide and the nearby residues in HLA-DQ8 are shown in ball and stick. The yellow portion in the insulin B chain has also been shown to form amyloid fibrils (PDB entry, 3HYD).

11-LVEALYL-17 (PDB entry, 3HYD[75]). Figure 1.10 shows an MHC–insulin B chain complex with the amyloid fibril-forming portion highlighted. Consistently, it has been observed that patients treated with a continuous subcutaneous infusion of insulin can develop allergies due to anti-insulin IgG antibodies raised against insulin fibrils present in the pump systems.[76,77] A rare human disease called injection site amyloidosis is also caused by insulin fibril formation at the sites of frequent injections.[75,77,78]

Metal ion binding sites can also be located near APRs in biologics.[79] It was recently observed that increasing molar concentrations of Fe^{3+} ions in a mAb solution lead to increased levels of oxidation and aggregation in the mAb. An examination of the mAb structure revealed several Fe^{3+} binding sites and their clusters throughout the structure, including the CDRs and the hinge region. In particular, co-localization of Fe^{3+} binding residues (Asp and Glu), aromatic residues (Trp, Tyr, and Phe), and APRs was observed in the CDRs of the mAb (Figure 1.11). Metal-catalyzed oxidation of the aromatic residues could potentially destabilize this mAb, leading to aggregation and potential immunogenicity.

1.7 ESTIMATING HIGH-CONCENTRATION-SOLUTION BEHAVIOR OF BIOLOGICAL CANDIDATES VIA COMPUTATION

Being able to deliver small volumes (~1 ml) of highly concentrated (100–150 mg/ml) solutions of biologic drugs is desirable from the perspective of patient convenience and compliance.[80] However, high viscosity can be a significant impediment to successful development of high-concentration biologic drug products.[81] Rapid identification of the biologic drug candidates whose development as high-concentration products could prove difficult is desirable. Computational tools to identify such

FIGURE 1.11 Co-localization of Fe^{3+} ion binding residues, aromatic residues, and APRs in CDRs of the Fv portion of a mAb is shown. The solvent-accessible Fe^{3+} ion binding residues, E46, D54, and D62 in the heavy chain, are depicted in ball-and-stick representation along with their proximal aromatic residues, W47, W52, W53, Y60, and Y106. Note the residues are numbered according to their position in the sequences. The following color code was used: heavy chain, green; light chain, red; and APR, yellow.

candidates are valuable at the early stages of drug discovery and formulation. To develop such tools, it is essential that experimental data available on different mAbs are consistent. That is, the biologic molecules are studied under the same conditions using identical experimental procedures as much as feasible. For example, Li et al.[44] recently obtained viscosity data on 11 therapeutic mAb candidates formulated in the same buffer that contained no added salt. All the viscosity measurements were performed using an Anton–Paar viscometer using the same procedures. Several molecular properties, such as net charge, pI, ξ-potential, hydrophobicity, and aggregation propensity, were computed from amino acid sequences and structural models for the Fv portions of these mAbs using the same computational methods. This facilitated statistical correlation analyses and building of molecular profiles. The analyses confirmed the hypothesis that molecular characteristics of the variable regions play important roles in determining high-concentration-solution behaviors of mAbs. This is because of the following reasons: The high-concentration-solution behavior of a mAb must arise from the nature of colloidal intermolecular interactions and self-associations it forms. The intermolecular interactions are determined by the amino acid sequence[82] and 3D structural features of the mAb. The colloidal intermolecular interactions involving only the constant regions of the human mAbs are likely to be the same for all mAb candidates. Therefore, variable regions (Fvs) that change from mAb to mAb also modulate their colloidal intermolecular interactions. The Fv portions of the mAbs that showed high viscosities at 150 mg/ml in the above-mentioned study had significantly negative net charge and negative ξ-potential at the formulation buffer pH, 5.8. The pI values of the Fv regions were also below formulation buffer pH. These observations helped devise mAb candidate screening profiles based on calculated values of net charge, ξ-potential, and pI of the variable regions (Z_{Fv}, ξ_{Fv}, and pI_{Fv}). Such profiles could be used for "red flagging" therapeutic mAb candidates, whose development as high-concentration drug products may be hindered by viscosity-related challenges. Figure 1.12 shows examples of these profiles by presenting electrostatic surfaces of the Fv regions in different mAbs.[44]

Multiscale simulations of solution behaviors of biologic candidates can yield insights into their colloidal intermolecular interactions. Such simulations require molecular systems that consist of several hundred to thousands of copies of the solute (biologics). To accelerate such simulations, molecular structural features of the biologic are coarse-grained (CG) and the solvent is only implicitly accounted for. Chaudhri et al. have pioneered the CG simulations of highly concentrated antibody solutions that yielded valuable insights into their self-association behavior.[83, 84] Figure 1.13 shows a CG model of the human IgG1 b12 antibody. In this model, each domain of the antibody is approximated by a spherical CG interaction bead. A single CG interaction bead was assigned to the center of mass for each structural domain of the b12 antibody. Mass and charge of each bead are the mass and net charge, respectively, of the domain to which the bead is assigned. Overall, each mAb molecule has 12 CG interaction beads. Because the variable domains are different in each mAb, the four CG interaction beads representing these domains vary from mAb to mAb.

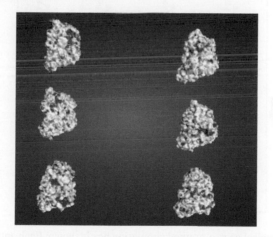

FIGURE 1.12 Examples of surface electrostatic maps for the Fv regions that fall in one of the three profiles deduced from the analyses of concentration-dependent viscosity behaviors of 11 mAbs[44] are shown. The top row shows examples of Fv regions that follow profile 1 (significantly positive net charge and ξ-potential for the Fv regions). Such mAbs typically have low viscosities even at high concentrations. The middle row shows electrostatic surfaces of the Fv regions follow profile 2 (net charge on the Fv region is nearly zero). Hydrophobic intermolecular interactions play more significant roles in such cases, and it is currently difficult to reliably predict concentration-dependent viscosity behavior of such mAbs. The Fv regions from mAbs that follow profile 3 (significant negative net charge and ξ-potential at the formulation pH and pI_{Fv} below the formulation buffer pH) are shown in the bottom row. The mAbs in this profile typically show high viscosities at high concentrations, as both electrostatic and hydrophobic intermolecular interactions promote reversible self-associations. In all these surface electrostatic maps, increasingly blue and red colors indicate increasingly positive and negative charges, respectively. (From Li et al., *Pharm. Res.*, 2014.[44])

FIGURE 1.13 A coarse-grained representation of human IgG1 b12 mAb is shown. Such coarse graining facilitates multiscale simulations of the solution behavior of antibodies.

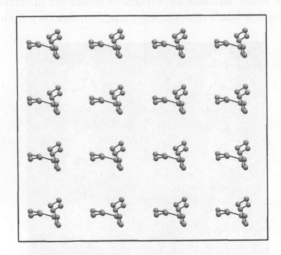

FIGURE 1.14 A typical coarse-grained simulation box is shown. Each antibody molecule is represented by a 12-bead coarse-grained model.

Figure 1.14 shows a typical CG simulation box. The CG simulations were used to compare diffusion behaviors of three mAbs (mAb 1, mAb 10, and mAb 11) from the above-mentioned data set of 11 mAbs at concentrations of 30 and 100 mg/ml.[44] Among the three, mAb 1 is well behaved (low viscosity at concentrations above 100 mg/ml), while mAbs 10 and 11 are poorly behaved (high viscosity at concentrations above 100 mg/ml). The simulations were performed using LAMMPS[85] and run in a vacuum with periodic boundary conditions at 300 K for 5 μs with a time step of 1 ps (10^{-12} s). These simulations showed that diffusion of mAbs 10 and 11 is significantly slower than that of mAb 1 at both concentrations. The electrostatic complementarity between Fab and Fc fragments increases due to an increasingly more negatively charged Fv region for mAbs 10 and 11. This resulted in increased self-association behavior for these mAbs. Molecular crowding at higher concentrations was also found to significantly impact the diffusion of the three mAbs.[44, 87]

1.8   CONCLUSIONS AND FUTURE DIRECTIONS

Biopharmaceutical informatics is an emerging field that seeks to advance applications of diverse computational tools such as informatics, data analyses, predictive modeling, and mathematical modeling toward discovery and development of biopharmaceutical drugs. Here, we have presented examples of how biopharmaceutical informatics can help improve our understanding of fundamental sequence–structural motifs that drive physicochemical degradations, immunogenicity, or high-concentration-solution behavior of biopharmaceuticals. This research enables development of rational design strategies for biologic products with improved molecular integrity[7, 86] and safety. Only minor tweaks in the protein sequences, that do not impact potency, may be required.[4, 7] Such design strategies are highly desirable to improve the cost effectiveness of biologics and prevent late stage drug failures.

Biopharmaceutical informatics is anticipated to be increasingly applied during biologic drug discovery and development in the foreseeable future. These applications will make biologic drug discovery and development cycles more efficient and nimble, leading to products with improved stability, safety, and ease-of-use profiles. It is expected that improvements in the chemistry, manufacturing, and control (CMC) aspects will lead to greater probability of success at translating discoveries into biologic drug products available in the market, thereby substantially reducing the drug development costs. This in turn will lead to greater utilization of biopharmaceuticals for treating human diseases and increased market share. Our increased collective experience with biologic drugs calls for creation of databases, both public and proprietary, that contain self-consistent experimental data on cell line expression, biophysical stability, immunogenicity, PK/PD, and drug safety, along with calculated molecular properties of biologic molecules. Such databases shall enable mining for desirable drug product characteristics and suggest strategies for incorporating them into the future biologic drug candidates. There is a need for development of reliable computational methods for simulating solution behavior of biologic drug molecules both *in vitro* and *in vivo*. Given the large size of biologic drug candidate molecules and our insufficient understanding of fundamental biochemistry of protein folding, protein–protein interactions, and aggregation, there are several opportunities for advances in both experimental and computational biophysical sciences. Such research needs to be prioritized, as it can be of much practical utility. For example, screening the amino acid sequences and preferably structural models in the antibody libraries and next-generation sequencing (NGS) outputs with respect to potential physicochemical degradation sites very early in the drug discovery projects can substantially reduce the risk of attrition in biologic drug discovery and development cycles. The ability to rapidly and reliably simulate conformational stability of biologic candidates in a wide variety of formulation buffers, pH values, and excipients can considerably reduce the need for pH, buffer, and excipient screening experiments during early formulation stability studies.

ACKNOWLEDGMENTS

The authors thank Pfizer BT for computational facilities. A postdoctoral fellowship for P.M.B. by Pfizer WRD is acknowledged. S.K. acknowledges numerous helpful discussions with colleagues both within and outside Pfizer on the topic of using computational tools for biologic drug development.

SUMMARY

This chapter describes various applications of computational tools and methods toward the development of biologic drug products. A new term, *biopharmaceutical informatics*, is introduced. Biopharmaceutical informatics is an emerging field where scientists seek to understand the challenges faced during the development of biologic drug products, such as recombinant human proteins, monoclonal antibodies and antibody drug conjugates, and so forth, in terms of their molecular sequence and structural properties using computational tools and methods. Applications of computational modeling and simulations during early stages of biologic drug discovery and formulation can facilitate developability risk assessments for the biopharmaceutical candidates. Such risk-based assessments can be used to prioritize candidates for development when multiple candidates are available. Alternatively, these assessments can be used to optimize a biopharmaceutical candidate for cost-effective drug development and improved safety via molecular redesign. The experimental data collected during multiple drug development projects can also be analyzed to find correlations with molecular attributes of the candidates, leading to greater understanding of desirable protein sequence–structural characteristics that potentially make biologic drug development efficient and cost-effective. It is anticipated that the process of biopharmaceutical drug development will become more efficient as these tools mature and their applications become more widespread in the biotechnology and pharmaceutical industry. Although this chapter focuses on the most successful class of biopharmaceuticals, namely, monoclonal antibodies, the tools and approaches described here can also be applied to other modalities.

REFERENCES

1. Reichert, J.M. Antibodies to watch in 2014. *MAbs*, 2014; 6(1): 5–14.
2. Zurdo, J. Developability assessment as an early de-risking tool for biopharmaceutical development. *Pharm. Bioprocess.*, 2013; 1(1): 29–50.
3. Scannell, J.W., et al. Diagnosing the decline in pharmaceutical R&D efficiency. *Nat. Rev. Drug Discov.*, 2012; 11(3): 191–200.
4. Wang, X., et al. Potential aggregation prone regions in biotherapeutics: A survey of commercial monoclonal antibodies. *MAbs*, 2009; 1(3): 254–67.
5. Wang, X., S.K. Singh, and S. Kumar. Potential aggregation-prone regions in complementarity-determining regions of antibodies and their contribution towards antigen recognition: A computational analysis. *Pharm. Res.*, 2010; 27(8): 1512–29.
6. Buck, P.M., et al. Computational methods to predict therapeutic protein aggregation. *Methods Mol. Biol.*, 2012; 899: 425–51.

7. Kumar, S., et al. Coupling of aggregation and immunogenicity in biotherapeutics: T- and B-cell immune epitopes may contain aggregation-prone regions. *Pharm. Res.*, 2011; 28(5): 949–61.

8. Buck, P.M., S. Kumar, and S.K. Singh. On the role of aggregation prone regions in protein evolution, stability, and enzymatic catalysis: Insights from diverse analyses. *PLoS Comp. Biol.*, 2013; 9(10): e1003291.

9. Vicini, P., and P.H. van der Graaf. Systems pharmacology for drug discovery and development: Paradigm shift or flash in the pan? *Clin. Pharmacol. Ther.*, 2013; 93(5): 379–81.

10. Danhof, M., et al. Mechanism-based pharmacokinetic-pharmacodynamic (PK-PD) modeling in translational drug research. *Trends Pharmacol. Sci.*, 2008; 29(4): 186–91.

11. Kumar, S., S.K. Singh, and M.M. Gromiha. Temperature dependent molecular adaptations in microbial proteins: Lessons for structure-based biotherapeutics design and development. In *Wiley Encyclopedia of Industrial Biotechnology*, ed. M. Flickinger. Wiley & Sons, 2010, pp. 4647–61.

12. Thangakani, A.M., et al. How do thermophilic proteins resist aggregation? *Proteins*, 2012; 80(4): 1003–15.

13. Motono, C., M.M. Gromiha, and S. Kumar. Thermodynamic and kinetic determinants of *Thermotoga* maritima cold shock protein stability: A structural and dynamic analysis. *Proteins*, 2008; 71(2): 655–69.

14. Kumar, S., and R. Nussinov. Different roles of electrostatics in heat and in cold: Adaptation by citrate synthase. *Chembiochem*, 2004; 5(3): 280–90.

15. Kumar, S., C.J. Tsai, and R. Nussinov. Thermodynamic differences among homologous thermophilic and mesophilic proteins. *Biochemistry*, 2001; 40(47): 14152–65.

16. Kumar, S., C.J. Tsai, and R. Nussinov. Factors enhancing protein thermostability. *Protein Eng.*, 2000; 13(3): 179–91.

17. Kumar, S., et al. Contribution of salt bridges toward protein thermostability. *J. Biomol. Struct. Dyn.*, 2000; 17(Suppl. 1): 79–85.

18. Harris, L.J., et al. Refined structure of an intact IgG2a monoclonal antibody. *Biochemistry*, 1997; 36(7): 1581–97.

19. Rose, P.W., et al. The RCSB Protein Data Bank: New resources for research and education. *Nucleic Acids Res.*, 2013; 41(database issue): D475–82.

20. Berman, H.M., et al. The Protein Data Bank. *Nucleic Acids Res.*, 2000; 28(1): 235–42.

21. Creighton, T.E. *Proteins: Structures and Molecular Properties.* 2nd ed. New York: W.H. Freeman, 1993.

22. McNally, E.J., and J.E. Hastedt, eds. *Protein Formulation and Delivery.* Drugs and the Pharmaceutical Sciences Series, 2nd ed., vol. 175. New York: Informa HealthCare, 2008, p. 351.

23. Bummer, P.M. Chemical considerations in protein and peptide stability. In *Protein Formulation and Delivery*, ed. E.J. McNally and J.E. Hastedt. Drugs and the Pharmaceutical Sciences Series, 2nd ed., vol. 175. New York: Informa Healthcare, 2008.

24. Kumar, S., X. Wang, and S.K. Singh. Identification and impact of aggregation-prone regions in proteins and therapeutic monoclonal antibodies. In *Aggregation of Therapeutic Proteins.* John Wiley & Sons, 2010, pp. 103–18.

25. Agrawal, N.J., et al. Aggregation in protein-based biotherapeutics: Computational studies and tools to identify aggregation-prone regions. *J. Pharm. Sci.*, 2011; 100(12): 5081–95.

26. Nelson, R., et al. Structure of the cross-beta spine of amyloid-like fibrils. *Nature*, 2005;435(7043): 773–78.

27. Sawaya, M.R., et al. Atomic structures of amyloid cross-beta spines reveal varied steric zippers. *Nature*, 2007; 447(7143): 453–57.

28. Thangakani, A.M., et al. Distinct position-specific sequence features of hexa-peptides that form amyloid-fibrils: Application to discriminate between amyloid fibril and amorphous beta-aggregate forming peptide sequences. *BMC Bioinformatics*, 2013; 14(Suppl. 8): S6.

29. Thangakani, A.M., et al. GAP: Towards almost hundred percent prediction for beta-strand mediated aggregating peptides with distinct morphologies. *Bioinformatics*, 2014.

30. Tsolis, A.C., et al. A consensus method for the prediction of "aggregation-prone" peptides in globular proteins. *PLoS One*, 2013; 8(1): e54175.

31. Conchillo-Sole, O., et al. AGGRESCAN: A server for the prediction and evaluation of "hot spots" of aggregation in polypeptides. *BMC Bioinformatics*, 2007; 8: 65.

32. Fernandez-Escamilla, A.M., et al. Prediction of sequence-dependent and mutational effects on the aggregation of peptides and proteins. *Nat. Biotechnol.*, 2004; 22(10): 1302–6.

33. Maurer-Stroh, S., et al. Exploring the sequence determinants of amyloid structure using position-specific scoring matrices. *Nat. Methods*, 2010; 7(3): 237–42.

34. Walsh, I., F. Seno, S.C.E. Tosatto, and A. Trovato. PASTA 2.0: An improved server for protein aggregation prediction. *Nucleic Acids Res.*, 42 (web server issue): 2014; W301-7.

35. Saphire, E.O., et al. Crystal structure of a neutralizing human IGG against HIV-1: A template for vaccine design. *Science*, 2001; 293(5532): 1155–59.

36. Tartaglia, G.G., et al. Prediction of aggregation rate and aggregation-prone segments in polypeptide sequences. *Protein Sci.*, 2005; 14(10): 2723–34.

37. Lopez de la Paz, M., and L. Serrano. Sequence determinants of amyloid fibril formation. *Proc. Natl. Acad. Sci. USA*, 2004; 101(1): 87–92.

38. Skolnick, J., H. Zhou, and M. Gao. Are predicted protein structures of any value for binding site prediction and virtual ligand screening? *Curr. Opin. Struct. Biol.*, 2013; 23(2): 191–97.

39. Mullins, J.G. Structural modelling pipelines in next generation sequencing projects. *Adv. Protein Chem. Struct. Biol.*, 2012; 89: 117–67.

40. Kuroda, D., et al. Computer-aided antibody design. *Protein Eng. Des. Sel.*, 2012; 25(10): 507–21.

41. Miller, A.K., et al. Characterization of site-specific glycation during process development of a human therapeutic monoclonal antibody. *J. Pharm. Sci.*, 2011; 100(7): 2543–50.

42. Chennamsetty, N., et al. Aggregation-prone motifs in human immunoglobulin G. *J. Mol. Biol.*, 2009; 391(2): 404–13.

43. Chennamsetty, N., et al. Design of therapeutic proteins with enhanced stability. *Proc. Natl. Acad. Sci. USA*, 2009; 106(29): 11937–42.

44. Li, L., et al. Concentration dependent viscosity of monoclonal antibody solutions: Explaining experimental behavior in terms of molecular properties. *Pharm. Res.*, 2014; in press.

45. Brummitt, R.K., et al. Non-native aggregation of an IgG1 antibody in acidic conditions: Part 1: Unfolding, colloidal interactions, and formation of high-molecular-weight aggregates. *J. Pharm. Sci.*, 2011; 100(6): 2087–103.

46. Brummitt, R.K., D.P. Nesta, and C.J. Roberts. Predicting accelerated aggregation rates for monoclonal antibody formulations, and challenges for low-temperature predictions. *J. Pharm. Sci.*, 2011.

47. Manning, M.C., et al. Stability of protein pharmaceuticals: An update. *Pharm. Res.*, 2010; 27(4): 544–75.

48. Nelson, R., and D. Eisenberg. Recent atomic models of amyloid fibril structure. *Curr. Opin. Struct. Biol.*, 2006; 16(2): 260–65.

49. Nelson, R., and D. Eisenberg. Structural models of amyloid-like fibrils. *Adv. Protein Chem.*, 2006; 73: 235–82.

50. Maas, C., et al. A role for protein misfolding in immunogenicity of biopharmaceuticals. *J. Biol. Chem.*, 2007; 282(4): 2229–36.

51. Sahin, E., et al. Comparative effects of pH and ionic strength on protein-protein interactions, unfolding, and aggregation for IgG1 antibodies. *J. Pharm. Sci.*, 2010; 99(12): 4830–48.

52. Kayser, V., et al. A screening tool for therapeutic monoclonal antibodies: Identifying the most stable protein and its best formulation based on thioflavin T binding. *Biotechnol. J.*, 2012; 7(1): 127–32.

53. Buck, P.M., S. Kumar, and S.K. Singh. Insights into the potential aggregation liabilities of the b12 Fab fragment via elevated temperature molecular dynamics. *Protein Eng. Des. Sel.*, 2013; 26(3): 195–206.

54. Wang, X., et al. Impact of deglycosylation and thermal stress on conformational stability of a full length murine igG_{2a} monoclonal antibody: Observations from molecular dynamics simulations. *Proteins*, 2013; 81(3): 443–60.

55. Buck, P.M., S. Kumar, and S.K. Singh. Consequences of glycan truncation on Fc structural integrity. *MAbs*, 2013; 5(6): 904–16.

56. Nelson, A.L., E. Dhimolea, and J.M. Reichert. Development trends for human monoclonal antibody therapeutics. *Nat. Rev. Drug Discov.*, 2010; 9(10): 767–74.

57. Baker, M.P., et al. Immunogenicity of protein therapeutics: The key causes, consequences and challenges. *Self Nonself*, 2010; 1(4): 314–322.

58. Singh, S.K. Impact of product-related factors on immunogenicity of biotherapeutics. *J. Pharm. Sci.*, 2011; 100(2): 354–87.

59. Jawa, V., et al. T-cell dependent immunogenicity of protein therapeutics: Preclinical assessment and mitigation. *Clin. Immunol.*, 2013; 149(3): 534–55.

60. De Groot, A.S., and L. Moise. Prediction of immunogenicity for therapeutic proteins: State of the art. *Curr. Opin. Drug Discov. Devel.*, 2007; 10(3): 332–40.

61. Bryson, C.J., T.D. Jones, and M.P. Baker. Prediction of immunogenicity of therapeutic proteins: Validity of computational tools. *Biodrugs*, 2010; 24(1): 1–8.

62. Koren, E., et al. Clinical validation of the "*in silico*" prediction of immunogenicity of a human recombinant therapeutic protein. *Clin. Immunol.*, 2007; 124(1): 26–32.

63. Rombach-Riegraf, V., et al. Aggregation of human recombinant monoclonal antibodies influences the capacity of dendritic cells to stimulate adaptive T-cell responses *in vitro*. *PLoS One*, 2014; 9(1): e86322.

64. Schafer, J.R., et al. Prediction of well-conserved HIV-1 ligands using a matrix-based algorithm, EpiMatrix. *Vaccine*, 1998; 16(19): 1880–84.

65. Ponomarenko, J., et al. ElliPro: A new structure-based tool for the prediction of antibody epitopes. *BMC Bioinformatics*, 2008; 9: 514.

66. van Schouwenburg, P.A., et al. Adalimumab elicits a restricted anti-idiotypic antibody response in autoimmune patients resulting in functional neutralisation. *Ann. Rheum. Dis.*, 2013; 72(1): 104–9.

67. Emini, E.A., et al. Induction of hepatitis A virus-neutralizing antibody by a virus-specific synthetic peptide. *J. Virol.*, 1985; 55(3): 836–39.

68. Haste Andersen, P., M. Nielsen, and O. Lund. Prediction of residues in discontinuous B-cell epitopes using protein 3D structures. *Protein Sci.*, 2006; 15(11): 2558–67.

69. Lee, H.Y., et al. Simulation and prediction of the adaptive immune response to influenza A virus infection. *J. Virol.*, 2009; 83(14): 7151–65.

70. Chen, X., T.P. Hickling, and P. Vicini. A mechanistic, multi-scale mathematical model of immunogenicity for therapeutic proteins: Part 2: Model applications. *CPT Pharmacometrics Syst. Pharmacol.*, 2014. 3: in press.

71. Sauerborn, M., et al. Immunological mechanism underlying the immune response to recombinant human protein therapeutics. *Trends Pharmacol. Sci.*, 2010; 31(2): 53–59.

72. Harding, F.A., et al. The immunogenicity of humanized and fully human antibodies: Residual immunogenicity resides in the CDR regions. *MAbs*, 2010; 2(3): 256–65.

73. Kumar, S., et al. Relationship between potential aggregation-prone regions and HLA-DR-binding T-cell immune epitopes: Implications for rational design of novel and follow-on therapeutic antibodies. *J. Pharm. Sci.*, 2012; 101(8): 2686–701.

74. Lee, K.H., K.W. Wucherpfennig, and D.C. Wiley. Structure of a human insulin peptide-HLA-DQ8 complex and susceptibility to type 1 diabetes. *Nat. Immunol.*, 2001; 2(6): 501–7.

75. Ivanova, M.I., et al. Molecular basis for insulin fibril assembly. *Proc. Natl. Acad. Sci. USA*, 2009; 106(45): 18990–95.

76. Fineberg, S.E., et al. Immunological responses to exogenous insulin. *Endocr. Rev.*, 2007; 28(6): 625–52.

77. Brange, J., et al. Toward understanding insulin fibrillation. *J. Pharm. Sci.*, 1997; 86(5): 517–25.

78. Dische, F.E., et al. Insulin as an amyloid-fibril protein at sites of repeated insulin injections in a diabetic patient. *Diabetologia*, 1988; 31(3): 158–61.

79. Kumar, S., S. Zhou, and S.K. Singh. Metal ion leachates and the physico-chemical stability of biotherapeutic drug products. *Curr. Pharm. Des.*, 2014; 20(8): 1173–81.

80. Galush, W.J., L.N. Le, and J.M. Moore. Viscosity behavior of high-concentration protein mixtures. *J. Pharm. Sci.*, 2012; 101(3): 1012–20.

81. Liu, J., et al. Reversible self-association increases the viscosity of a concentrated monoclonal antibody in aqueous solution. *J. Pharm. Sci.*, 2005; 94(9): 1928–40.

82. Yadav, S., et al. Establishing a link between amino acid sequences and self-associating and viscoelastic behavior of two closely related monoclonal antibodies. *Pharm. Res.*, 2011; 28(7): 1750–64.

83. Chaudhri, A., et al. Coarse-grained modeling of the self-association of therapeutic monoclonal antibodies. *J. Phys. Chem. B*, 2012; 116(28): 8045–57.

84. Chaudhri, A., et al. The role of amino acid sequence in the self-association of therapeutic monoclonal antibodies: Insights from coarse-grained modeling. *J. Phys. Chem. B*, 2013; 117(5): 1269–79.

85. Plimpton, S.J. Fast parallel algorithms for short-range molecular dynamics. *J. Comp. Phys.*, 1995; 117: 1–19.

86. Barbosa, M.D.F.S., et al. Biosimilars and biobetters as tools for understanding and mitigating immunogenicity of biotherapeutics. *Drug Discov. Today*, 2012; 17(23–24): 1282–1288.

87. Buck, P.M., Chaudhari, A., Kumar, S. and Singh, S.K. Highly viscous antibody solutions are a consequence of network formation caused by domain–domain electrostatic complementarities: Insights from coarse-grained simulations. *Molec. Pharmaceut.*, 2015; 12(1): 127–29.

2 Computational Methods in the Optimization of Biologic Modalities

*Surjit B. Dixit**

CONTENTS

2.1 INTRODUCTION

Proteins and their complexes can be appreciated as molecular machines with an elaborate three-dimensional (3D) structure, carrying out particular tasks along the complex biological pathways effective in the physiology of the organism.[1] Such a

---

* Zymeworks, Inc., Vancouver, British Columbia, Canada

perspective of proteins—their structure, interactions, and function—is an essential element of most modern drug discovery efforts. The developments in biological and biochemical sciences in the last century have fostered drug development activity away from empirical discovery and toward a rational knowledge- and data-driven approach.[2] Albeit, it is thought provoking that a serendipitous find remains a valuable element of drug discovery[3] and is an indicator of the complexity of the biological organism. Work across the areas of genomics and molecular biology combined with biochemistry, crystallography, and physical chemistry of proteins is providing the necessary knowledge base to advance a mechanistic perspective of protein action and its role in human diseases.

As we learn more about proteins and their behavior, the concept of structure in the structure–function relationship has expanded beyond what is typically observed in x-ray crystallography, the most robust technique to generate detailed structural insight on proteins. There is a growing appreciation of proteins as conformationally dynamic entities.[4] For example, protein–ligand recognition, a central event in the mechanistic pathway, is better appreciated as an event involving the capture of a preferred conformation from an ensemble of geometries available to the interacting proteins, rather than the rigid lock-and-key-like fit of associating partners proposed by Emil Fischer in 1894.[5,6] Intrinsic conformational change, induced by binding partners, and aspects of the kinetics and free energy cost associated with such conformational change are being recognized as critical components of recognition.[7] Thus, along the structure–function axis of protein activity, the system is best appreciated by taking into consideration structure, dynamic, and energetic aspects as well. Conformational dynamics leads to the contemplation of proteins as ensembles of conformational substates on a free energy landscape with barriers and dynamic transitions between them. Characterization of conformational diversity intrinsic to the protein and the equilibria between the conformational substates is important for appreciating factors contributing to the protein's stability and binding affinity. In this regard, *in silico* protein modeling and simulation combine computational algorithms with empirical data, sequence, and structural information to develop a detailed and realistic appreciation of the system and its dynamics. This allows one to forge critical and novel physicochemical insights, expanding into the paradigm of dynamic structure–function relationship.[8,9] Molecular modeling and computational simulations present a unique advantage in studying these dynamic and thermodynamic aspects of proteins and their complexes. The growing value of computational biomolecular modeling and simulation is emphasized by the 2013 Nobel Prize awarded to Martin Karplus, Michael Levitt, and Arieh Warshel for their work in this field.[10] As the field of drug discovery and development is expanding from the domain of small molecules to protein therapeutics,[11] *in silico* rational protein engineering is also expanding its role in biologic development as a critical tool that can complement other means of experimental explorations.

Use of rational protein engineering methods can be valuable for optimizing therapeutic protein modalities addressing a range of issues, including stability, efficacy, pharmacokinetics, pharmacodynamics, and expression productivity. While traditional quality by design (QbD) efforts have focused on process control and optimization in the development of therapeutics to achieve product quality, there have been calls to expand QbD to the entire drug development process.[12] A structure-guided rationale for the engineered design of therapeutic proteins could provide valuable scientific insight aimed at de-risking their development, the philosophy at the core of QbD.[13] This chapter will review the uses of computational methods in the rational structure and knowledge-driven approach to drug discovery and development, with emphasis on engineering the modality of protein-based therapeutic action and activity.

2.2 COMPUTATIONAL TECHNOLOGIES FOR PROTEIN THERAPEUTIC OPTIMIZATION

While the field of computer-aided drug design has a very well-established role in small-molecule drug (SMD) discovery,[14] the field of *in silico* biologic design is growing to establish its place. Figure 2.1 presents various pieces of technology necessary to efficiently use a computational platform for rational structure-guided biologic engineering, with the goal of predicting mutations that achieve the biophysical or functional properties of interest. Accurately modeling the proteins and their variant analogs, capturing diverse aspects of their conformational flexibility and interactions with other proteins and molecular partners, understanding the energy criteria that control and define these properties, and developing software and hardware to make this approach tractable is a multidisciplinary effort. The components in the figure are represented as puzzle pieces to emphasize the fact that while there might be extensive and detailed work addressing each piece, it is challenging to bring the diverse pieces together into an integrated technology platform.

2.2.1 MODEL BUILDING AND STRUCTURAL ANALYSIS

The molecular modeling infrastructure for preparing the 3D structure and performing analysis of the model or simulation output, is a core piece of the technology. Modeling and simulation are highly susceptible to the adage "garbage in, garbage out," and in this regard, structure preparation addresses the artifacts of a typical crystallographically derived structure and reads on the quality management aspects of modeling.[15, 16] Assigning protonation states, especially for systems that show pH-dependent activity, is an important element of model quality.[17] The ability to recognize and explicitly model critical water molecules and ions on the protein surface or interior is another factor that can impact model quality.[18, 19] The literature is rich with examples where proteins have been shown to employ bound water molecules as extensions of their amino acid side chains to achieve specific functionality. Comparative modeling to obtain novel protein structure is a critical technique in this tool chain and will be addressed in more detail in the context of antibody structure modeling later in this chapter.

Structural analysis using molecular modeling infrastructure can bring to light a number of relevant features that could impact properties of the protein and contribute to the knowledge toward discovery and optimization steps. Figure 2.2 presents

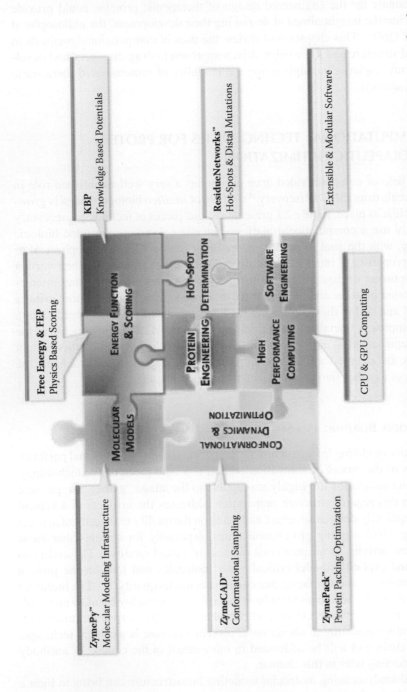

FIGURE 2.1 Components of a computational platform for molecular optimization of protein therapeutics. The figure presents areas of molecular simulation addressed at Zymeworks in order to develop a tool chain that can be employed for rational structure and computational modeling-guided engineering of candidate protein therapeutics. The development and application of the computational platform for biologic optimization is a multidisciplinary exercise calling upon expertise in the areas of chemistry, physics, structural biology, mathematics, data mining, and computer science.

a list of features in the protein structure that help guide the structure–function relational understanding. Details of the individual methods for protein structure analysis can be found in other places.[20,21] Machine learning approaches based on this data-rich analysis can be applied to draw novel insights on protein structure–function relationships. The importance of individual or groups of residues on the basis of such detailed structural analysis can be used to computationally define hotspots or locations in the protein that may be advantageous or detrimental to protein engineering. Good quality high-resolution crystal structures and the use of novel room temperature x-ray crystallography provide direct evidence that atoms in the structure can occupy multiple positions, thereby presenting multiple alternate conformational states. There is leading information suggesting that the allosteric correlations between these conformational states of residues are able to form functionally relevant residue networks within the protein structure.[22,23] Developments in theoretical molecular simulations and their applications are capable of providing such intricate details about protein structure dynamics and its contribution to functional activity.

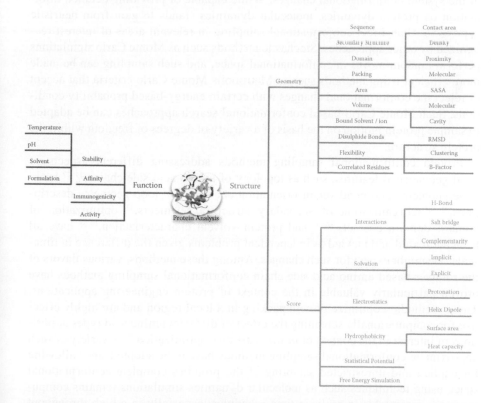

FIGURE 2.2 A "mindmap" of problems and approaches in relation to protein structure and model analysis. While protein optimization is typically directed at stability or affinity improvements, the approach would involve delving into structure or geometry aspects, along with information on their energetic aspects.

2.2.2 SIMULATION OF CONFORMATIONAL CHANGE AND DYNAMICS

Modeling a protein and the structural changes associated with introduced mutations requires use of appropriate sampling approaches to understand the potential
conformations adopted by the protein and its mutants. The use of simulations for
modeling protein conformation change helps expand the utility of the structural
information derived by methods such as crystallography. The added information
is in terms of insights about the system in altered states, effected by either conditions extrinsic or intrinsic to the protein. From an algorithmic standpoint, protein
conformational sampling methods can be classified as deterministic, stochastic,
or combined deterministic and stochastic approximations.[24] Deterministic simulations involve algorithms that evolve the system in a predictable manner. In a deterministic approach such as molecular dynamics, atoms or particles in the system
are simulated on the basis of Newton's laws of motion, with the particle motion
being determined by the forces acting on it as a result of interactions with other
particles in the system. Such an effective force and the resultant displacement are
computed iteratively during short time steps in order to predict the time evolution
of the system conformational changes. While capable of providing detailed information on protein dynamics, molecular dynamics stands to gain from heuristic
approaches that direct conformational sampling in relevant areas of interest recognized by other approaches. Stochastic methods such as Monte Carlo simulations
perform random moves in conformational space, and such sampling can be made
more efficient using methods such as Metropolis Monte Carlo criteria that accept
or reject the conformational changes with certain energy-based probability conditions. Such Monte Carlo-based conformational search approaches can be adapted
to sample protein motion on the basis of a variety of degrees of freedom within the
protein structure.

Protein conformational sampling methods addressing different aspects of
local geometrical features, such as topology of amino acid side chains,[24-27] backbone geometry and bond vector orientation changes,[28-30] loop geometry descrip-
tions,[31, 32] reorganization of secondary structural elements,[33] optimization of
domain–domain contacts,[34-37] and protein–solvent characterization,[19, 38] have all
been addressed and tackled as independent problems given the difference in timescale one might expect for such changes. Among these methods, various flavors of
the rotamer-based amino acid side chain conformational sampling methods have
proven particularly valuable in the context of protein engineering applications.
These methods reoptimize protein packing in a local region and are highly effective for computationally screening the effect of different amino acid types at positions of interest in the context of *in silico* affinity optimization.[26] A variety of such
approximate conformational sampling methods have to be adopted since allowing
for explicit and unrestricted sampling of the protein's complete conformational
space using techniques such as molecular dynamics simulations remains computationally intractable in realistic time schedules, especially in a high-throughput
manner for the hundreds of thousands of mutations one would screen in a protein
engineering effort. This is slowly changing with the growing effort to adopt highly
specialized high-performance computing hardware and graphic processing units

(GPUs) to solve such iterative numerical problems.[39,40] There is a complex link between protein structure, the nature of mutations, and their impact on stability, as a result of changes to the local interactions, flexibilities, and altered folding propensity.[41] Protein conformation modeling and molecular dynamics simulations provide valuable tools to evaluate such questions, to dissect and develop insight in regard to the structural aspects of the protein and its mutation.

2.2.3 ENERGY FUNCTION AND SCORING

The other critical factor that defines the reliability of protein simulation is the nature of the energy functions that model the interactions of particles in the protein and its environment. Typical energy functions for protein simulation are based on effective pair potentials derived from high-level *ab initio* quantum mechanical calculations, as well as by fitting to data derived from molecular analyses such as spectroscopy, crystallography, and solvation energies of small molecules that function as amino acid analogs. The physics-based potential energy function for protein simulations, popularly referred to as the force field, typically includes functional terms to capture the topological features of the protein, such as bonds, bond angles, and torsional angles. The energy associated with bond length and angle changes is described as a harmonic function. The torsional term is presented as a limited Fourier expansion to describe the stereochemical effect of rotation about bonds. Further, pair-wise non-bonded interaction terms comprise functions to model the ubiquitous but weak van der Waals (dispersion–repulsion) interactions, typically modeled by the Lennard–Jones equation, and the charge–charge (electrostatic) interactions modeled by the Coulomb equation.[42–45] More details and functional forms of these energy terms are presented in Table 2.1. The van der Waals interactions have a short range, which suggests one could approximate it by cutting off (i.e., avoiding) interactions beyond a certain interatomic distance. On the other hand, the electrostatic effects are much more long range. As an alternative to explicit treatment of the numerous solvent molecules (water and ion) in the protein environment, which becomes computationally demanding, the thermodynamic effects of solvation can be modeled fairly accurately on the basis of continuum model approximations of electrostatics that represent the dielectric shielding effect of a solvent. In this regard, the Poisson–Boltzmann approach to treat the solvent environment is particularly valuable for estimating electrostatic effects on solvation and binding. The treatment of hydrophobic effects, which relates to the entropic effect of solvent rearrangement around the non-polar components of the protein surface, is usually modeled on the basis of changes in the solvent-accessible surface area of the protein.[46]

The accurate scoring and ranking of individual analogs or mutants relative to each other or to the parent protein is one of the most challenging facets of the computer-aided approach. To put this in perspective, the stability of a folded protein is typically about 5–10 kcal/mol more favorable than that of a reference unfolded state, while the strength of a well-formed interaction such as one hydrogen bond could be as much as 2–9 kcal/mol.[47] As a result, even small inaccuracies in the description of these critical interactions could make simulation results quantitatively incorrect.

TABLE 2.1
Energy Terms in a Typical Force Field Used to Compute Pair-Wise Interactions in a Protein

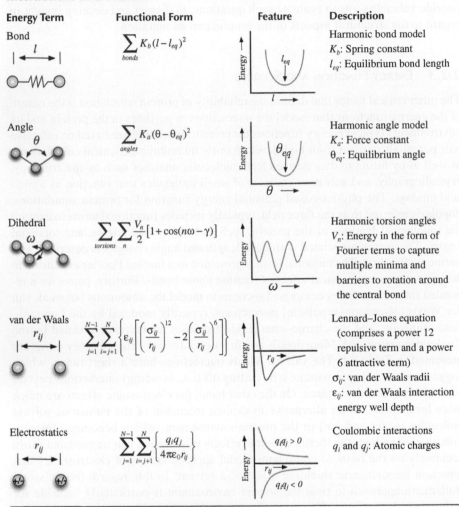

Energy Term	Functional Form	Feature	Description
Bond $\|\xleftarrow{l}\rightarrow\|$	$\sum_{bonds} K_b (l - l_{eq})^2$		Harmonic bond model K_b: Spring constant l_{eq}: Equilibrium bond length
Angle θ	$\sum_{angles} K_a (\theta - \theta_{eq})^2$		Harmonic angle model K_a: Force constant θ_{eq}: Equilibrium angle
Dihedral ω	$\sum_{torsions} \sum_n \frac{V_n}{2} [1 + \cos(n\omega - \gamma)]$		Harmonic torsion angles V_n: Energy in the form of Fourier terms to capture multiple minima and barriers to rotation around the central bond
van der Waals $\|\xleftarrow{r_{ij}}\rightarrow\|$	$\sum_{j=1}^{N-1} \sum_{i=j+1}^{N} \left\{ \varepsilon_{ij} \left[\left(\frac{\sigma_{ij}^*}{r_{ij}} \right)^{12} - 2 \left(\frac{\sigma_{ij}^*}{r_{ij}} \right)^6 \right] \right\}$		Lennard–Jones equation (comprises a power 12 repulsive term and a power 6 attractive term) σ_{ij}: van der Waals radii ε_{ij}: van der Waals interaction energy well depth
Electrostatics $\|\xleftarrow{r_{ij}}\rightarrow\|$	$\sum_{j=1}^{N-1} \sum_{i=j+1}^{N} \left\{ \frac{q_i q_j}{4\pi\varepsilon_0 r_{ij}} \right\}$		Coulombic interactions q_i and q_j: Atomic charges

Rigorous free energy and potential of mean force simulation methods that predict the thermodynamic free energy differences between states of the system have been developed, but the inherent approximations in the energy functions and the computational cost of achieving convergence in these simulations make their regular use extremely challenging.[48] Residue or atomic pairing probabilities derived from the database of high-resolution 3D protein structures have been used as a means to calibrate and develop statistical potentials (also known as knowledge-based potentials) relevant to proteins.[49, 50] Such knowledge-based potentials intrinsically represent the effect of all the thermodynamic factors leading to the folded state of the protein.

In light of the fact that these computational algorithms are trying to emulate the extremely complex physics behind the molecular architecture and interactions with fairly simple effective pair potentials, there is plenty of opportunity to refine and optimize these methods so as to employ them in a predictive manner. An efficient way to draw value from computational methods, keeping in mind their limitations, is by being involved in a design cycle involving computational and structure-guided modeling to propose a structure–function hypothesis, followed by experimental evaluation of the hypothesis. Iterative feedback and training of the model with experimental data help refine the hypotheses for subsequent optimization.[51]

2.3 APPLICATION OF COMPUTATIONAL METHODS FOR PROTEIN THERAPEUTIC OPTIMIZATION

2.3.1 Target Discovery and Computational Methods

Central to the rational approach to drug discovery is the identification and extensive characterization of a biomolecule uniquely associated with the disease as a *target* for therapeutic treatment. The target is typically a protein exhibiting an altered expression profile in the diseased state, often with functional characteristics contributing to disease progression. It may be involved in a signal transduction pathway belonging to the category of receptors, enzymes, ion channels, growth factors, cytokines, nucleic acid binding proteins, or other factors that interact with these signaling elements. A drug candidate typically binds and/or modulates the activity of the target so as to impact the disease condition.

Among the many factors contributing to the growing significance of biologics to the pharmaceutical industry is the limitation in number of protein targets that can be effectively treated with small-molecule drugs (SMDs), that is, are druggable.[52] Genome and protein structural bioinformatics is playing a lead role in gaining a high-level picture of druggable protein space in the genome for target identification. It has been pointed out that limited SMD utility originates in the inadequate differentiation of structural elements in targets, differentiation necessary to allow small drug-like molecules to bind with very high affinity and specificity (druggability). Biologics, particularly antibodies, can effectively address these issues related to specific targeting. Recent statistics on clinical trial phase transitions suggest that the likelihood of approval for biologics is nearly twice that of SMDs.[53]

In the context of target identification, there is growing realization that the reality of complex biological systems is not conducive to the reductionist one-disease, one-target, one-drug modality. In spite of the tremendous utility of these mechanistic models for drug development, one cannot forget that the biology of drug action happens at the systems level via its effect on signaling cascades that involve multiple proteins working collectively.[54] There is significant impact of the drug on elements outside of the target to which it binds, and hence the target is a biochemical system rather than just a protein. This introduces complex toxicity issues, making the translational aspect of target validation extremely challenging. This also opens up the opportunity for data- and information-driven drug discovery and optimization approaches. The emphasis on target validation feeds into the strong interest to develop biobetters and best-in-class biologics, an area that is well served by rational drug discovery and optimization efforts. Biobetters, also referred to as biosuperiors, are potential next-generation therapeutics comprised of versions of known biotherapeutics against known targets. Biobetters are usually engineered to improve functional or physicochemical properties with notable impact on efficacy.[55] A good number of therapeutics served by computational optimizations fall in the category of biobetters.

Systems biology is a growing avenue of bioinformatics with applications in the appreciation of targets and development of novel biologics.[56] A systems approach to the numerous molecular players in the biochemical networks that are active in the cellular environment can help gain a deeper understanding of complex diseases such as cancer. Modeling such biological networks relies on empirical data derived from a variety of high-throughput "omics" techniques combined with sophisticated mathematical and statistical models of these dynamic networks. Analysis of these models can help generate new predictive insights into the biological system and opportunities to apply that knowledge toward drug discovery activity. Schoeberl and colleagues[57] have employed computational models of ErbB signaling pathways to capture the activity of various known ligands and inhibitors in this system and used the information to identify targets and features of interest in a therapeutic candidate. Systems biology could play a valuable role in the development of personalized medicine and in the appreciation of resistance mechanisms evolved by disease-causing cells.[58,59]

2.3.2 MODALITIES OF BIOLOGIC ACTION

The use of biologics can be broadly classified into three therapeutic groups based on their modalities of action.[60] The first therapeutic group is comprised of exogenous or recombinantly derived enzymes or regulatory proteins used to treat patients with deficient or abnormal endogenous protein. Recombinant cytokines or growth factors can be used to augment natural activity in certain disease conditions. The second class involves use of antibodies or other specific target binding proteins that may interfere with the activity of the target protein or pathway it is involved in or mediate the delivery of other compounds or proteins. The antibody is a very versatile molecule involved in numerous functionally relevant molecular interactions that can be important from a therapeutic design perspective (Figure 2.3). The emerging field of engineered bispecific antibodies represents a novel category under the class of target binding proteins wherein the drug is able to concurrently bind two different targets in order to achieve its activity. In the subsequent sections, we will review the use of in silico protein engineering techniques to optimize therapeutics in these two broad classes of biologics. The third class of biologics is comprised of protein vaccines. While there are a number of established in silico techniques supporting this third class of biologics, such as in the design of vaccines, identification of T-cell epitopes, and so forth, this avenue is not covered in this chapter.

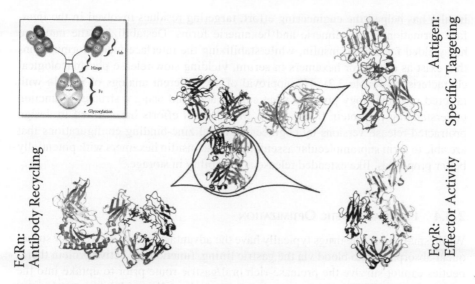

FIGURE 2.3 The antibody is a versatile multifunctional macromolecule capable of various functionally important protein–protein interactions involving its different regions. The figure presents representative structures of the antibody domains (subunits) with various other proteins, in particular, complexes of the Fc domain with FcRn and FcγR proteins, as well as a representative antigen–Fab complex structure involving interactions via the CDRs at the tip of the variable domains (Protein Data Bank [PDB] IDs: 1HZH, 1A1X, 1E4K). The availability of such detailed structural information at the atomic level provides opportunity for structure-guided optimization of these interactions. The inset figure presents a schematic representation of the antibody.

2.3.3 OPTIMIZATION OF INSULIN

The most prominent example of a regulatory protein being employed as a therapeutic agent is insulin. The optimization and development of insulin and insulin analogs for diabetes management exemplify the opportunities available for rationally engineered biological therapeutics toward achieving better modalities of action. Insulin is involved in functions related to glucose metabolism and control (homeostasis), deficient to various degrees in different types of diabetes patients. Active insulin in humans is a post-translationally modified protein comprising two chains (A and B), 21 and 30 amino acids long, respectively, cross-linked by three cysteine disulfide bonds. The crystallographic structure of insulin was among the first proteins to be solved by x-ray diffraction.[61] Insulin can exist as a monomer, dimer, or hexamer in solution. The hexamer configuration includes bound Zn^{2+} ions and functions as a store of insulin in the pancreatic cells, while the dissociated form is biologically active. In order to recapitulate the physiologic pattern of insulin secretion in the therapeutic setting and achieve glucose homeostasis, various protein-engineered analogs of insulin that alter its therapeutic modality, particularly its pharmacokinetic properties, have been developed.[62] The available structural information on

insulin has helped the engineering effort, targeting residues involved in the interface formation of the dimeric and hexameric forms. Destabilizing the interface
has yielded fast-acting insulin, while stabilizing the interface yields formulations
that exist as dimers or hexamers in serum, yielding slow-release pharmacological
characteristics (Table 2.2). The approval of six different analogs of insulin with
tailored pharmacology portrays what can be achieved using a structure–function
understanding of protein therapeutics. More recent efforts have aimed to design
protracted-release versions by engineering novel zinc-binding configurations that
are able to form supramolecular assemblies of the insulin hexamers with potentially
better properties, like extended release and stability in storage.[63]

2.3.4 Pharmacokinetic Optimization

While small-molecule drugs typically have the advantage of oral delivery for subsequent absorption into blood via the gastric lining, functionally active protein therapeutics cannot survive the protease-rich oral/gastric route prior to uptake into the
circulatory system. Protein therapeutics are typically delivered by parenteral routes
(subcutaneous or intravenous injection), and for this reason, frequent dosing is not
a very convenient option to maintain necessary serum concentration. Significant
effort has been put into developing rational protein engineering methods that can
improve the pharmacokinetic (PK) properties of therapeutic proteins, and hence
improve their bioavailability and convenience of use. Once in the serum, the fate
of the protein therapeutic is a function of its size, as well as physical and chemical
properties, such as charge, hydrophobicity, and shape. Filtration by the kidneys for
proteins smaller than 60 kDa, receptor-mediated clearance, and an immunogenic
response with the development of anti-drug antibodies are prominent routes for
removal from the circulatory system.[64] A number of small therapeutically relevant
proteins, including insulin, human growth hormone (rHGH), colony-stimulating
factor (CSF), erythropoietin (EPO), interferons α-2a and α-2b, and human inter-
leukin (IL-11), have all been shown to have a short plasma half-life when used as
therapeutics. Increasing the size and hydrodynamic radius of the therapeutic protein
provides a means to rationally improve the serum half-life. One of the approaches
to achieve a biobetter version of protein therapeutic in this regard has been to
conjugate the protein with an inert polymer, such as polyethylene glycol (PEG).
PEGylated versions of recombinant G-CSF and interferons α-2a and α-2b, and
adenosine deaminase are approved for clinical use and show improved PK properties. The potential impact of such conjugation on target binding and pharmacodynamics activity is an obvious follow-up concern in such an effort. Structure-guided
approaches to engineer the PEGylation site without unfavorably impacting therapeutic efficacy can be undertaken.[65,66]

In the case of EPO, scientists at Amgen developed a glycoengineered version
that comprises five amino acid substitutions to introduce two new N-glycosylation
sites by introducing the consensus site N-X-S/T sequence, in addition to the three
N-glycosylation and one O-glycosylation sites present in the parent recombinant
protein. These substitutions result in a version of EPO with five carbohydrate chains

TABLE 2.2

Insulin Analogs, Their Engineering and Modalities of Action

Insulin Analog	Year and Manufacturer	Description	Effect of Modification	Duration of Action
NPH insulin	1950 Eli Lilly & Co.	Porcine/bovine insulin		
Humulin	1982 Genentech/Eli Lilly & Co.	Recombinant human insulin		
Lispro/Humalog®	1996 Eli Lilly & Co.	Pro_B30 → Lys; Lys_B28 → Pro	Impairs dimerization	Short (3–4 h)
Aspart/NovoLog®	2000 Novo-Nordisk	Pro_B28 → Asp	Impairs dimerization	Short (3–5 h)
Glargine/Lantus®	2000 Sanofi-Aventis	Insert Arg_B31-Arg_B32; Asp_A21 → Gly	Shifts pI and oligomer solubility	Long (~24 h)
Glulisine/Apidra®	2004 Sanofi-Aventis	Asn_B3 → Lys; Lys_B29 → Glu	Impairs oligomerization	Short (4 h)
Detemir/Levemir®	2006 Novo-Nordisk	Lys_B29 tethered to fatty acid	Binds to albumin	Long (up to 24 h)
Degludec/Tresiba®	2013 Nono-Nordisk	Delete Thr_B30; Glu linked to Lys_B29; thapsic acid ligated to Glu	Multihexamer assembly	Long-acting

and up to 22 sialic acid residues when produced recombinantly. As a result of the addition of extra carbohydrate chains and negative charge on the molecule due to the sialic acid, the serum half-life of the molecule is improved almost threefold, and hence shows better efficacy.[67] At the time this work was carried out, detailed structural information was not available, and hence the researchers screened several dozen analogs with the consensus site to recognize two functionally active positions that were subsequently combined in the same sequence to achieve the analog of interest with two new carbohydrate chains. From a structural perspective, the introduction of the glycosylation site would require selecting positions where the introduced carbohydrate moiety would remain solvent exposed and not significantly impact the local protein conformation. Structure-guided approaches to efficiently introduce novel glycosylation sites with desired improvements in the serum half-life of other therapeutic candidates have been attempted.[68, 69]

2.3.5   FcRn Receptor and Biologics

Designing fusion of therapeutic protein candidates with the antibody Fc is another strategy to improve their serum half-life. Alternately, fusion to human serum albumin, or peptides or other small molecules that bind to albumin, can enhance the serum half-life of therapeutic candidates. The antibody Fc and albumin interact with the FcRn receptors on the endothelial surface in a pH-dependent fashion, binding at acidic pH and releasing at physiological or neutral pH. In the process, antibody Fc and albumin escape lysosomal degradation following pinocytosis, the fate of most proteins in serum.[70–72] Structural analysis coupled with experimental screening has been employed to guide the development of structure–function models of this interaction and further optimize this fascinating rescue mechanism.[73] Some of these molecular events are extremely subtle and it can be challenging to model the overall effect, calling for high accuracy with regard to the simulation of conformational change and coupled changes in the protonation state and binding. A recent study indicated that functionally active mutations could be quite distant from the Fc-FcRn interface.[74] Such information is guiding the design and engineering of variants with altered pharmacokinetic properties with the goal of potentially employing them to develop biosuperior versions of blockbuster antibodies with longer serum persistence. On the other hand, in some therapeutic designs, amino acid mutations could be engineered to manipulate FcRn binding such that the therapeutic or therapeutic–target complex is preferentially sorted for degradation.[75]

2.3.6   Fcγ Receptor (FcγR) Affinity Modulation

Given the broad utility of Fc in both antibodies and Fc fusion molecules, a great deal of effort has gone into engineering the Fc for tailored functional activity. In particular, the ability of the antibody to bind the target (cell) and recruit immune effector cells via the Fc to mediate target killing is an important modality of antibody-based therapeutic action. The FcγRs present on immune cells can be engaged by antibody

Fc in an immune complex to induce effector activity, such as antibody-dependent cellular cytotoxicity (ADCC) or antibody-dependent cellular phagocytosis (ADCP), against the target cell.[76] Similarly, Fc interactions with the C1q protein of the complement system can induce complement-dependent cytotoxicity (CDC) activity. The immune effector cells express a variety of FcγRs on their cell surfaces that are notably homologous in the extracellular region recognized by the Fc domain. These receptors differ in the intracellular immunoreceptor tyrosine kinase motif they contain, which can play either an activating (ITAM) or inhibiting (ITIM) signaling role. In humans, the FcγRI (CD64), FcγRIIa (CD32a), and FcγRIIIa (CD16a) contain the ITAM motif and induce an activating immune response, while FcγRIIb (CD32b) has an ITIM motif capable of negatively regulating the B-cell immune response. Another category of the Fcγ receptor, FcγRIIIb lacks the tyrosine kinase motif, but instead has a glycophosphatidyl-inositol (GPI) anchor. Depending on the type of immune cells relevant to a disease condition and their Fcγ receptor expression level, as well as the relative affinities of the Fc for each of these receptors, the degree of engagement of immune cells and their response can be modulated. A number of structure-guided antibody engineering efforts have been undertaken, with the goal of achieving Fc analogs that modulate immune effector activity.[77-79] Engineering the Fc to selectively bind the FcγRIIIa or FcγRIIa has yielded antibody analogs with increased activator/cell killing activity, while the FcγRIIb has been targeted to achieve an inhibitory response.[80] Given the extensive use of the Fc fusion strategy to primarily improve the PK properties of the therapeutic candidate, structure-guided strategies have also been adopted to engineer mutations to knock out the native effector activity induced by Fc-FcγR interactions.

2.3.7 Fc Fusion and Dimerization

Apart from the FcRn- and FcγR-based activity rendered by an Fc-containing molecule, engineering of therapeutic proteins or peptides as Fc fusions has also been undertaken to utilize the homodimerizing feature of the Fc and the range of antagonistic or agonistic activity these fusions can provide.[81] It has been shown that in the case of abatacept, a fusion of CTLA-4 with Fc, cross-arm binding as a dimer allows for more avid binding to its targets, CD80 and CD86. Belatacept, a rationally engineered variant of abatacept with two additional mutations, binds close to fourfold more avidly to CD86 and twofold more avidly to CD80, with a tenfold increase in potency.[82]

2.3.8 Bispecific Antibodies

Beyond its highly specific target recognition features, the ability of an antibody to concurrently interact with multiple target/receptor proteins is critical for its utility as a novel therapeutic agent relative to SMDs. As a case in point, rituximab promotes the formation of immune complexes via the simultaneous interaction of its Fab domains with the B-cell antigen CD20, while the Fc domain interacts with immune effector cells, thereby bridging the cells and facilitating an antibody-dependent cytotoxicity response.[83] The promise of achieving novel modalities of action with multispecific protein therapeutics capable of mediating contact between two or more targets or epitopes has been the driver for development of engineered therapeutic formats referred to as bispecifics.[84] The design of bispecific molecular formats or scaffolds has been a playground for protein engineering efforts, resulting in what has been referred to as a zoo of molecular formats that have undergone various levels of evaluation.[85] Transforming these diverse molecular formats into a real therapeutic candidate calls into question the developability of the engineered solutions. From a molecular platform or scaffold standpoint, developability refers to the easy usability of the scaffold in a plug-and-play manner in different systems in order to test and screen for bispecific activity with regard to the targets of interest. From a therapeutic candidate perspective, going beyond efficacy in the form of PK and pharmacodynamics (PD) of the molecule, developability addresses various other CMC aspects of product development, like manufacturability, quality control, and cost of goods, in order to achieve a product that is homogeneous, reproducible, analytically tractable, easy to formulate, and stable at the requisite drug concentration. Detailed evaluation of the fundamental biophysical properties of the protein early in development provides a great deal of insight in this regard prior to initialization of process development.[86–88]

2.3.9 HETERODIMERIC Fc ENGINEERING

The intrinsic tendency of the Fc domain to homodimerize can be altered by engineering to obtain a heterodimeric Fc that is the basis for a bispecific antibody scaffold.[89] The major advantage of such a design is that it could retain the native tertiary structure and other interaction propensities of the Fc. The knobs-into-holes (KiH) strategy employed by Carter, Presta, and coworkers at the C_H3 interface of an Fc involved introduction of a knob mutation on one chain by mutating a smaller residue with a larger one (e.g., T366W), while a hole to accommodate the knob was created in the complementary surface of the other chain with larger to smaller residue mutations (e.g., T366S, L368A, and Y407V).[90] Such a design biases the system toward preferentially forming the heterodimeric species. However, this design resulted in significant stability loss (~15°C) of the C_H3 domain relative to the wild-type Fc, which was subsequently optimized by phage display.[91] In contrast to the steric complementarity approach in the KiH designs, Gunasekaran and coworkers[92] employed a structure-guided electrostatic charge inversion (CI) design strategy to achieve selective heterodimerization of the Fc. Based on a sequence- and structure-guided engineering approach, Davis and coworkers[93] have designed a strand exchange engineered domain (SEED) C_H3 that is comprised of alternating segments of human IgA and IgG C_H3 sequences, yielding preferentially associating complementary heterodimers. Following the trend for the KiH design, heterodimeric C_H3 domains derived from the CI and SEEDbody approaches resulted in a thermal stability of ~68°C. There are advantages to developing therapeutics that address such fundamental biophysical stability loss.[87, 94, 95] Iterative structure and computational modeling–guided protein engineering combined with biophysical characterization has been employed to design mutations in the Fc that allow for exquisite pairing specificity of the two designed complementary Fc chains while retaining native antibody-like stability and developability attributes.[51] Improvements in high-throughput biophysical characterization and screening technologies can streamline such a rational optimization approach.[96]

2.3.10 ANTIBODY VARIABLE DOMAIN MODELING

Phage and other display technology approaches are the most widely used techniques to address antigen binding specificity and affinity maturation requirements in the context of antibody development for therapeutic application.[97] Computer-aided antibody design approaches are just beginning to make progress in this regard, with a limited number of success stories to show.[98] Traditionally, a great deal of effort has been put into using homology information to model antibody structure, the details of which are employed to model variants with improvements in therapeutically relevant features, such as affinity and stability.[99] Early on, structural modeling of antibody Fab was central to the humanization process.[100] This process was guided by the understanding that sequence and geometry of the complementarity determining regions (CDRs), along with a limited number of anchor residues in the framework, were critical for antigen binding. Such humanized antibodies of mouse origin constitute a good fraction of the antibodies currently licensed for clinical use in humans.

Sequence- and structure-guided homology modeling of antibody Fab structures is the area of antibody structure modeling with the broadest following in the antibody engineering community.[101] Modeling the geometry of the CDRs and the angle between the constant and variable domains are big challenges in accurately predicting antibody structure. Of the six CDR loops that define the basic hyper-variable contact space with the antigen, modeling the third loop in the heavy chain (H3), which can be fairly diverse in length, sequence and conformation, typically contributes the most toward antigen binding and is one of the most challenging aspects of antibody Fab structure modeling. The geometry of the remaining loops, which tend to be shorter, can be modeled effectively on the basis of sequence homology and their canonical geometries in other known antibody structures. Use of refined energy functions and tailored Ramachandran maps as a knowledge base to guide conformational sampling is driving the improvements in this H3 loop modeling effort.[102–104] Elements of the H3 loop geometry are also coupled to the orientation of the V_L/V_H domains in the Fab, the other critical variable in the prediction of antibody structures.[105] A bioinformatics-based appreciation of antibody structural elements is also critical in the design and construction of synthetic antibody libraries.[106]

In modeling the antibody–antigen complexes, their docking configuration and binding affinity are a bigger challenge.[107] In contrast to about 400 $Å^2$ of surface area buried upon protein–ligand binding for a typical SMD,[108] the antibody–antigen interface typically buries about 1200–2000 $Å^2$.[109] While this large surface area accessed upon complex formation enables the exquisite specificity conferred by antibodies, it renders accurate modeling of the interface for structure-guided antibody engineering quite complex. The flexible nature of the loops constituting the interface and its entropic contributions adds to the challenge of accurately modeling binding and affinity in these complexes.[110] There is a growing appreciation that distal allosteric effects could be at play in an antibody structure and its action.[111] Computational approaches have utility in developing further insight into such properties. As we recognize and develop new modalities to use antibodies as therapeutics, tailored antigen binding characteristics will be critical to rationally modify the functional activity of these therapeutic agents.[112] Structure-guided selection of residues for mutagenesis can be followed by experimental affinity maturation techniques to accelerate development of optimal variants.[113] Comparative modeling can also help better appreciate the sequence–structure–function relationships in the context of antibody–antigen binding. Early in the development of this field, a structure-guided approach was published to reengineer a humanized antibody to achieve species cross-reactivity to facilitate non-human primate study in the preclinical drug development process.[114]

2.3.11 Drug Delivery System

There has been a great deal of interest to materialize Paul Ehrlich's vision of a therapeutic magic bullet that would very specifically deliver disease-curative toxins to the target cell.[115] Such agents are expected to present better safety profiles, and thus improve the therapeutic index of the treatment, than non-targeted cytotoxic agents that have a systemic effect. For a number of years, antibodies with their highly specific targeting capability have been considered Trojan horses to deliver conjugated drug molecules to the target disease cells, and the field is showing rejuvenated interest with the recent approval of brentuximab vedotin and ado-trastuzumab emtansine.[116] In case of both these therapeutics, the antibody–drug conjugates (ADCs) show better activity than their parent antibodies, SGN-30 and trastuzumab, respectively. The conjugated molecules are able to expand the modalities of therapeutic action beyond what is achieved by the parent antibody. SGN-30 has been tested in phase II clinical trials with modest efficacy outcome,[117] and trastuzumab treatment, although revolutionary, has high rates of primary and treatment-emergent resistance impeding long-term outcomes.[118] With the goal that toxicity challenges can be managed by engineering and design of the ADC, conjugating a toxic payload with a highly specific targeting agent such as an antibody can potentially open up the utility of both antibody and small drug-like molecules that do not show much promise by themselves.

The receptor targeted by the antibody, the drug used in conjugation, the linker employed to conjugate the drug to the antibody, and the location of the conjugation site on the antibody are all important factors that determine the PK, PD, and toxicity of the ADC.[119] Achieving a homogenous product with respect to the number of drug molecules conjugated and retaining stable biophysical properties of the conjugated antibody, properties that are correlated to the therapeutic window, are critical elements in the challenge of ADC development.[120] A number of toxin molecules, such as tubulin binders and DNA intercalators, used in ADCs are highly hydrophobic in nature, requiring serious consideration of the potential destabilizing effect these might have on the conjugated antibody and the complexity of producing the conjugated material.[121] Structure-based modeling has been used to recognize positions that could be conducive for cysteine mutations, subsequently used for conjugation of the toxin molecule in antibodies referred to as Thio-Mabs.[122] In this study it was shown that the local conformational and biophysical properties at the site of conjugation impact drug release and overall activity of the ADC. Molecular dynamics studies have been attempted on these Thio-Mabs to appreciate the impact of mutations and the conjugation of small molecules on local conformation dynamics.[123] Recognizing sites for engineering mutations where the conjugation of drug molecule could be favorable calls for a detailed structure–function understanding of the drug conjugation site and the chemistries of drug conjugation and release. Modeling and simulation studies to appreciate such details would be particularly valuable in the context of novel protein engineering strategies that aim to introduce non-standard amino acids in the protein chain in order to achieve controlled and site-specific conjugation of the toxin to the antibody.[124]

Along similar lines, there has been a long-standing interest to conjugate protein toxins derived from plants and other species to antibody and the therapeutic use of these engineered proteins, referred to as immunoconjugates.[125, 126] Structure-guided approaches can aid in engineering the protein toxin molecule for various stability- and immunogenicity-related properties.[127, 128] Similarly, cytokines as key modulators of the immune system have been considered as payload for targeted action and present novel engineering opportunities.[129]

2.4 CONCLUSIONS

The utility of rational knowledge- and computational modeling-driven design and optimization of proteins for use as therapeutics is expanding with the surge of information derived from the adoption of high-throughput techniques to appreciate natural biological activity at the molecular level. Concurrent improvements in high-performance computing hardware and software design are opening the opportunities to adopt such technologies for the optimization and application of novel molecular modeling and simulation technologies to tackle problems that are hard to address via traditional approaches. The effective development and use of theoretical computational modeling approaches for therapeutic protein optimization calls for a collaborative effort between computational sciences, chemistry, biology, and experimental evaluation. Computational modeling–guided protein engineering approaches help optimize both pharmacodynamic and pharmacokinetic aspects of therapeutic candidates. Such a rational data-driven optimization procedure is particularly valuable to refine the modality of therapeutic action and in the development of best-in-class candidate solutions, addressing aspects of both molecular stability and specific binding.

ACKNOWLEDGMENT

I would like to thank Dr. Steve Seredick for his comments on the manuscript.

REFERENCES

1. Alberts B. 1998. The cell as a collection of protein machines: Preparing the next generation of molecular biologists. *Cell* 92, 291–94.
2. Stevens E. 2014. *Medicinal Chemistry: The Modern Drug Discovery Process*. Prentice Hall, NJ.
3. Hargrave-Thomas E., Yu B., and Reynisson J. 2012. Serendipity in anticancer drug discovery. *World J Clin Oncol* 3, 1–6.
4. Henzler-Wildman K. and Kern D. 2007. Dynamic personalities of proteins. *Nature* 450, 964–72.
5. Koshland D. E., Jr. and Neet K. E. 1968. The catalytic and regulatory properties of enzymes. *Annu Rev Biochem* 37, 359–410.
6. Kumar S., Ma B., Tsai C. J., Sinha N., and Nussinov R. 2000. Folding and binding cascades: Dynamic landscapes and population shifts. *Protein Sci* 9, 10–19.
7. Zhou H. X. 2010. From induced fit to conformational selection: A continuum of binding mechanism controlled by the timescale of conformational transitions. *Biophys J* 98, L15–17.
8. Brooks C. L., Karplus M., and Pettitt B. M. 1988. *Proteins: A Theoretical Perspective of Dynamics, Structure, and Thermodynamics*. John Wiley, New York.
9. Durrant J. D. and McCammon J. A. 2011. Molecular dynamics simulations and drug discovery. *BMC Biol* 9, 71.
10. Thiel W. and Hummer G. 2013. Nobel 2013 chemistry: Methods for computational chemistry. *Nature* 504, 96–97.
11. EvaluatePharma. 2014. *EvaluatePharma® World Preview 2014, Outlook to 2020*. EvaluatePharma. Boston, MA.
12. Martin-Moe S., Lim F. J., Wong R. L., Sreedhara A., Sundaram J., and Sane S. U. 2011. A new roadmap for biopharmaceutical drug product development: Integrating development, validation, and quality by design. *J Pharm Sci* 100, 3031–43.
13. Swann P. G., Tolnay M., Muthukkumar S., Shapiro M. A., Rellahan B. L., and Clouse K. A. 2008. Considerations for the development of therapeutic monoclonal antibodies. *Curr Opin Immunol* 20, 493–99.
14. Sliwoski G., Kothiwale S., Meiler J., and Lowe E. W., Jr. 2014. Computational methods in drug discovery. *Pharmacol Rev* 66, 334–95.
15. Chen V. B., Arendall W. B., 3rd, Headd J. J., Keedy D. A., Immormino R. M., Kapral G. J., Murray L. W., Richardson J. S., and Richardson D. C. 2010. MolProbity: All-atom structure validation for macromolecular crystallography. *Acta Crystallogr D Biol Crystallogr* 66, 12–21.
16. Sondergaard C. R., Garrett A. E., Carstensen T., Pollastri G., and Nielsen J. E. 2009. Structural artifacts in protein-ligand x-ray structures: Implications for the development of docking scoring functions. *J Med Chem* 52, 5673–84.
17. Li H., Robertson A. D., and Jensen J. H. 2005. Very fast empirical prediction and rationalization of protein pKa values. *Proteins* 61, 704–21.
18. Chaplin M. 2006. Do we underestimate the importance of water in cell biology? *Nat Rev Mol Cell Biol* 7, 861–66.
19. Jiang L., Kuhlman B., Kortemme T., and Baker D. 2005. A "solvated rotamer" approach to modeling water-mediated hydrogen bonds at protein-protein interfaces. *Proteins* 58, 893–904.

20. de Beer T. A., Berka K., Thornton J. M., and Laskowski R. A. 2014. PDBsum additions. *Nucleic Acids Res* 42, D292–96.
21. Leach A. R. 2001. *Molecular Modelling: Principles and Applications.* Prentice Hall.
22. Fraser J. S., van den Bedem H., Samelson A. J., Lang P. T., Holton J. M., Echols N., and Alber T. 2011. Accessing protein conformational ensembles using room-temperature x-ray crystallography. *Proc Natl Acad Sci USA* 108, 16247–52.
23. Kohn J. E., Afonine P. V., Ruscio J. Z., Adams P. D., and Head-Gordon T. 2010. Evidence of functional protein dynamics from x-ray crystallographic ensembles. *PLoS Comput Biol* 6, e1000911.
24. Schlick T. 2010. *Molecular Modeling and Simulation: An Interdisciplinary Guide.* Springer Science + Business Media, New York.
25. Desmet J., Spriet J., and Lasters I. 2002. Fast and accurate side-chain topology and energy refinement (FASTER) as a new method for protein structure optimization. *Proteins* 48, 31–43.
26. Gordon D. B., Hom G. K., Mayo S. L., and Pierce N. A. 2003. Exact rotamer optimization for protein design. *J Comput Chem* 24, 232–43.
27. Shapovalov M. V. and Dunbrack R. L., Jr. 2011. A smoothed backbone-dependent rotamer library for proteins derived from adaptive kernel density estimates and regressions. *Structure* 19, 844–58.
28. Cahill M., Cahill S., and Cahill K. 2002. Proteins wriggle. *Biophys J* 82, 2665–70.
29. Davis I. W., Arendall W. B., 3rd, Richardson D. C., and Richardson J. S. 2006. The backrub motion: How protein backbone shrugs when a sidechain dances. *Structure* 14, 265–74.
30. Mandell D. J. and Kortemme T. 2009a. Backbone flexibility in computational protein design. *Curr Opin Biotechnol* 20, 420–28.
31. Canutescu A. A. and Dunbrack R. L., Jr. 2003. Cyclic coordinate descent: A robotics algorithm for protein loop closure. *Protein Sci* 12, 963–72.
32. Coutsias E. A., Seok C., Jacobson M. P., and Dill K. A. 2004. A kinematic view of loop closure. *J Comput Chem* 25, 510–28.
33. Li X., Jacobson M. P., and Friesner R. A. 2004. High-resolution prediction of protein helix positions and orientations. *Proteins* 55, 368–82.
34. Huang P. S., Love J. J., and Mayo S. L. 2007. A de novo designed protein protein interface. *Protein Sci* 16, 2770–74.
35. Karaca E., Melquiond A. S., de Vries S. J., Kastritis P. L., and Bonvin A. M. 2010. Building macromolecular assemblies by information-driven docking: Introducing the HADDOCK multibody docking server. *Mol Cell Proteomics* 9, 1784–94.
36. Lippow S. M. and Tidor B. 2007. Progress in computational protein design. *Curr Opin Biotechnol* 18, 305–11.
37. Mandell D. J. and Kortemme T. 2009b. Computer-aided design of functional protein interactions. *Nat Chem Biol* 5, 797–807.
38. Petukhov M., Cregut D., Soares C. M., and Serrano L. 1999. Local water bridges and protein conformational stability. *Protein Sci* 8, 1982–89.
39. Shaw D. E., Maragakis P., Lindorff-Larsen K., Piana S., Dror R. O., Eastwood M. P., Bank J. A., Jumper J. M., Salmon J. K., Shan Y., and Wriggers W. 2010. Atomic-level characterization of the structural dynamics of proteins. *Science* 330, 341–46.
40. Sweet J. C., Nowling R. J., Cickovski T., Sweet C. R., Pande V. S., and Izaguirre J. A. 2013. Long timestep molecular dynamics on the graphical processing unit. *J Chem Theory Comput* 9, 3267–81.
41. Tokuriki N., Stricher F., Schymkowitz J., Serrano L., and Tawfik D. S. 2007. The stability effects of protein mutations appear to be universally distributed. *J Mol Biol* 369, 1318–32.

42. Cornell W. D., Cieplak P., Bayly C. I., Gould I. R., Merz K. M., Ferguson D. M., Spellmeyer D. C., Fox T., Caldwell J. W., and Kollman P. A. 1995. A second generation force field for the simulation of proteins, nucleic acids, and organic molecules. *J Am Chem Soc* 117, 5179–97.

43. Hornak V., Abel R., Okur A., Strockbine B., Roitberg A., and Simmerling C. 2006. Comparison of multiple Amber force fields and development of improved protein backbone parameters. *Proteins* 65, 712–25.

44. Jorgensen W. L. and Tirado-Rives J. 2005. Potential energy functions for atomic-level simulations of water and organic and biomolecular systems. *Proc Natl Acad Sci USA* 102, 6665–70.

45. Zhu X., Lopes P. E., and Mackerell A. D., Jr. 2012. Recent developments and applications of the CHARMM force fields. *Wiley Interdiscip Rev Comput Mol Sci* 2, 167–85.

46. Baker N. A. 2004. Poisson-Boltzmann methods for biomolecular electrostatics. *Methods Enzymol* 383, 94–118.

47. Fersht A. 1999. *Structure and Mechanism in Protein Science: A Guide to Enzyme Catalysis and Protein Folding*. W.H. Freeman & Company, NY.

48. Wereszczynski J. and McCammon J. A. 2012. Statistical mechanics and molecular dynamics in evaluating thermodynamic properties of biomolecular recognition. *Q Rev Biophys* 45, 1–25.

49. Vajda S., Sippl M., and Novotny J. 1997. Empirical potentials and functions for protein folding and binding. *Curr Opin Struct Biol* 7, 222–28.

50. Zhang J. and Zhang Y. 2010. A novel side-chain orientation dependent potential derived from random-walk reference state for protein fold selection and structure prediction. *PLoS One* 5, e15386.

51. Spreter Von Kreudenstein T., Lario P. I., and Dixit S. B. 2014. Protein engineering and the use of molecular modeling and simulation: The case of heterodimeric Fc engineering. *Methods* 65, 77–94.

52. Hopkins A. L. and Groom C. R. 2002. The druggable genome. *Nat Rev Drug Discov* 1, 727–30.

53. Hay M., Thomas D. W., Craighead J. L., Economides C., and Rosenthal J. 2014. Clinical development success rates for investigational drugs. *Nat Biotechnol* 32, 40–51.

54. Imming P., Sinning C., and Meyer A. 2006. Drugs, their targets and the nature and number of drug targets. *Nat Rev Drug Discov* 5, 821–34.

55. Beck A. 2011. Biosimilar, biobetter and next generation therapeutic antibodies. *mAbs* 3, 107–10.

56. Wang E. 2010. *Cancer Systems Biology*. CRC Press, Boca Raton, FL.

57. Schoeberl B., Pace E. A., Fitzgerald J. B., Harms B. D., Xu L., Nie L., Linggi B., Kalra A., Paragas V., Bukhalid R., Grantcharova V., Kohli N., West K. A., Leszczyniecka M., Feldhaus M. J., Kudla A. J., and Nielsen U. B. 2009. Therapeutically targeting ErbB3: A key node in ligand-induced activation of the ErbB receptor-PI3K axis. *Sci Signal* 2, ra31.

58. Csermely P., Korcsmaros T., Kiss H. J., London G., and Nussinov R. 2013. Structure and dynamics of molecular networks: A novel paradigm of drug discovery: A comprehensive review. *Pharmacol Ther* 138, 333–408.

59. Kell D. B. 2013. Finding novel pharmaceuticals in the systems biology era using multiple effective drug targets, phenotypic screening and knowledge of transporters: Where drug discovery went wrong and how to fix it. *FEBS J* 280, 5957–80.

60. Leader B., Baca Q. J., and Golan D. E. 2008. Protein therapeutics: A summary and pharmacological classification. *Nat Rev Drug Discov* 7, 21–39.

61. Adams M. J., Blundell T. L., Dodson E. J., Dodson G. G., Vijayan M., Baker E. N., Harding M. M., Hodgkin D. C., Rimmer B., and Sheat S. 1969. Structure of rhombohedral 2 zinc insulin crystals. *Nature* 224, 491–95.

62. Morello C. M. 2011. Pharmacokinetics and pharmacodynamics of insulin analogs in special populations with type 2 diabetes mellitus. *Int J Gen Med* 4, 827–35.
63. Pandyarajan V. and Weiss M. A. 2012. Design of non-standard insulin analogs for the treatment of diabetes mellitus. *Curr Diab Rep* 12, 697–704.
64. Vugmeyster Y., Xu X., Theil F. P., Khawli L. A., and Leach M. W. 2012. Pharmacokinetics and toxicology of therapeutic proteins: Advances and challenges. *World J Biol Chem* 3, 73–92.
65. Grace M. J., Lee S., Bradshaw S., Chapman J., Spond J., Cox S., Delorenzo M., Brassard D., Wylie D., Cannon-Carlson S., Cullen C., Indelicato S., Voloch M., and Bordens R. 2005. Site of pegylation and polyethylene glycol molecule size attenuate interferon-alpha antiviral and antiproliferative activities through the JAK/STAT signaling pathway. *J Biol Chem* 280, 6327–36.
66. Zhang C., Yang X. L., Yuan Y. H., Pu J., and Liao F. 2012. Site-specific PEGylation of therapeutic proteins via optimization of both accessible reactive amino acid residues and PEG derivatives. *Biodrugs* 26, 209–15.
67. Egrie J. C. and Browne J. K. 2002. Development and characterization of darbepoetin alfa. *Oncology* (Williston Park) 16, 13–22.
68. Perlman S., van den Hazel B., Christiansen J., Gram-Nielsen S., Jeppesen C. B., Andersen K. V., Halkier T., Okkels S., and Schambye H. T. 2003. Glycosylation of an N-terminal extension prolongs the half-life and increases the *in vivo* activity of follicle stimulating hormone. *J Clin Endocrinol Metab* 88, 3227–35.
69. Stork R., Zettlitz K. A., Muller D., Rether M., Hanisch F. G., and Kontermann R. E. 2008. N-Glycosylation as novel strategy to improve pharmacokinetic properties of bispecific single-chain diabodies. *J Biol Chem* 283, 7804–12.
70. Andersen J. T. and Sandlie I. 2009. The versatile MHC class I-related FcRn protects IgG and albumin from degradation: Implications for development of new diagnostics and therapeutics. *Drug Metab Pharmacokinet* 24, 318–32.
71. Roopenian D. C. and Akilesh S. 2007. FcRn: The neonatal Fc receptor comes of age. *Nat Rev Immunol* 7, 715–25.
72. Ward E. S. and Ober R. J. 2009. Chapter 4 multitasking by exploitation of intracellular transport functions. *Adv Immunol* 103, 77–115.
73. Yeung Y. A., Leabman M. K., Marvin J. S., Qiu J., Adams C. W., Lien S., Starovasnik M. A., and Lowman H. B. 2009. Engineering human IgG1 affinity to human neonatal Fc receptor: Impact of affinity improvement on pharmacokinetics in primates. *J Immunol* 182, 7663–71.
74. Monnet C., Jorieux S., Souyris N., Zaki O., Jacquet A., Fournier N., Crozet F., de Romeuf C., Bouayadi K., Urbain R., Behrens C. K., Mondon P., and Fontayne A. 2014. Combined glyco- and protein-Fc engineering simultaneously enhance cytotoxicity and half-life of a therapeutic antibody. *mAbs* 6, 422–36.
75. Kenanova V. E., Olafsen T., Salazar F. B., Williams L. E., Knowles S., and Wu A. M. 2010. Tuning the serum persistence of human serum albumin domain III:diabody fusion proteins. *Protein Eng Des Sel* 23, 789–98.
76. Nimmerjahn F. and Ravetch J. V. 2007. Antibodies, Fc receptors and cancer. *Curr Opin Immunol* 19, 239–45.
77. Lazar G. A., Dang W., Karki S., Vafa O., Peng J. S., Hyun L., Chan C., Chung H. S., Eivazi A., Yoder S. C., Vielmetter J., Carmichael D. F., Hayes R. J., and Dahiyat B. I. 2006. Engineered antibody Fc variants with enhanced effector function. *Proc Natl Acad Sci USA* 103, 4005–10.

78. Shields R. L., Namenuk A. K., Hong K., Meng Y. G., Rae J., Briggs J., Xie D., Lai J., Stadlen A., Li B., Fox J. A., and Presta L. G. 2001. High resolution mapping of the binding site on human IgG1 for Fc gamma RI, Fc gamma RII, Fc gamma RIII, and FcRn and design of IgG1 variants with improved binding to the Fc gamma R. *J Biol Chem* 276, 6591–604.

79. Strohl W. R. 2009. Optimization of Fc-mediated effector functions of monoclonal antibodies. *Curr Opin Biotechnol* 20, 685–91.

80. Mimoto F., Katada H., Kadono S., Igawa T., Kuramochi T., Muraoka M., Wada Y., Haraya K., Miyazaki T., and Hattori K. 2013. Engineered antibody Fc variant with selectively enhanced FcgammaRIIb binding over both FcgammaRIIa(R131) and FcgammaRIIa(H131). *Protein Eng Des Sel* 26, 589–98.

81. Czajkowsky D. M., Hu J., Shao Z., and Pleass R. J. 2012. Fc-fusion proteins: New developments and future perspectives. *EMBO Mol Med* 4, 1015–28.

82. Larsen C. P., Pearson T. C., Adams A. B., Tso P., Shirasugi N., Strobert E., Anderson D., Cowan S., Price K., Naemura J., Emswiler J., Greene J., Turk L. A., Bajorath J., Townsend R., Hagerty D., Linsley P. S., and Peach R. J. 2005. Rational development of LEA29Y (belatacept), a high-affinity variant of CTLA4-Ig with potent immunosuppressive properties. *Am J Transplant* 5, 443–53.

83. Weiner G. J. 2010. Rituximab: Mechanism of action. *Semin Hematol* 47, 115–23.

84. Segal D. M., Weiner G. J., and Weiner L. M. 1999. Bispecific antibodies in cancer therapy. *Curr Opin Immunol* 11, 558–62.

85. Kontermann R. 2012. Dual targeting strategies with bispecific antibodies. *mAbs* 4, 182–97.

86. Buchanan A., Clementel V., Woods R., Harn N., Bowen M. A., Mo W., Popovic B., Bishop S. M., Dall'acqua W., Minter R., Jermutus L., and Bedian V. 2013. Engineering a therapeutic IgG molecule to address cysteinylation, aggregation and enhance thermal stability and expression. *mAbs* 5, 255–62.

87. Demarest S. J. and Glaser S. M. 2008. Antibody therapeutics, antibody engineering, and the merits of protein stability. *Curr Opin Drug Discov Devel* 11, 675–87.

88. Yang X., Xu W., Dukleska S., Benchaar B., Mengisen S., Antochshuk V., Cheung J., Mann L., Babadjanova Z., Rowand J., Gunawan R., McCampbell A., Beaumont M., Meininger D., Richardson D., and Ambrogelly A. 2013. Developability studies before initiation of process development: Improving manufacturability of monoclonal antibodies. *mAbs* 5.

89. Carter P. 2001. Bispecific human IgG by design. *J Immunol Methods* 248, 7–15.

90. Merchant A. M., Zhu Z., Yuan J. Q., Goddard A., Adams C. W., Presta L. G., and Carter P. 1998. An efficient route to human bispecific IgG. *Nat Biotechnol* 16, 677–81.

91. Atwell S., Ridgway J. B., Wells J. A., and Carter P. 1997. Stable heterodimers from remodeling the domain interface of a homodimer using a phage display library. *J Mol Biol* 270, 26–35.

92. Gunasekaran K., Pentony M., Shen M., Garrett L., Forte C., Woodward A., Ng S. B., Born T., Retter M., Manchulenko K., Sweet H., Foltz I. N., Wittekind M., and Yan W. 2010. Enhancing antibody Fc heterodimer formation through electrostatic steering effects: Applications to bispecific molecules and monovalent IgG. *J Biol Chem* 285, 19637–46.

93. Davis J. H., Aperlo C., Li Y., Kurosawa E., Lan Y., Lo K. M., and Huston J. S. 2010. SEEDbodies: Fusion proteins based on strand-exchange engineered domain (SEED) C_H3 heterodimers in an Fc analogue platform for asymmetric binders or immunofusions and bispecific antibodies. *Protein Eng Des Sel* 23, 195–202.

94. Manning M. C., Chou D. K., Murphy B. M., Payne R. W., and Katayama D. S. 2010. Stability of protein pharmaceuticals: An update. *Pharm Res* 27, 544–75.

95. Von Kreudenstein T. S., Escobar-Carbrera E., Lario P. I., D'Angelo I., Brault K., Kelly J., Durocher Y., Baardsnes J., Woods R. J., Xie M. H., Girod P. A., Suits M. D., Boulanger M. J., Poon D. K., Ng G. Y., and Dixit S. B. 2013. Improving biophysical properties of a bispecific antibody scaffold to aid developability: Quality by molecular design. *mAbs* 5, 646–54.

96. Alsenaidy M. A., Jain N. K., Kim J. H., Middaugh C. R., and Volkin D. B. 2014. Protein comparability assessments and potential applicability of high throughput biophysical methods and data visualization tools to compare physical stability profiles. *Front Pharmacol* 5, 39.

97. Carter P. J. 2006. Potent antibody therapeutics by design. *Nat Rev Immunol* 6, 343–57.

98. Kuroda D., Shirai H., Jacobson M. P., and Nakamura H. 2012. Computer-aided antibody design. *Protein Eng Des Sel* 25, 507–21.

99. Caravella J. A., Wang D., Glaser S. M., and Lugovskoy A. 2010. Structure-guided design of antibodies. *Curr Comput Aided Drug Des.*, 6, 128–38.

100. Presta L. G. 2006. Engineering of therapeutic antibodies to minimize immunogenicity and optimize function. *Adv Drug Deliv Rev* 58, 640–56.

101. Teplyakov A., Luo J., Obmolova G., Malia T. J., Sweet R., Stanfield R. L., Kodangattil S., Almagro J. C., and Gilliland G. L. 2014. Antibody modeling assessment II. Structures and models. *Proteins* 82, 1563–82.

102. Mandell D. J., Coutsias E. A., and Kortemme T. 2009. Sub-angstrom accuracy in protein loop reconstruction by robotics-inspired conformational sampling. *Nat Methods* 6, 551–52.

103. North B., Lehmann A., and Dunbrack R. L., Jr. 2011. A new clustering of antibody CDR loop conformations. *J Mol Biol* 406, 228–56.

104. Zhu K. and Day T. 2013. *Ab initio* structure prediction of the antibody hypervariable H3 loop. *Proteins* 81, 1081–89.

105. Abhinandan K. R. and Martin A. C. 2010. Analysis and prediction of V_H/V_L packing in antibodies. *Protein Eng Des Sel* 23, 689–97.

106. Knappik A., Ge L., Honegger A., Pack P., Fischer M., Wellnhofer G., Hoess A., Wolle J., Pluckthun A., and Virnekas B. 2000. Fully synthetic human combinatorial antibody libraries (HuCAL) based on modular consensus frameworks and CDRs randomized with trinucleotides. *J Mol Biol* 296, 57–86.

107. Sivasubramanian A., Sircar A., Chaudhury S., and Gray J. J. 2009. Toward high-resolution homology modeling of antibody Fv regions and application to antibody-antigen docking. *Proteins* 74, 497–514.

108. Lijiness M. S., Vieth M., and Erickson J. 2004. Molecular properties that influence oral drug-like behavior. *Curr Opin Drug Discov Devel* 7, 470–77.

109. Janin J., Bahadur R. P., and Chakrabarti P. 2008. Protein-protein interaction and quaternary structure. *Q Rev Biophys* 41, 133–80.

110. Kastritis P. L. and Bonvin A. M. 2013. On the binding affinity of macromolecular interactions: Daring to ask why proteins interact. *J R Soc Interface* 10, 20120835.

111. Sela-Culang I., Alon S., and Ofran Y. 2012. A systematic comparison of free and bound antibodies reveals binding-related conformational changes. *J Immunol* 189, 4890–99.

112. Rudnick S. I., Lou J., Shaller C. C., Tang Y., Klein-Szanto A. J., Weiner L. M., Marks J. D., and Adams G. P. 2011. Influence of affinity and antigen internalization on the uptake and penetration of Anti-HER2 antibodies in solid tumors. *Cancer Res* 71, 2250–59.

113. Barderas R., Desmet J., Timmerman P., Meloen R., and Casal J. I. 2008. Affinity maturation of antibodies assisted by *in silico* modeling. *Proc Natl Acad Sci USA* 105, 9029–34.

114. Werther W. A., Gonzalez T. N., O'Connor S. J., McCabe S., Chan B., Hotaling T., Champe M., Fox J. A., Jardieu P. M., Berman P. W., and Presta L. G. 1996. Humanization of an anti-lymphocyte function-associated antigen (LFA)-1 monoclonal antibody and reengineering of the humanized antibody for binding to rhesus LFA-1. *J Immunol* 157, 4986–95.

115. Strebhardt K. and Ullrich A. 2008. Paul Ehrlich's magic bullet concept: 100 years of progress. *Nat Rev Cancer* 8, 473–80.

116. Lambert J. M. 2013. Drug-conjugated antibodies for the treatment of cancer. *Br J Clin Pharmacol* 76, 248–62.

117. Katz J., Janik J. E., and Younes A. 2011. Brentuximab Vedotin (SGN-35). *Clin Cancer Res* 17, 6428–36.

118. Gradishar W. J. 2013. Emerging approaches for treating HER2-positive metastatic breast cancer beyond trastuzumab. *Ann Oncol* 24, 2492–500.

119. Strop P., Liu S. H., Dorywalska M., Delaria K., Dushin R. G., Tran T. T., Ho W. H., Farias S., Casas M. G., Abdiche Y., Zhou D., Chandrasekaran R., Samain C., Loo C., Rossi A., Rickert M., Krimm S., Wong T., Chin S. M., Yu J., Dilley J., Chaparro-Riggers J., Filzen G. F., O'Donnell C. J., Wang F., Myers J. S., Pons J., Shelton D. L., and Rajpal A. 2013. Location matters: Site of conjugation modulates stability and pharmacokinetics of antibody drug conjugates. *Chem Biol* 20, 161–67.

120. Sievers E. L. and Senter P. D. 2013. Antibody-drug conjugates in cancer therapy. *Annu Rev Med* 64, 15–29.

121. Chari R. V., Miller M. L., and Widdison W. C. 2014. Antibody-drug conjugates: An emerging concept in cancer therapy. *Angew Chem Int Ed Engl* 53, 3796–827.

122. Shen B. Q., Xu K., Liu L., Raab H., Bhakta S., Kenrick M., Parsons-Reponte K. L., Tien J., Yu S. F., Mai E., Li D., Tibbitts J., Baudys J., Saad O. M., Scales S. J., McDonald P. J., Hass P. E., Eigenbrot C., Nguyen T., Solis W. A., Fuji R. N., Flagella K. M., Patel D., Spencer S. D., Khawli L. A., Ebens A., Wong W. L., Vandlen R., Kaur S., Sliwkowski M. X., Scheller R. H., Polakis P., and Junutula J. R. 2012. Conjugation site modulates the *in vivo* stability and therapeutic activity of antibody-drug conjugates. *Nat Biotechnol* 30, 184–89.

123. Kortkhonjia E., Brandman R., Zhou J. Z., Voelz V. A., Chorny I., Kabakoff B., Patapoff T. W., Dill K. A., and Swartz T. E. 2013. Probing antibody internal dynamics with fluorescence anisotropy and molecular dynamics simulations. *mAbs* 5, 306–22.

124. Panowksi S., Bhakta S., Raab H., Polakis P., and Junutula J. R. 2014. Site-specific antibody drug conjugates for cancer therapy. *mAbs* 6, 34–45.

125. Chaddock J. A. and Acharya K. R. 2011. Engineering toxins for 21st century therapies. *FEBS J* 278, 899–904.

126. Govindan S. V. and Goldenberg D. M. 2012. Designing immunoconjugates for cancer therapy. *Expert Opin Biol Ther* 12, 873–90.

127. Kurnikov I. V., Kyrychenko A., Flores-Canales J. C., Rodnin M. V., Simakov N., Vargas-Uribe M., Posokhov Y. O., Kurnikova M., and Ladokhin A. S. 2013. pH-triggered conformational switching of the diphtheria toxin T-domain: The roles of N-terminal histidines. *J Mol Biol* 425, 2752–64.

128. Reiter Y., Brinkmann U., Webber K. O., Jung S. H., Lee B., and Pastan I. 1994. Engineering interchain disulfide bonds into conserved framework regions of Fv fragments: Improved biochemical characteristics of recombinant immunotoxins containing disulfide-stabilized Fv. *Protein Eng* 7, 697–704.

129. Vazquez-Lombardi R., Roome B., and Christ D. 2013. Molecular engineering of therapeutic cytokines. *Antibodies* 2, 426–51.

3 Understanding, Predicting, and Mitigating the Impact of Post-Translational Physicochemical Modifications, Including Aggregation, on the Stability of Biopharmaceutical Drug Products

Neeraj J. Agrawal and *Naresh Chennamsetty†*

CONTENTS

* Amgen, Inc., Thousand Oaks, California
† Bristol-Myers Squibb Company, New Brunswick, New Jersey

Biopharmaceutical molecules such as monoclonal antibodies (mAbs) undergo a series of complex processing steps to obtain the final product. Specifically, protein-based biotherapeutics can undergo steps such as production, harvest, purification, refolding, freeze–thaw, drying, formulation, filling, nebulization, and shipping to obtain the final product. These processing steps and factors such as high concentrations, variable temperatures, pH extremes, varying ionic strength, agitation, light, shear stresses, air–liquid interface, and a variety of solid–liquid interfaces subject the drug molecules to many stresses and cause them to degrade.[1] Advances in analytical chemistry have identified many degradation pathways that occur in protein therapeutics over time. These pathways, which depend not only on the above-mentioned factors, but also on the protein sequence–structure, generate either physical instability or chemical instability, or both. The physical degradation pathway includes unfolding, dissociation, denaturation, precipitation, and aggregation, while the chemical degradation pathway includes oxidation, deamidation, aspartate isomerization, and peptide bond hydrolysis. Quite often, protein degradation pathways are synergistic; that is, a chemical degradation event triggers a physical degradation event, such as when oxidation of a labile residue in the protein is followed by protein aggregation. These degradations not only lead to protein drug product heterogeneity, but also might lead to immunogenicity when administered to patients, reduced target binding, altered pharmacokinetics, and so forth.[2,3]

In this chapter, the current understanding and computational tools for prediction of degradation, such as aggregation-, oxidation-, and deamidation-prone sites, are discussed. Since, among these pathways, protein aggregation is one of the most common pathways of degradation, a large section of this chapter is devoted to computational tools for addressing protein aggregation.

3.1 PROTEIN AGGREGATION

Individual protein molecules can come together in several different ways to form oligomers/polymers resulting in soluble aggregates, subvisible and visible particulates, fibers, and precipitates.[4] Protein aggregation has been observed both *in vivo* (e.g., formation of amyloids in human diseases such as Alzheimer's, Parkinson's, and Huntington's and aggregation during expression of recombinant protein) and *in vitro* (e.g., during manufacturing, storage, and handling of protein products). Since protein aggregation, both *in vivo* and *in vitro*, is of significant concern, a large number of experimental and computational studies have been performed on the topic of protein aggregation and broadly they identify five general mechanisms of protein aggregation:[5] (1) native aggregation, that is, aggregation of native protein; (2) non-native aggregation, that is, aggregation of conformationally altered protein; (3) aggregation of chemically modified protein; (4) nucleation-mediated aggregation; and (5) surface-induced aggregation. Historically, aggregation of conformationally altered protein via an amyloid-like aggregation pathway has been well studied in protein science due to its importance in neurodegenerative diseases. Amyloid formation by globular protein under native conditions (conditions under which proteins are initially folded) requires a global or partial (or local) unfolding of the protein. Recently, protein aggregation via the native pathway, which does not require partial or complete unfolding of the protein, has also been studied for biopharmaceuticals.

In general, computational studies on protein aggregation can be divided into two broad categories: (1) studies to understand the aggregation mechanisms and (2) studies to predict potential aggregation-prone regions (APRs) in proteins.

3.1.1 COMPUTATIONAL TOOLS TO UNDERSTAND THE AGGREGATION MECHANISM

A number of computational studies have been performed using molecular modeling techniques such as coarse-grained models and atomistic models to elucidate the mechanisms of protein aggregation and the factors that drive protein aggregation.[6–9] Most of these studies have focused on small peptides known to aggregate via the amyloid pathway, though recent studies involving full-length protein aggregation via the non-amyloid pathway have also emerged.[10–12]

Coarse-grained lattice models have been frequently used to study protein aggregation.[13–19] In one simulation study, simple two-dimensional hydrophobic polar (HP) lattice models were used to study the effect of denaturant and protein concentration on aggregation.[13,20] These simulations reported that aggregation begins via association of the partially folded intermediates, and the native HP toy protein with hydrophobic beads exposed to the surface is susceptible to aggregation, especially when solution conditions favor partially unfolded conformations. Hence, strengthening hydrophobic interactions within the molecules could suppress aggregation by stabilizing their native states. In another simulation study, more detailed interaction parameters for each amino acid were used to study protein aggregation. Cellmer et al. used renormalized Miyazawa–Jernigan potentials, which are knowledge-based potentials, for representing interactions among the amino acids of a 64-mer three-dimensional (3D) lattice protein with a specific sequence.[16] Their simulations showed that the higher protein concentration led to destabilization of folding landscapes and increased preference for the misfolded states. These misfolded states were accompanied by increased interchain interactions, which appear to compensate for the loss of intrachain interactions.

Atomistic simulations have also been used to study protein aggregation. A fully atomistic molecular dynamics (MD) simulation performed on AGAAAAGA, a highly amyloidogenic short peptide from Syrian hamster prion protein (ShPrP) (residues 113–120), and its analog AAAAAAAA (A8) showed that β-sheet oligomers containing six to eight strands are stable.[21] Another MD study, performed using an intermediate resolution model to represent peptides, reported that poly-alanine peptides can undergo spontaneous aggregation to form amyloid fibrils.[14] These atomistic simulations have also been used to explore the effect of sequence context on the seeding oligomer size and conformation in aggregation of amyloidogenic short peptides.[22–24] For example, atomistic simulation along with the high-temperature MD simulations correctly reproduced antiparallel β-sheet orientations for $A\beta_{16–22}$ (KLVFFAE), as seen in the solid-state nuclear magnetic resonance (NMR) experiments.[23]

Atomistic simulations have also been performed to study the effect of environmental factors such as the pH of the peptide aggregation. MD simulations on $A\beta_{1–28}$ and $A\beta_{1–40}$ peptides at low, medium, and neutral pH in explicit solvent conditions showed that both peptides become highly flexible near their isoelectric point. $A\beta_{1-40}$ showed greater flexibility than $A\beta_{1-28}$ due to the hydrophobic tail.[25] The authors concluded that conformational flexibility, determined by both hydrophobicity and charge effect, is the main mechanistic determinant of aggregation propensity. Another set of atomistic simulations has been performed to elucidate the relative contributions of polypeptide backbone and amino acid side chains toward initiation and propagation of aggregates. Several studies have demonstrated that though polypeptide backbone drives amyloidosis, the specific side chain sequences can affect the kinetics and thermodynamics of aggregation.[26–28]

3.1.2 COMPUTATIONAL TOOLS TO PREDICT AGGREGATION-PRONE REGIONS

A plethora of experimental and computational studies have shown that a few (sequence- or structure-based) specific regions on proteins drive protein aggregation.[29–35] These regions, termed aggregation-prone regions (APRs), tend to have unique features with respect to charge, hydrophobicity, aromaticity, and secondary structural preference that can be used to delineate them from the rest of the protein. Both bioinformatics and phenomenological models have been developed that identify unique features of APRs using a training data set and use this knowledge to predict unknown APRs in proteins. Most of these computational methods use only the protein sequence as input to identify short APRs of five to nine residues capable of forming amyloid-like fibrils. Other methods based on pattern recognition, 3D protein structure, and molecular simulations are also emerging.[34–42] A summary of available *in silico* tools for identification of APRs in proteins is provided in Table 3.1, and below we discuss three methods—TANGO, PAGE, and SAP—in more detail since applications of these three methods for engineering biotherapeutic molecules with lower aggregation propensity have been published in the literature.[4,11]

3.1.2.1 TANGO and PAGE

Both TANGO and PAGE are sequence-based APR prediction bioinformatics tools. While these tools have been extensively validated against amyloidogenic proteins, a few recent studies have also shown successful application of these tools against biotherapeutic molecules such as antibodies.[4] TANGO is a statistical mechanics-based, algorithm and it calculates aggregation scores based on the principles of β-sheet formation.[53] It assumes that the core regions of the nucleating β-aggregating regions are fully buried in the aggregate core and satisfy their hydrogen bond potential. This method also considers competing conformational states such as the α-helix, β-strand, turn, random coil, and β-aggregates. Apart from the protein sequence, this method also accounts for environmental factors such as pH, protein concentration, and ionic strength in the calculation of the aggregation propensity score. Like TANGO, PAGE is also a sequence-based tool, and it predicts aggregation propensity and the absolute aggregation rate by sliding a small window of five to nine residues along the protein sequence.[55] Aggregation scores calculated by PAGE are based on aromaticity, β-strand propensity, charge, solubility, and the average polar/non-polar accessible surface area of each residue in a given window.

TABLE 3.1
Computational Tools to Predict APRs in Peptides and Proteins

Tool	Input	Brief Description	Description of Validation
AGGRESCAN[43]	Sequence	Intracellular aggregation propensity for mutants of $A\beta_{42}$ peptide (mutants were generated by single point mutation in residues 17–21). The algorithm was parameterized on $A\beta_{42}$ peptide mutants.	Experimental data on 24 fibrillar deposition-linked polypeptides were used to validate this method.
AMYLPRED[44]	Sequence	Uses consensus among five different methods to predict APRs.	Validated using experimentally known amyloidogenic short stretches in 18 proteins involved in amyloidoses.
FoldAmyloid[45]		Expected probability of hydrogen bond formation and expected packing density are used to identify amyloidogenic peptides.	Validated on a data set of 407 peptides (144 amyloidogenic and 263 non-amyloidogenic peptides).
GAP[46]	Sequence	Energy potential based on distinct preferences of adjacent and alternate position residue pairs in amyloid fibrils and amorphous β-aggregating hexapeptides used for machine learning. Trained on 139 amyloid fibril-forming hexapeptides and 168 amorphous β-aggregating hexapeptides.	Validated on 428 non-redundant hexa- and longer amyloid fibril-forming peptides.
Packing density[47]	Sequence	A number of neighboring residues are used as a measure of packing density for a given amino acid. The average packing density for each amino acid was derived from protein crystal structures.	Validated using 12 amyloid-forming peptides and proteins.
Pafig[42]	Sequence	Support vector machine learning-based method trained on 2452 hexapeptides.	—
PAGE[48]	Sequence	Aggregation propensity calculated based on aromaticity, β-strand propensity, and charge. The algorithm was parameterized using peptides found in diseases causing amyloidogenic proteins.	Experimental data on a number of diseases causing amyloidogenic proteins were used to validate.

(Continued)

TABLE 3.1 (CONTINUED)
Computational Tools to Predict APRs in Peptides and Proteins

Tool	Input	Brief Description	Description of Validation
PASTA[49,50]	Sequence	Pair-wise interaction potentials for a pair of residues to be found facing each other in a cross-β-motif. The interaction potentials were determined from a data set of 500 high-resolution globular protein crystal structures.	Experimental data on 179 peptides used to validate TANGO were used in this case too.
ProA-SVM/ ProA-RF[51]	Sequence	Physicochemical features important to protein aggregation identified by support vector machine and random forest methods using a data set on 354 peptides.	Standard 10-fold and leave-one-protein-out cross-validation methods used on the 354-peptide data set.
SALSA[52]	Sequence	Chou–Fasman β-strand propensity. The algorithm was parameterized using a protein data set.	α-Synuclein, Aβ, and Tau were used to validate this approach.
SAP[11]	Structure	SAP finds the effective dynamically exposed hydrophobicity of a certain patch on the protein surface. High SAP values are indicative of aggregation-prone regions.	mAbs, which were redesigned based on the SAP prediction, showed reduced aggregation propensity in experiments.
TANGO[53]	Sequence	Statistical mechanics-based method. Takes into account physicochemical principles behind β-sheet formation. The algorithm was parameterized using short aggregating and non-aggregating peptides.	Experimental data on a set of 179 peptides were used to validate this method.
Waltz[54]	Sequence	A combination of physicochemical properties of β-sheet formation with position specific substitution matrices. The algorithm was parameterized using data on 200 aggregating and non-aggregating hexapeptides.	Waltz correctly predicted 7 out of 12 peptides in the sup35 N-terminal domain to be amyloidogenic.
Zyggregator[55–57]/ AggreSolve	Sequence	Relative propensities for folding and aggregation in a given sequence region. The algorithm was parameterized using a data set of short peptides.	A number of known amyloidogenic peptides were used to validate.
3D Profile[37]	Sequence	Molecular modeling-based method that evaluates compatibility of a sequence with the crystal structure of hexapeptide NNQQNY. The algorithm uses the Rosetta energy function.	Strongly predicted peptides were shown to form fibrils in experimental studies.

3.1.2.2 SAP

Unlike TANGO and PAGE, SAP (i.e., Spatial Aggregation Propensity) is a structure-based APR prediction tool. SAP identifies dynamically exposed hydrophobic patches on the protein surface as APRs. Using a protein 3D structure as an input, this method identifies hydrophobic patches, that is, clusters of exposed hydrophobic residues. Since this method is based on the protein structure, an ensemble of structures generated via molecular dynamics simulation can also be used in SAP to identify hydrophobic patches that become exposed during normal protein fluctuations. The SAP-identified APRs are not the amyloidogenic sequence patterns; in fact, these APRs consist of residues that may or may not be contiguous in an amino acid sequence.

3.1.3 COMPUTATIONAL STRATEGIES LEADING TO STABLE BIOTHERAPEUTICS

Identification of APR via computational tools can be leveraged to design biotherapeutic molecules that are stable against aggregation. These computational methods can be used to either engineer proteins that are stable against aggregation or identify the most stable protein from a pool of proteins. The computational tools can also be used to design excipients for mitigation of protein aggregation.

3.1.3.1 Aggregation Mitigation via Protein Engineering

Identification of APR via computational tools can be leveraged to design biotherapeutic molecules that are stable against aggregation by disruption of these APRs. However, an obvious concern with performing mutation to reduce aggregation is that the mutation can affect the protein binding affinity to its partner and protein biophysical properties such as thermal stability. For example, any mutation in the complementarity determining regions (CDRs) of an antibody performed for increasing the antibody's aggregation stability can potentially affect the antibody's binding affinity for its antigen.

So far, only a handful of independent studies have been published on identifying APR in biotherapeutic molecules such as mAbs. Apart from these published studies, there is some evidence that commercially available computational tools from Lonza, Accelrys, Molecular Operations Environment (MOE), and Schrödinger can also identify APRs in mAbs. In one study, sequence-based tools such as TANGO and PAGE were used to identify several APRs in mAbs.[58] Many of these APRs coincided with the antigen binding region of the mAb, and hence a rational structure-based approach was recommended for disruption of these potential APRs in the CDR loops.[59] In another study, SAP, which is a structure-based tool, was used to identify several APR in mAbs.[11] Figures 3.1 and 3.2 show the APR predicted by TANGO/PAGE and SAP, respectively, for a full-length human IgG1 mAb (Protein Data Bank [PDB] ID: 1HZH[60]). A few of the APRs are predicted by both TANGO/PAGE and SAP. For example, APR 302-VVSVLTVL-309 predicted by TANGO/PAGE in the Fc region overlaps with the APR containing L309 predicted by SAP. Mutation of L309 to a Lys residue can be performed to disrupt this APR, and thus to stabilize the mAb against aggregation. In fact, a L309K variant of an IgG1 mAb showed enhanced stability against aggregation under accelerated aggregation experiments.[11] Additional variants generated by mutating SAP-predicted APR also

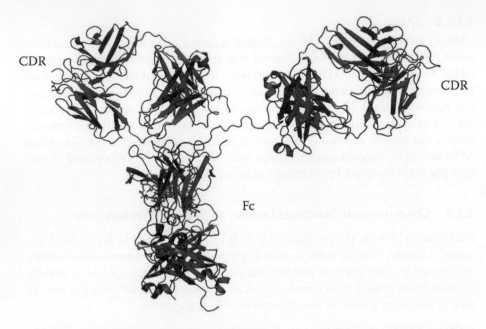

FIGURE 3.1   The TANGO/PAGE-predicted APR (red) for antibody b12 (PDB ID: 1HZH[60]). (From X. Wang et al., *mAbs* 1(3):254–267, 2009.)

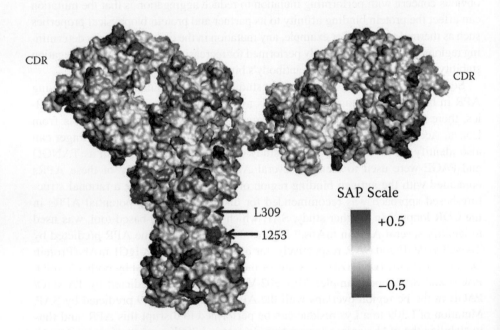

FIGURE 3.2   The SAP-predicted APR (red) for antibody b12 (PDB ID: 1HZH[60]). (From N. Chennamsetty et al., *Proc Natl Acad Sci USA* 106(29):11937–11942, 2009.)

showed enhanced stability. However, a few of these variants affected the binding of antibody to its partners. For example, though variant I253K showed very low levels of aggregation, it showed poor binding to protein A, which is used for purification of antibodies. Similarly, for another IgG1 antibody, while mutations of SAP-predicted residues in the CDR improved the stability of these variants, most of the variants lost their binding to the antigen.

3.1.3.2 Aggregation Mitigation via Screening for Stable Proteins

Identification of APR via computational tools can also be leveraged for ranking of proteins according to their aggregation propensity, and thus identification of stable proteins. This strategy can be used, instead of engineering, when the APR is present in a region, which is important for protein function. For example, as discussed in the previous section, engineering of an APR present in the CDR of an antibody often leads to loss of binding to the antigen. In such a case, the APR identification computational tools can be applied to perform fast developability (or manufacturability) screening of therapeutic mAbs coming out of the discovery phase. In a published study based on a limited number of mAbs, the SAP tool was extended to rank these antibodies according to their aggregation propensity. This extended version, termed the developability index (DI), combined the effect of exposed hydrophobic patches (as measured by the SAP tool) and the protein net charge for screening the proteins according to their aggregation propensity. Such computational tools can be used to identify stable and unstable drug candidates from the mAbs coming out of the discovery phase. The stable candidates can be moved forward to the development, while the unstable candidates can be either discarded or engineered for stability.

3.1.3.3 Aggregation Mitigation via Excipient Design

The excipients, such as sugars, polyols, amino acids, and polymers, aid in stabilizing the protein solutions, especially when the protein itself cannot be modified to decrease aggregation. Such a situation occurs when the aggregation-prone regions are also involved in antigen binding (e.g., CDRs in antibodies could be responsible for aggregation as well as antigen binding[61]). Different proteins require a different combination of excipients for stabilization. These excipients are typically selected using trial-and-error methods and empirically derived heuristics. This is both tedious and expensive. Computational modeling can aid in the rational selection of excipients and can give a mechanistic understanding of how different excipients work in inhibiting aggregation.

The dominant influence of the excipient arises from whether it is attracted or repelled from the protein surface.[62] Thus, if the concentration of the excipient in the local domain around a protein differs from the concentration in the bulk solution, significant changes in the thermodynamic properties of the protein will arise that influence solubility and conformational stability.[62,63] Such behavior is quantified using the preferential interaction coefficient, Γ_{23}, which is a measure of the preference an excipient has for the protein surface.[64,65] Γ_{23} is defined by the following expression:

$$\Gamma_{23} = \left(\frac{\partial m_3}{\partial m_2}\right)_{T,P,\mu_3} = -\left(\frac{\partial \mu_2}{\partial \mu_3}\right)_{T,P,m_2} \tag{3.1}$$

where m, T, P, and μ represent molal concentration, temperature, pressure, and chemical potential, respectively, and the subscripts indicate solution components in Scatchard notation: water (subscript 1), protein (subscript 2), or excipient (subscript 3).[66] Additives with a positive Γ_{23} are preferentially bound to the protein surface due to an increase in the concentration of the co-solute in the local domain, and this favorable interaction, as indicated by Equation 3.1, lowers the chemical potential of the protein. The opposite is true for excipients with a negative Γ_{23}, which are preferentially excluded from the surface of the protein. The preferential binding or exclusion can arise from either non-specific interactions with the protein surface (e.g., volume exclusion or perturbation of surface free energy) or specific interactions with the protein surface (e.g., electrostatic interactions, hydrogen bonding, cation-π interactions with aromatic residues, hydrophobic interactions, and solvophobic effects).[67]

Computational all-atom models were used to calculate the preferential interaction coefficient for several protein and excipient combinations.[67–71] For instance, the preferential interaction coefficient was used to explain the molecular mechanism behind the synergistic effect of the excipient mixture of arginine and glutamic acid on protein aggregation.[71] Here, computational modeling showed that the synergistic effect is related to the relative increase in the number of arginine and glutamic acid molecules around the protein in the equimolar mixtures due to additional hydrogen bonding interactions between the excipients on the surface of the protein when both excipients are present. The presence of these additional molecules around the protein leads to enhanced crowding, which suppresses the protein association. Thus, computer models based on a preferential interaction coefficient can give estimates for the effect of excipients as well as insight into the molecular mechanisms involved.

3.2 PROTEIN CHEMICAL MODIFICATIONS

A majority of biopharmaceuticals approved or in clinical trials bear some form of post-translational modification (PTM), which can profoundly affect protein properties relevant to their therapeutic application.[72] These PTMs include oxidation, deamidation, glycation, glycosylation, disulfide scrambling, and so forth. These PTMs could pose significant challenges in the development of therapeutic proteins and impact stability, potency, pharmacokinetic and pharmacodynamics parameters, and safety and immunogenicity.[73] Experimental detection of PTM sites is often labor-intensive and limited by the availability of enzymatic reactions. Here we discuss the computational modeling tools developed for identifying the potential PTMs and degradation sites based on protein sequence or structure.

3.2.1 OXIDATION

Oxidation is a major problem for therapeutic proteins,[74–77] which can adversely impact their safety and efficacy.[78–81] Oxidation can occur at a number of residues, including methionine, cysteine, tryptophan, tyrosine, and histidine residues.[82,83] The most liable residue among them is methionine, which typically oxidizes into methionine sulfoxide (Figure 3.3).[84,85] Proteins often have multiple methionine residues,

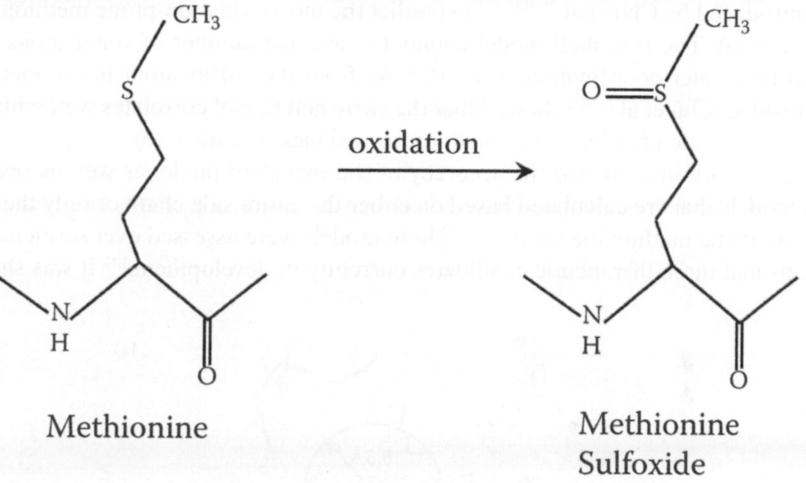

CH$_3$

S

oxidation →

CH$_3$

O=S

N
H

O

Methionine

N
H

O

Methionine
Sulfoxide

FIGURE 3.3   The oxidation of methionine into methionine sulfoxide.

each of which may oxidize to a different extent. The experimental approaches to assess the most oxidation prone residues, such as peptide mapping and liquid chromatography–tandem mass spectrometry (LC/MS-MS), are both tedious and expensive.[86,87] Computational modeling can complement experiments and aid in predicting the most oxidation-prone residues.

There are multiple sources for methionine oxidation, such as metal catalysts, peroxides, or exposure to light.[86–88] The site of oxidation can differ based on the source of oxidation. Metal-catalyzed oxidation typically occurs close to the metal binding pockets within the protein (called site-specific oxidation), whereas photo- or peroxide-induced oxidation can occur anywhere in the protein (called non-site-specific oxidation). In the case of metal-catalyzed oxidation, the metal typically binds at the Gly, Asp, His, or Cys residues. Once bound, reactive oxygen species (ROS) is generated at the metal binding cite that can oxidize either the bound ligands (such as His or Cys) or other oxidation-prone residues close by (such as Met). A variety of computational models have been developed to predict the metal-ion binding sites in proteins.[89–93] Once the binding site is predicted with these models, structural modeling can be used to identify the Met, His, or Cys residues close to the binding site that could be prone to oxidation.

Peroxides are another major source of oxidation, especially for the methionine residues within proteins. These peroxides can come from excipients, containers, or tubing used in either formulation or delivery.[74,94,95] Peroxide oxidation has been shown to be influenced by the protein structural features, such as solvent exposure of methionine and protein conformational stability.[96–98] Accordingly, the solvent-accessible area (SAA) is the most commonly used model to predict the most oxidation-prone methionine. This model calculates the solvent-exposed surface area of the side chain of a methionine residue.[99] Typically, a static protein conformation (an x-ray or homology-modeled structure) is used in calculating SAA because of its simplicity and ease of use. Another model, called the two-shell water coordination number, was introduced by Chu et al.[96,100–103] to predict the most oxidation-prone methionines (Figure 3.4a). The two-shell model counts the average number of water molecules within two water coordination shells (5.5 Å) from the sulfur atom in the methionine residue. Chu et al.[96,102] showed that the two-shell model correlates well with the experimental rate of oxidation of methionine residues (Figure 3.4b).

Recent work investigated the accuracy of the two-shell model as well as several SAA models that are calculated based on either the entire side chain or only the sulfur atom in the methionine residue.[104] These models were assessed over seven model proteins and three therapeutic candidates currently in development.[104] It was shown

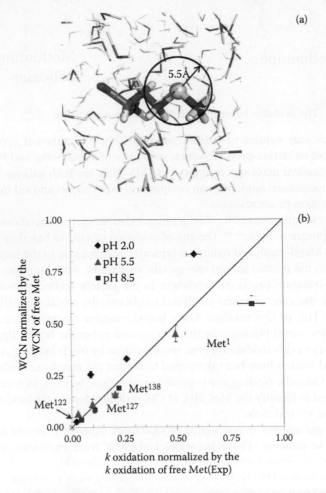

FIGURE 3.4 (a) A representation of the two-shell water coordination number (2SWCN) of the sulfur atom of a methionine amino acid. Red lines are water molecules.[102] (b) Correlation between the 2SWCNs of the methionine sulfur atoms (cutoff radius 5.5 Å) and the rates of oxidation of methionine residues of G-CSF at different pH values. The rates of oxidation of methionine residues and methionine sulfur WCNs are normalized to their values for the free methionine amino acid.[102]

that both the two-shell and the SAA models based on simulations perform very well in predicting oxidation-prone methionines due to peroxides. In addition, the models based on simulation were much more accurate than the ones based on static structure (such as an x-ray). These simulation models not only gave accurate ranking of relative propensity for oxidation, but also gave good semiquantitative agreement with the experimental data. Thus, computational modeling can be used to identify the most oxidation-prone methionines, which enables researchers to improve protein stability through direct mutagenesis or formulation optimization.[80]

3.2.2 DEAMIDATION

Asparagine (Asn) residues within protein pharmaceuticals can undergo deamidation during manufacture and storage. The deamidated species could lead to protein structural changes, aggregation, change in pharmacokinetics, loss of activity, and potential immunogenicity.[105–107] Glutamine (Gln) residues can get deamidated as well, but the Gln deamidation rate is 100-fold slower than that of Asn, and is therefore typically not a concern for protein pharmaceuticals.[108,109] In neutral or basic pH conditions (pH > 7), deamidation of Asn proceeds through a cyclic intermediate called succinimide. This intermediate then hydrolyzes to form a mixture of aspartic acid and iso aspartic acid in a 1:3 ratio[110] (Figure 3.5). Under acidic conditions (pH < 7), deamidation is acid catalyzed, leading directly to aspartic acid. The transition states for the network of reactions involved in deamidation as well as their pH dependence have been characterized using *ab initio* modeling.[111]

The rate of deamidation varies with both intrinsic (primary, secondary, tertiary structure, etc.) and extrinsic (pH, temperature, buffer composition, etc.) factors. A significant work in determining the effect of primary sequence on deamidation comes from Robinson and Robinson.[112,113] They studied the deamidation rates of Asn in pentapeptides (Gly-Xxx-Asn-Yyy-Gly) and showed that the halftimes of Asn deamidation are mainly dependent on the residue following Asn. Asn residues followed by Gly, His, Ser, or Ala have the most propensities for deamidation. This is

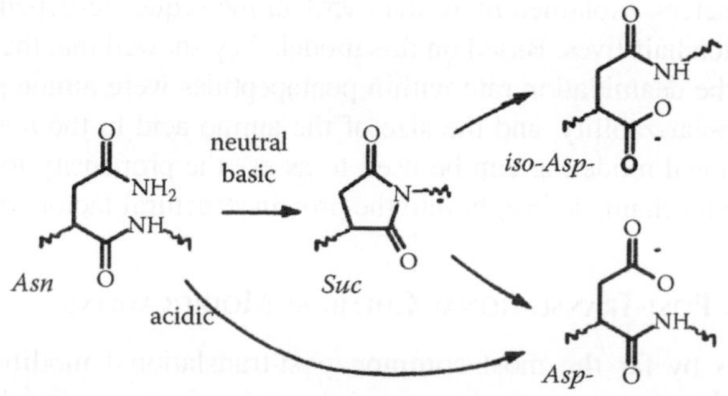

FIGURE 3.5 The intermediate (succinimide) and final (aspartic acid, iso-aspartic acid) products of asparagine deamidation. (Taken from B. Peters and B.L. Trout, *Biochemistry* 45(16):5384–5392, 2006.)

attributed to the small size of residues, such as Gly, Ser, or Ala, which offer less steric hindrance to intramolecular cyclization.

Deamidation rates in proteins are typically reduced compared to peptides because of the decrease in flexibility. Furthermore, other factors can affect deamidation in proteins, such as interactions with neighboring residues, including hydrogen bonds and steric hindrance to cyclization. Deamidation in proteins is typically assessed experimentally using tryptic peptide mapping, which can give a quantitative estimate of the extent of deamidation. However, this procedure is quite tedious. Therefore, computational modeling tools have been developed to predict the extent of deamidation from a protein structure. One such model was developed using the sequence-determined deamidation rate of Asn in pentapeptides, as well as several protein structure factors.[114,115] In this model, the deamidation coefficient, C_D, is defined as[114,115]

$$C_D = (0.01)(t_{1/2})(e^{f(c_m, c_{Sn}, s_n)})$$

where $t_{1/2}$ is the pentapeptide primary structure half-life, C_m is a structure proportionality factor, C_{Sn} is the 3D structure coefficient for the nth structure observation, S_n is that observation, and

$$f(C_m, C_{Sn}, S_n) = C_m [(CS_1)(S_1) + (CS_2)(S_2) + (CS_3)(S_3) - (CS_{4,5})(S_4)/ \\ (S_5) + (CS_6)(S_6) + (CS_7)(S_7) + (CS_8)(S_8) + (CS_9)(S_9) + \\ (CS_{10})(1 - S_{10}) + (CS_{11})(5 - S_{11}) + (CS_{12})(5 - S_{12})]$$

The structure observations, S_n, were selected as those most likely to impede deamidation, including hydrogen bonds, positions with respect to α-helical and β-sheet regions, and peptide inflexibilities. This modeling approach reliably predicted the relative Asn deamidation rates in agreement with the experimental data for 23 proteins.[114,115] Kosky et al.[116] used a multivariate regression technique called the projection to latent structures (PLS) to establish mathematical relationships between a set of protein physicochemical properties and the deamidation halftimes of the amino acid sequences from Robinson and Robinson[112] (Table 3.1). They showed that the PLS models with the reduced propensity scales, combined with the hydrogen exchange rates and flexibility parameters, explained more than 95% of the sequence-dependent variation in the deamidation half-lives. Based on this model, they showed that the principal factors governing the deamidation rate within pentapeptides were amide proton acidity, hydrophilicity, polarizability, and the size of the amino acid in the $n + 1$ position.[116] Thus, computational modeling can be used to assess the propensity for deamidation and in gaining a mechanistic insight into the protein structural factors involved.

3.2.3 Other Post-Translational Chemical Modifications

Glycosylation is by far the most common post-translational modification (PTM) associated with biopharmaceutical products that influences protein folding, binding, trafficking, pharmacokinetics, and stabilization.[117] Glycosylation is the enzymatic addition of sugars at specific sites on the protein and is typically essential for protein function. The most common glycosylations are N- and O-linked glycosylation.

N-linked glycosylation occurs by the addition of a sugar moiety to a nitrogen atom on the side chain of the asparagine residue. O-Linked glycosylation occurs by the addition of the sugar to the hydroxyl group of serine or threonine. However, not all Asn, Ser, or Thr residues in a protein are glycosylated. Therefore, it is useful to characterize which sites are glycosylated because it can affect protein stability and binding. Experimental detection of occupied glycosylation sites is an expensive and laborious process.[118] A number of computational models and web-based servers were developed for predicting the glycosylation sites, as well as analyzing their structure.[119-123] For instance, a model called NGlycPred was developed to predict the N-linked glycosylation sites based on the structural as well as residue pattern information using the Random Forest algorithm.[121] Structural modeling of an IgG1 antibody showed that its glycosylation covers a region consisting of several hydrophobic residues that could potentially cause aggregation if exposed.[124] In addition, molecular dynamics simulations were used to understand the effect of glycosylation.[125,126] Simulations for the IgG1 and IgG2 class of antibodies showed that sugar truncation or removal can cause deformation in the protein tertiary and quaternary structure, potentially affecting protein binding and aggregation.[125,126] Thus, computational modeling can complement experiments in predicting the site of glycosylation, as well as in understanding its effect on protein structure and stability.

Glycation is another PTM, similar to glycosylation, and it also involves the addition of a sugar molecule to the protein. However, while glycosylation is an enzymatic addition at specific sites on the protein, glycation is a non-enzymatic haphazard process that typically impairs the protein function. Glycation occurs at the primary amine sites on the protein, either the lysine side chain or the amino N-terminus. Glycation leads to formation of a Schiff base and then a ketoamine, both of which can oxidize and rearrange to form advanced glycation end products (AGEs).[127] Glycation occurs naturally for proteins in vivo during the normal aging processes, and AGEs are more prevalent in disease states such as diabetes.[128] In the case of therapeutic proteins, glycation can occur either during circulation in vivo or during the cell culture production and manufacturing processes, where they are exposed to sugars.[129-131] Glycation has been shown to affect binding in some cases,[132] while not in others,[129] and it has the potential to alter the structure, function, and stability of therapeutic proteins. Therefore, it is essential to characterize the site and extent of glycation during therapeutic protein development to mitigate risks associated with product quality and safety. It was shown that glycation is site dependent; certain sites with lysine residues are more prone to glycation than others.[130,131] A number of factors were observed to be influencing glycation, such as pKa and solvent exposure of lysine, and the proximity of specific residues, including aspartic acid or other lysines.[131,133,134] Computational modeling and analysis of a therapeutic protein showed that the high glycation observed at a certain site was due to the spatial proximity of an aspartic acid residue.[131] Another computational analysis showed that highly solvent-exposed lysines were indeed highly glycated.[135] A model called NetGlycate was developed using machine learning algorithms that predict the site of glycation with reasonable accuracy.[134]

Disulfide scrambling is a PTM that occurs when the covalent disulfide bonds between Cys residue pairs within a protein get interchanged.[136] This phenomenon is predominantly observed in the IgG2 and IgG4 classes of therapeutic antibodies, but relatively less in the IgG1 class.[137–142] Disulfide scrambling can alter protein conformation and subunit associations and can cause product heterogeneity, potentially leading to loss of drug efficacy, unfolding, aggregation, and immunogenicity.[73,136] It was shown to cause therapeutic IgG4s to exchange Fab arms with endogenous human IgG4s, and it may affect their pharmacokinetics and pharmacodynamics.[140] Computational modeling can aid in gaining a fundamental understanding of protein structure–function relationships, including an understanding of the reasons behind the variation in the covalent disulfide-bonded structure of proteins.[143] Atomistic molecular dynamics simulations were used to analyze the repeated close approach of sulfur atoms from Cys residue pairs to indicate potential for disulfide exchange.[143–145] For instance, simulation of an IgG2 antibody showed that the hinge region dynamics can place the Cys residue pairs in a large number of spatially proximal positions, leading to several potential non-canonical disulfide bonding schemes, leading to disulfide scrambling.[143] Thus, computational modeling can be used to assess the potential for disulfide scrambling and the resulting structural consequences. In addition to the models discussed above, there are several computational models and web-based servers available for predicting sites of other PTMs, such as phosphorylation, acetylation, and methylation[146–149] (Figure 3.6). Thus, the diverse computational

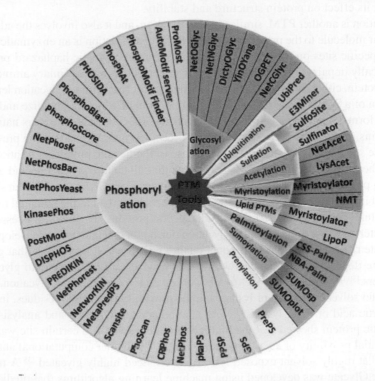

FIGURE 3.6 Computational models available as web-based servers for predicting protein sites with PTMs. The figure segregates various tools as per the type of PTM. Each type of PTM is presented with the most popular predictors. (Taken from K.S. Kamath et al., *J Proteomics* 75(1):127–144, 2011.)

models and tools described in this report can complement experiments and aid in identifying the sites of potential PTMs within the protein sequence and structure.

3.3 CONCLUSIONS

In this chapter, we have presented a comprehensive survey of computational methods and tools used in identifying various physicochemical degradation sites in proteins. Due to their speed and relatively low cost compared to experiments, these predictive tools for identifying degradation-prone sites can prove very useful in biotherapeutic discovery and development. For example, these predictive tools can be used very early on in the discovery process to engineer out the degradation-prone sites. The rational engineering of proteins guided by the computational tool has a much higher chance of success in reducing degradation-prone sites than a trial-and-error approach. When engineering of these sites is not feasible, these tools can be used to rank a large number of candidates based on their degradation liabilities. Only a handful of screened candidates, which show low degradation liabilities, can be experimentally studied for identification of a lead candidate for further development. A rational approach for addressing the post-translational modification in proteins at the early stage of protein discovery in the biopharmaceutical industry can lead to more developable or manufacturable therapeutic candidates. Addressing these modifications in the early stages of drug discovery and development has the potential to reduce the costs and time associated with bringing a protein-based drug to the patients.

REFERENCES

1. Cromwell MEM, Hilario E, Jacobson F. 2006. Protein aggregation and bioprocessing. *AAPS J* 8(3):E572–E579.
2. Kumar S, Singh SK, Wang X, Rup B, Gill D. 2011. Coupling of aggregation and immunogenicity in biotherapeutics: T- and B-cell immune epitopes may contain aggregation prone regions. *Pharm Res* 28:949–961.
3. Rosenberg AS. 2006. Effects of protein aggregates: An immunologic perspective. *AAPS J* 8(3):E501–E507.
4. Wang X, Das TK, Singh SK, Kumar S. 2009. Potential aggregation prone regions in biotherapeutics: A survey of commercial monoclonal antibodies. *mAbs* 1(3):254–267.
5. Philo JS, Arakawa T. 2009. Mechanisms of protein aggregation. *Curr Pharm Biotechnol* 10(4):348–351.
6. Cellmer T, Bratko D, Prausnitz JM, Blanch HW. 2007. Protein aggregation *in silico*. *Trends Biotechnol* 25(6):254–261.
7. Ma B, Nussinov R. 2006. Simulations as analytical tools to understand protein aggregation and predict amyloid conformation. *Curr Opin Chem Biol* 10(5):445–452.
8. Ma B, Nussinov R. 2002. Stabilities and conformations of Alzheimer's beta-amyloid peptide oligomers (Abeta 16–22, Abeta 16–35, and Abeta 10–35): Sequence effects. *Proc Natl Acad Sci USA* 99(22):14126–14131.
9. Clarke OJ, Parker MJ. 2009. Identification of amyloidogenic peptide sequences using a coarse-grained physicochemical model. *J Comput Chem* 30(4):621–630.
10. Chennamsetty N, Helk B, Voynov V, Kayser V, Trout BL. 2009. Aggregation-prone motifs in human immunoglobulin G. *J Mol Biol* 391(2):404–413.

11. Chennamsetty N, Voynov V, Kayser V, Helk B, Trout BL. 2009. Design of therapeutic proteins with enhanced stability. *Proc Natl Acad Sci USA* 106(29):11937–11942.
12. Chennamsetty N, Voynov V, Kayser V, Helk B, Trout BL. 2010. Prediction of aggregation prone regions of therapeutic proteins. *J Phys Chem B* 114(19):6614–6624.
13. Gupta P, Hall CK, Voegler AC. 1998. Effect of denaturant and protein concentrations upon protein refolding and aggregation: A simple lattice model. *Protein Sci* 7(12):2642–2652.
14. Nguyen HD, Hall CK. 2004. Molecular dynamics simulations of spontaneous fibril formation by random-coil peptides. *Proc Natl Acad Sci USA* 101(46):16180–16185.
15. Cellmer T, Bratko D, Prausnitz JM, Blanch H. 2005. The competition between protein folding and aggregation: Off-lattice minimalist model studies. *Biotechnol Bioeng* 89(1):78–87.
16. Cellmer T, Bratko D, Prausnitz JM, Blanch H. 2005. Protein-folding landscapes in multichain systems. *Proc Natl Acad Sci USA* 102(33):11692–11697.
17. Cellmer T, Bratko D, Prausnitz JM, Blanch HW. 2007. Protein aggregation *in silico*. *Trends Biotechnol* 25(6):254–261.
18. Bratko D, Blanch HW. 2001. Competition between protein folding and aggregation: A three-dimensional lattice-model simulation. *J Chem Phys* 114(1):561–569.
19. Bratko D, Cellmer T, Prausnitz JM, Blanch HW. 2006. Effect of single-point sequence alterations on the aggregation propensity of a model protein. *J Am Chem Soc* 128(5):1683–1691.
20. Zhang L, Lu D, Liu Z. 2008. How native proteins aggregate in solution: A dynamic Monte Carlo simulation. *Biophys Chem* 133(1–3):71–80.
21. Ma B, Nussinov R. 2002. Molecular dynamics simulations of alanine rich beta-sheet oligomers: Insight into amyloid formation. *Protein Sci* 11(10):2335–2350.
22. Tsai H-H, Reches M, Tsai C-J, Gunasekaran K, Gazit E, Nussinov R. 2005. Energy landscape of amyloidogenic peptide oligomerization by parallel-tempering molecular dynamics simulation: Significant role of Asn ladder. *Proc Natl Acad Sci USA* 102(23):8174–8179.
23. Ma B, Nussinov R. 2002. Stabilities and conformations of Alzheimer's b-amyloid peptide oligomers ($A\beta_{16-22}$, $A\beta_{16-35}$, and $A\beta_{10-35}$): Sequence effects. *Proc Natl Acad Sci USA* 99(22):14126–14131.
24. Chen H-F. 2009. Aggregation mechanism investigation of the GIFQINS cross-[beta] amyloid fibril. *Comput Biol Chem* 33(1):41–45.
25. Valerio M, Colosimo A, Conti F, Giuliani A, Grottesi A, Manetti C, Zbilut JP. 2005. Early events in protein aggregation: Molecular flexibility and hydrophobicity/charge interaction in amyloid peptides as studied by molecular dynamics simulations. *Proteins Struct Funct Bioinformatics* 58(1):110–118.
26. Baumketner A, Shea JE. 2005. Free energy landscapes for amyloidogenic tetrapeptides dimerization. *Biophys J* 89(3):1493–1503.
27. Tjernberg L, Hosia W, Bark N, Thyberg J, Johansson J. 2002. Charge attraction and beta propensity are necessary for amyloid fibril formation from tetrapeptides. *J Biol Chem* 277(45):43243–43246.
28. Wolf MG, Jongejan JA, Laman JD, de Leeuw SW. 2008. Quantitative prediction of amyloid fibril growth of short peptides from simulations: Calculating association constants to dissect side chain importance. *J Am Chem Soc* 130(47):15772–15773.
29. de Groot N, Pallares I, Aviles F, Vendrell J, Ventura S. 2005. Prediction of "hot spots" of aggregation in disease-linked polypeptides. *BMC Struct Biol* 5(1):18.
30. Chiti F, Taddei N, Baroni F, Capanni C, Stefani M, Ramponi G, Dobson CM. 2002. Kinetic partitioning of protein folding and aggregation. *Nat Struct Mol Biol* 9(2):137–143.

31. Ventura S, Zurdo J, Narayanan S, Parreño M, Mangues R, Reif B, Chiti F, Giannoni E, Dobson CM, Aviles FX, Serrano L. 2004. Short amino acid stretches can mediate amyloid formation in globular proteins: The Src homology 3 (SH3) case. *Proc Natl Acad Sci USA* 101(19):7258–7263.
32. Ivanova MI, Sawaya MR, Gingery M, Attinger A, Eisenberg D. 2004. An amyloid-forming segment of b2-microglobulin suggests a molecular model for the fibril. *Proc Natl Acad Sci USA* 101(29):10584–10589.
33. Monsellier E, Ramazzotti M, de Laureto PP, Tartaglia G-G, Taddei N, Fontana A, Vendruscolo M, Chiti F. 2007. The distribution of residues in a polypeptide sequence is a determinant of aggregation optimized by evolution. *Biophys J* 93(12):4382–4391.
34. Chennamsetty N, Helk B, Voynov V, Kayser V, Trout BL. 2009. Aggregation-prone motifs in human immunoglobulin G. *J Mol Biol* 391(2):404–413.
35. Chennamsetty N, Voynov V, Kayser V, Helk B, Trout BL. 2009. Design of therapeutic proteins with enhanced stability. *Proc Natl Acad Sci USA* 106(29):11937–11942.
36. Zhang Z, Chen H, Lai L. 2007. Identification of amyloid fibril-forming segments based on structure and residue-based statistical potential. *Bioinformatics* 23(17):2218–2225.
37. Thompson MJ, Sievers SA, Karanicolas J, Ivanova MI, Baker D, Eisenberg D. 2006. The 3D profile method for identifying fibril-forming segments of proteins. *Proc Natl Acad Sci USA* 103(11):4074–4078.
38. Cecchini M, Curcio R, Pappalardo M, Melki R, Caflisch A. 2006. A molecular dynamics approach to the structural characterization of amyloid aggregation. *J Mol Biol* 357(4):1306–1321.
39. Vitalis A, Wang X, Pappu RV. 2008. Atomistic simulations of the effects of polyglutamine chain length and solvent quality on conformational equilibria and spontaneous homodimerization. *J Mol Biol* 384(1):279–297.
40. Ma B, Nussinov R. 2006. Simulations as analytical tools to understand protein aggregation and predict amyloid conformation. *Curr Opin Cheml Biol* 10(5):445–452.
41. Clarke OJ, Parker MJ. 2009. Identification of amyloidogenic peptide sequences using a coarse-grained physicochemical model. *J Comput Chem* 30(4):621–630.
42. Tian J, Wu N, Guo J, Fan Y. 2009. Prediction of amyloid fibril-forming segments based on a support vector machine. *BMC Bioinformatics* 10(Suppl 1):S45.
43. Conchillo-Sole O, de Groot N, Aviles F, Vendrell J, Daura X, Ventura S. 2007. AGGRESCAN: A server for the prediction and evaluation of "hot spots" of aggregation in polypeptides. *BMC Bioinformatics* 8(1):65.
44. Frousios KK, Iconomidou VA, Karletidi C-MK, Hamodrakas SJ. 2009. Amyloidogenic determinants are usually not buried. *BMC Struct Biol* 9(1):44.
45. Garbuzynskiy SO, Lobanov MY, Galzitskaya OV. 2010. FoldAmyloid: A method of prediction of amyloidogenic regions from protein sequence. *Bioinformatics* 26(3):326–332.
46. Thangakani AM, Kumar S, Nagarajan R, Velmurugan D, Gromiha MM. 2014. GAP: Towards almost 100 percent prediction for beta-strand-mediated aggregating peptides with distinct morphologies. *Bioinformatics*, 30(14):1983–90.
47. Galzitskaya OV, Garbuzynskiy SO, Lobanov MY. 2006. Prediction of amyloidogenic and disordered regions in protein chains. *PLoS Comput Biol* 2(12):e177.
48. Tartaglia GG, Cavalli A, Pellarin R, Caflisch A. 2005. Prediction of aggregation rate and aggregation-prone segments in polypeptide sequences. *Protein Sci* 14(10):2723–2734.
49. Trovato A, Chiti F, Maritan A, Seno F. 2006. Insight into the structure of amyloid fibrils from the analysis of globular proteins. *PLoS Comput Biol* 2(12):e170.
50. Trovato A, Seno F, Tosatto SCE. 2007. The PASTA server for protein aggregation prediction. *Protein Eng Des Sel* 20(10):521–523.
51. Fang Y, Gao S, Tai D, Middaugh CR, Fang J. 2013. Identification of properties important to protein aggregation using feature selection. *BMC Bioinformatics* 14:314.

52. Zibaee S, Makin OS, Goedert M, Serpell LC. 2007. A simple algorithm locates beta-strands in the amyloid fibril core of alpha-synuclein, Abeta, and tau using the amino acid sequence alone. *Protein Sci* 16(5):906–918.

53. Fernandez-Escamilla A-M, Rousseau F, Schymkowitz J, Serrano L. 2004. Prediction of sequence-dependent and mutational effects on the aggregation of peptides and proteins. *Nat Biotechnol* 22(10):1302–1306.

54. Maurer-Stroh S, Debulpaep M, Kuemmerer N, de la Paz ML, Martins IC, Reumers J, Morris KL, Copland A, Serpell L, Serrano L, Schymkowitz JWH, Rousseau F. 2010. Exploring the sequence determinants of amyloid structure using position-specific scoring matrices. *Nat Methods* 7(3):237–242.

55. Tartaglia GG, Pawar AP, Campioni S, Dobson CM, Chiti F, Vendruscolo M. 2008. Prediction of aggregation-prone regions in structured proteins. *J Mol Biol* 380(2):425–436.

56. DuBay KF, Pawar AP, Chiti F, Zurdo J, Dobson CM, Vendruscolo M. 2004. Prediction of the absolute aggregation rates of amyloidogenic polypeptide chains. *J Mol Biol* 341(5):1317–1326.

57. Tartaglia GG, Vendruscolo M. 2008. The Zyggregator method for predicting protein aggregation propensities. *Chem Soc Rev* 37(7):1395–1401.

58. Wang X, Das TK, Singh SK, Kumar S. 2009. Potential aggregation prone regions in biotherapeutics: A survey of commercial monoclonal antibodies. *mAbs* 1(3):254–267.

59. Wang X, Singh SK, Kumar S. 2010. Potential aggregation-prone regions in complementarity determining regions of antibodies and their contribution towards antigen recognition: A computational analysis. *Pharm Res* 27:1512–1529.

60. Saphire EO, Parren PW, Pantophlet R, Zwick MB, Morris GM, Rudd PM, Dwek RA, Stanfield RL, Burton DR, Wilson IA. 2001. Crystal structure of a neutralizing human IGG against HIV-1: A template for vaccine design. *Science* 293(5532):1155–1159.

61. Chennamsetty N, Voynov V, Kayser V, Helk B, Trout BL. 2009. Design of therapeutic proteins with enhanced stability. *Proc Natl Acad Sci USA*, 106(29):11937–42.

62. Timasheff SN. 1993. The control of protein stability and association by weak interactions with water: How do solvents affect these processes? *Annu Rev Biophys Biomol Struct* 22(1):67–97.

63. Arakawa T, Timasheff SN. 1985. [3]Theory of protein solubility. In HW Wyckoff, ed., *Methods in Enzymology*. Academic Press, New York, pp. 49–77.

64. Record Jr MT, Anderson CF. 1995. Interpretation of preferential interaction coefficients of nonelectrolytes and of electrolyte ions in terms of a two-domain model. *Biophys J* 68(3):786–794.

65. Anderson CF, Felitsky DJ, Hong J, Thomas Record Jr M. 2002. Generalized derivation of an exact relationship linking different coefficients that characterize thermodynamic effects of preferential interactions. *Biophys Chem* 101–102(0):497–511.

66. Scatchard G. 1946. Physical chemistry of protein solutions. I. Derivation of the equations for the osmotic pressure. *J Am Chem Soc* 68(11):2315–2319.

67. Shukla D, Schneider CP, Trout BL. 2011. Molecular level insight into intra-solvent interaction effects on protein stability and aggregation. *Adv Drug Deliv Rev* 63(13):1074–1085.

68. Baynes BM, Trout BL. 2003. Proteins in mixed solvents: A molecular-level perspective. *J Phys Chem B* 107(50):14058–14067.

69. Shukla D, Shinde C, Trout BL. 2009. Molecular computations of preferential interaction coefficients of proteins. *J Phys Chem B* 113(37):12546–12554.

70. Vagenende V, Yap MGS, Trout BL. 2009. Mechanisms of protein stabilization and prevention of protein aggregation by glycerol. *Biochemistry* 48(46):11084–11096.

71. Shukla D, Trout BL. 2011. Understanding the synergistic effect of arginine and glutamic acid mixtures on protein solubility. *J Phys Chem B* 115(41):11831–11839.

72. Walsh G, Jefferis R. 2006. Post-translational modifications in the context of therapeutic proteins. *Nat Biotechnol* 24(10):1241–1252.
73. Correia IR. 2010. Stability of IgG isotypes in serum. *MAbs* 2(3):221–232.
74. Carpenter JF, Manning MC. 2002. *Rational Design of Stable Protein Formulations: Theory and Practice.* Berlin: Springer.
75. Cleland JL, Powell MF, Shire SJ. 1993. The development of stable protein formulations: A close look at protein aggregation, deamidation, and oxidation. *Crit Rev Ther Drug Carrier Syst* 10(4):307–377.
76. Wang W. 1999. Instability, stabilization, and formulation of liquid protein pharmaceuticals. *Int J Pharm* 185(2):129–188.
77. Nguyen Tue H. 1994. *Oxidation Degradation of Protein Pharmaceuticals: Formulation and Delivery of Proteins and Peptides.* American Chemical Society, Washington, D.C., pp. 59–71.
78. Torosantucci R, Schöneich C, Jiskoot W. 2014. Oxidation of therapeutic proteins and peptides: Structural and biological consequences. *Pharm Res* 31(3):541–553.
79. Pan H, Chen K, Chu L, Kinderman F, Apostol I, Huang G. 2009. Methionine oxidation in human IgG2 Fc decreases binding affinities to protein A and FcRn. *Protein Sci* 18(2):424–433.
80. Kim YH, Berry AH, Spencer DS, Stites WE. 2001. Comparing the effect on protein stability of methionine oxidation versus mutagenesis: Steps toward engineering oxidative resistance in proteins. *Protein Eng* 14(5):343–347.
81. Li S, Schöneich C, Borchardt RT. 1995. Chemical instability of protein pharmaceuticals: Mechanisms of oxidation and strategies for stabilization. *Biotechnol Bioeng* 48(5):490–500.
82. Stadtman E. 1993. Oxidation of free amino acids and amino acid residues in proteins by radiolysis and by metal-catalyzed reactions. *Annu Rev Biochem* 62(1):797–821.
83. Hovorka SW, Schöneich C. 2001. Oxidative degradation of pharmaceuticals: Theory, mechanisms and inhibition. *J Pharm Sci* 90(3):253–269.
84. Vogt W. 1995. Oxidation of methionyl residues in proteins: Tools, targets, and reversal. *Free Radical Biol Med* 18(1):93–105.
85. Lu HS, Fausset PR, Narhi LO, Horan T, Shinagawa K, Shimamoto G, Boone TC. 1999. Chemical modification and site-directed mutagenesis of methionine residues in recombinant human granulocyte colony-stimulating factor: Effect on stability and biological activity. *Arch Biochem Biophys* 362(1):1–11.
86. Manning MC, Chou DK, Murphy BM, Payne RW, Katayama DS. 2010. Stability of protein pharmaceuticals: An update. *Pharm Res* 27(4):544–575.
87. Gao J, Yin DH, Yao Y, Sun H, Qin Z, Schöneich C, Williams TD, Squier TC. 1998. Loss of conformational stability in calmodulin upon methionine oxidation. *Biophys J* 74(3):1115–1134.
88. Levine RL, Mosoni L, Berlett BS, Stadtman ER. 1996. Methionine residues as endogenous antioxidants in proteins. *Proc Natl Acad Sci USA* 93(26):15036–15040.
89. Goyal K, Mande SC. 2008. Exploiting 3D structural templates for detection of metal-binding sites in protein structures. *Protein Struct Funct Bioinformatics* 70(4):1206–1218.
90. Levy R, Edelman M, Sobolev V, 2009. Prediction of 3D metal binding sites from translated gene sequences based on remote-homology templates. *Protein Struct Funct Bioinformatics* 76(2):365–374.
91. Lin C-T, Lin K-L, Yang C-H, Chung I-F, Huang C-D, Yang Y-S. Protein Metal Binding Residue Prediction Based on Neural Networks, in *Neural Information Processing*, Calcutta, India, Pal NR, Kasabov N, Mudi RK, Pal S, Parui, SK, Eds., Springer-Verlag, Berlin, Germany, 1316–21.

92. Passerini A, Punta M, Ceroni A, Rost B, Frasconi P. 2006. Identifying cysteines and histidines in transition-metal-binding sites using support vector machines and neural networks. *Protein Struct Funct Bioinformatics* 65(2):305–316.

93. Schymkowitz JW, Rousseau F, Martins IC, Ferkinghoff-Borg J, Stricher F, Serrano L. 2005. Prediction of water and metal binding sites and their affinities by using the Fold-X force field. *Proc Natl Acad Sci USA* 102(29):10147–10152.

94. Kerwin BA. 2008. Polysorbates 20 and 80 used in the formulation of protein biotherapeutics: Structure and degradation pathways. *J Pharm Sci* 97(8):2924–2935.

95. Wasylaschuk WR, Harmon PA, Wagner G, Harman AB, Templeton AC, Xu H, Reed RA. 2007. Evaluation of hydroperoxides in common pharmaceutical excipients. *J Pharm Sci* 96(1):106–116.

96. Chu J-W, Yin J, Wang DI, Trout BL. 2004. Molecular dynamics simulations and oxidation rates of methionine residues of granulocyte colony-stimulating factor at different pH values. *Biochemistry* 43(4):1019–1029.

97. Thirumangalathu R, Krishnan S, Bondarenko P, Speed-Ricci M, Randolph TW, Carpenter JF, Brems DN. 2007. Oxidation of methionine residues in recombinant human interleukin-1 receptor antagonist: Implications of conformational stability on protein oxidation kinetics. *Biochemistry* 46(21):6213–6224.

98. Pan B, Abel J, Ricci MS, Brems DN, Wang DI, Trout BL. 2006. Comparative oxidation studies of methionine residues reflect a structural effect on chemical kinetics in rhG-CSF. *Biochemistry* 45(51):15430–15443.

99. Lee B, Richards FM. 1971. The interpretation of protein structures: Estimation of static accessibility. *J Mol Biol* 55(3):379–400.

100. Chu J-W, Trout BL. 2004. On the mechanisms of oxidation of organic sulfides by H2O2 in aqueous solutions. *J Am Chem Soc* 126(3):900–908.

101. Chu J-W, Yin J, Wang DI, Trout BL. 2004. A structural and mechanistic study of the oxidation of methionine residues in hPTH (1–34) via experiments and simulations. *Biochemistry* 43(44):14139–14148.

102. Chu J-W, Yin J, Brooks BR, Wang DI, Ricci MS, Brems DN, Trout BL. 2004. A comprehensive picture of non-site specific oxidation of methionine residues by peroxides in protein pharmaceuticals. *J Pharm Sci* 93(12):3096–3102.

103. Chu J-W, Brooks BR, Trout BL. 2004. Oxidation of methionine residues in aqueous solutions: Free methionine and methionine in granulocyte colony-stimulating factor. *J Am Chem Soc* 126(50):16601–16607.

104. Chennamsetty N, Quan Y, Nashine V, Sadineni V, Lyngberg O, Krystek S. Modeling the oxidation of methionine residues in proteins. *J Pharm Sci*, 104:1246–55.

105. Huang L, Lu J, Wroblewski VJ, Beals JM, Riggin RM. 2005. *In vivo* deamidation characterization of monoclonal antibody by LC/MS/MS. *Anal Chem* 77(5):1432–1439.

106. Johnson BA, Shirokawa JM, Hancock WS, Spellman MW, Basa LJ, Aswad DW. 1989. Formation of isoaspartate at two distinct sites during *in vitro* aging of human growth hormone. *J Biol Chem* 264(24):14262–14271.

107. Doyle HA, Zhou J, Wolff MJ, Harvey BP, Roman RM, Gee RJ, Koski RA, Mamula MJ. 2006. Isoaspartyl post-translational modification triggers anti-tumor T and B lymphocyte immunity. *J Biol Chem* 281(43):32676–32683.

108. Robinson AB, Scotchler JW, McKerrow JH. 1973. Rates of nonenzymic deamidation of glutaminyl and asparaginyl residues in pentapeptides. *J Am Chem Soc* 95(24):8156–8159.

109. Robinson AB, Rudd CJ. 1974. Deamidation of glutaminyl and asparaginyl residues in peptides and proteins. *Curr Topics Cell Regul* 8(0):247–295.

110. Capasso S, Di Donato A, Esposito L, Sica F, Sorrentino G, Vitagliano L, Zagari A, Mazzarella L. 1996. Deamidation in proteins: The crystal structure of bovine pancreatic ribonuclease with an isoaspartyl residue at position 67. *J Mol Biol* 257(3):492–496.

111. Peters B, Trout BL. 2006. Asparagine deamidation: pH-dependent mechanism from density functional theory. *Biochemistry* 45(16):5384–5392.
112. Robinson NE, Robinson AB. 2001. Molecular clocks. *Proc Natl Acad Sci USA* 98(3):944–949.
113. Robinson NE, Robinson AB. 2004. *Molecular Clocks: Deamidation of Asparaginyl and Glutaminyl Residues in Peptides and Proteins.* Cave Junction, OR: Althouse Press.
114. Robinson NE, Robinson AB. 2001. Prediction of protein deamidation rates from primary and three-dimensional structure. *Proc Natl Acad Sci USA* 98(8):4367–4372.
115. Robinson N, Robinson A. 2001. Deamidation of human proteins. *Proc Natl Acad Sci USA* 98(22):12409–12413.
116. Kosky AA, Dharmavaram V, Ratnaswamy G, Manning MC. 2009. Multivariate analysis of the sequence dependence of asparagine deamidation rates in peptides. *Pharm Res* 26(11):2417–2428.
117. Walsh G. 2010. Post-translational modifications of protein biopharmaceuticals. *Drug Discov Today* 15(17–18):773–780.
118. Zaia J. 2008. Mass spectrometry and the emerging field of glycomics. *Chem Biol* 15(9):881–892.
119. Frank M, Schloissnig S. 2010. Bioinformatics and molecular modeling in glycobiology. *Cell Mol Life Sci* 67(16):2749–2772.
120. Mazola Y, Chinea G, Musacchio A. 2011. Integrating bioinformatics tools to handle glycosylation. *PLoS Comput Biol* 7(12):e1002285.
121. Chuang G-Y, Boyington JC, Joyce MG, Zhu J, Nabel GJ, Kwong PD, Georgiev I. 2012. Computational prediction of N-linked glycosylation incorporating structural properties and patterns. *Bioinformatics* 28(17):2249–2255.
122. Chauhan JS, Rao A, Raghava GPS. 2013. *In silico* platform for prediction of N-, O- and C-glycosites in eukaryotic protein sequences. *PLoS ONE* 8(6):e67008.
123. Blom N, Sicheritz-Pontén T, Gupta R, Gammeltoft S, Brunak S. 2004. Prediction of post-translational glycosylation and phosphorylation of proteins from the amino acid sequence. *Proteomics* 4(6):1633–1649.
124. Kayser V, Chennamsetty N, Voynov V, Forrer K, Helk B, Trout BL. 2011. Glycosylation influences on the aggregation propensity of therapeutic monoclonal antibodies. *Biotechnol J* 6(1):38–44.
125. Buck PM, Kumar S, Singh SK. 2013. Consequences of glycan truncation on Fc structural integrity. *mAbs* 5(6):904–916.
126. Wang X, Kumar S, Buck PM, Singh SK. 2013. Impact of deglycosylation and thermal stress on conformational stability of a full length murine igG2a monoclonal antibody: Observations from molecular dynamics simulations. *Proteins Struct Funct Bioinformatics* 81(3):443–460.
127. Rabbani N, Thornalley P. 2012. Glycation research in amino acids: A place to call home. *Amino Acids* 42(4):1087–1096.
128. Brownlee M. 2000. Negative consequences of glycation. *Metabolism* 49(2, Suppl 1):9–13.
129. Goetze AM, Liu YD, Arroll T, Chu L, Flynn GC. 2012. Rates and impact of human antibody glycation *in vivo*. *Glycobiology* 22(2):221–234.
130. Miller AK, Hambly DM, Kerwin BA, Treuheit MJ, Gadgil HS. 2011. Characterization of site-specific glycation during process development of a human therapeutic monoclonal antibody. *J Pharm Sci* 100(7):2543–2550.
131. Zhang B, Yang Y, Yuk I, Pai R, McKay P, Eigenbrot C, Dennis M, Katta V, Francissen KC. 2008. Unveiling a glycation hot spot in a recombinant humanized monoclonal antibody. *Anal Chem* 80(7):2379–2390.
132. Dolhofer-Bliesener R, Gerbitz KD. 1990. Impairment by glycation of immunoglobulin G Fc fragment function. *Scand J Clin Lab Invest* 50(7):739–746.

133. Venkatraman J, Aggarwal K, Balaram P. 2001. Helical peptide models for protein glycation: Proximity effects in catalysis of the Amadori rearrangement. *Chem Biol* 8(7):611–625.

134. Johansen MB, Kiemer L, Brunak S. 2006. Analysis and prediction of mammalian protein glycation. *Glycobiology* 16(9):844–853.

135. Lapolla A, Tonani R, Fedele D, Garbeglio M, Senesi A, Seraglia R, Favretto D, Traldi P. 2002. Non-enzymatic glycation of IgG: An *in vivo* study. *Horm Metab Res* 34(5):260–264.

136. Liu H, May K. 2012. Disulfide bond structures of IgG molecules: Structural variations, chemical modifications and possible impacts to stability and biological function. *mAbs* 4(1):17–23.

137. Wypych J, Li M, Guo A, Zhang Z, Martinez T, Allen MJ, Fodor S, Kelner DN, Flynn GC, Liu YD, Bondarenko PV, Ricci MS, Dillon TM, Balland A. 2008. Human IgG2 antibodies display disulfide-mediated structural isoforms. *J Biol Chem* 283(23):16194–16205.

138. Dillon TM, Ricci MS, Vezina C, Flynn GC, Liu YD, Rehder DS, Plant M, Henkle B, Li Y, Deechongkit S, Varnum B, Wypych J, Balland A, Bondarenko PV. 2008. Structural and functional characterization of disulfide isoforms of the human IgG2 subclass. *J Biol Chem* 283(23):16206–16215.

139. van der Neut Kolfschoten M, Schuurman J, Losen M, Bleeker WK, Martínez-Martínez P, Vermeulen E, den Bleker TH, Wiegman L, Vink T, Aarden LA, De Baets MH, van de Winkel JGJ, Aalberse RC, Parren PWHI. 2007. Anti-inflammatory activity of human IgG4 antibodies by dynamic Fab arm exchange. *Science* 317(5844):1554–1557.

140. Labrijn AF, Buijsse AO, van den Bremer ET, Verwilligen AY, Bleeker WK, Thorpe SJ, Killestein J, Polman CH, Aalberse RC, Schuurman J, van de Winkel JG, Parren PW. 2009. Therapeutic IgG4 antibodies engage in Fab-arm exchange with endogenous human IgG4 *in vivo*. *Nat Biotechnol* 27(8):767–771.

141. Zhang B, Harder AG, Connelly HM, Maheu LL, Cockrill SL. 2009. Determination of Fab–hinge disulfide connectivity in structural isoforms of a recombinant human immunoglobulin G2 antibody. *Anal Chem* 82(3):1090–1099.

142. Wang Y, Lu Q, Wu SL, Karger BL, Hancock WS. 2011. Characterization and comparison of disulfide linkages and scrambling patterns in therapeutic monoclonal antibodies: Using LC-MS with electron transfer dissociation. *Anal Chem* 83(8):3133–3140.

143. Wang X, Kumar S, Singh SK. 2011. Disulfide scrambling in IgG2 monoclonal antibodies: Insights from molecular dynamics simulations. *Pharm Res* 28(12):3128–3144.

144. Schmid N, Bolliger C, Smith LJ, van Gunsteren WF. 2008. Disulfide bond shuffling in bovine α-lactalbumin: MD simulation confirms experiment. *Biochemistry* 47(46):12104–12107.

145. Allison JR, Moll G-P, van Gunsteren WF. 2010. Investigation of stability and disulfide bond shuffling of lipid transfer proteins by molecular dynamics simulation. *Biochemistry* 49(32):6916–6927.

146. Ivanisenko V, Afonnikov D, Kolchanov N. 2008. Web-based computational tools for the prediction and analysis of post-translational modifications of proteins. In C Kannicht, ed., *Post-translational Modifications of Proteins*. Humana Press, Totowa, NJ, pp. 363–384.

147. Xue Y, Gao X, Cao J, Liu Z, Jin C, Wen L, Yao X, Ren J. 2010. A summary of computational resources for protein phosphorylation. *Curr Protein Peptide Sci* 11(6):485–496.

148. Charpilloz C, Veuthey AL, Chopard B, Falcone JL. 2014. Motifs tree: A new method for predicting post-translational modifications. *Bioinformatics*, 30(14):1974–82.

149. Kamath KS, Vasavada MS, Srivastava S. 2011. Proteomic databases and tools to decipher post-translational modifications. *J Proteomics* 75(1):127–144.

4 Preclinical Immunogenicity Risk Assessement of Biotherapeutics

Tim D. Jones, Anette C. Karle,†*
*and Matthew P. Baker**

CONTENTS

4.1 INTRODUCTION

A surprising aspect of the development of human proteins for use as therapeutics has been the observation that many of them elicit anti-drug antibodies (ADAs) in patients. The consequences of ADA responses in the clinic vary from mild to severe. Antibodies may be transient or non-neutralizing, in which case the effects may not be adverse, although they may impact therapeutic clearance, either positively

* ¹Antitope Ltd. (an Abzena company), Babraham Research Campus, Babraham, Cambridge, United Kingdom
† ²Novartis Pharma AG, Integrated Biologics Profiling Unit, Immunogenicity Risk Assessment, Werk Klybeck, Basel, Switzerland

or negatively,[1, 2] leading to additional dosing requirements.[3] Where antibodies are neutralizing, this can limit efficacy, as in the cases of granulocyte macrophage colony-stimulating factor (GM-CSF)[4] and IFNβ.[5] In some cases, the antibodies can also recognize the endogenous protein with serious consequences for the patient, as in the case of erythropoietin (EPO), associated with pure red cell aplasia during the treatment of anaemia,[6, 7] or thrombopoietin (TPO), associated with immune thrombocytopenic purpura (ITP) and thrombocytopenia.[8] The problem of ADAs has also been a central focus of the development of therapeutic antibody technology, which has striven to reduce the frequency of ADAs in patients by increasing the humanness of candidate therapeutics. One of the first antibodies to enter clinical use was the fully murine antibody muromonab (for allograft rejection), which was immunogenic in a high proportion of patients.[9, 10] Subsequently, chimeric antibodies were produced that contained murine variable domains fused to human antibody constant domains (e.g., infliximab and abciximab), but ADAs were still produced at high frequency.[11–13] The humanness of therapeutic antibodies has been further increased by humanization technologies (whereby complementarity determining regions [CDRs] of murine antibodies, together with selected framework residues, are grafted onto human germline antibody sequences[14]) and by the selection of fully human antibodies from phage display libraries or isolation from mice transgenic for human antibody genes.[15] While there is a general trend for these technologies to reduce the frequencies of ADAs in the clinic, they do not routinely deliver therapeutic antibodies with low immunogenicity profiles,[16] as exemplified by the fully human antibody adalimumab, where the development of ADAs is linked to lack of clinical response.[17]

The immunological response of patients to therapeutics in clinical development may result in failure of the drug to reach market, as in the case of TPO and a humanized antibody against the A33 tumor antigen.[18] Immunogenicity testing is therefore regarded as a key component for the demonstration of clinical safety and efficacy.[19] It is unlikely that regulatory approval would be granted for a biologic without a clinical assessment of immunogenicity, namely, the production of ADAs, and FDA guidance is provided on methods for detecting ADAs[20] and for risk mitigation.[21] Evidence, as outlined above, has shown us that humanness is not a predictor for the development of ADAs, and there are many factors that influence whether or not ADAs develop.[21] These may be related to the patient, disease indication, route of administration, or the nature and level of purity of the therapeutic,[22] all of which can contribute to the breaking of immunological tolerance to human therapeutic proteins. Regardless of the contributing factors, as will be discussed below, central to the process of eliciting an immune response is the presence of human leukocyte antigen (HLA) class II restricted CD4+ T-cell epitopes within the protein sequence, which drive the process of ADA development (Figure 4.1), culminating in the formation of high-affinity isotype switched antibodies that bind to B-cell epitopes on the surface of the therapeutic protein.[23] Therefore, computational methods for assessing immunogenicity risk have focused on the prediction of T and B-cell epitopes within the therapeutic protein as rapid and cost-effective tools either for lead selection or for reengineering proteins for epitope removal. This chapter focuses on

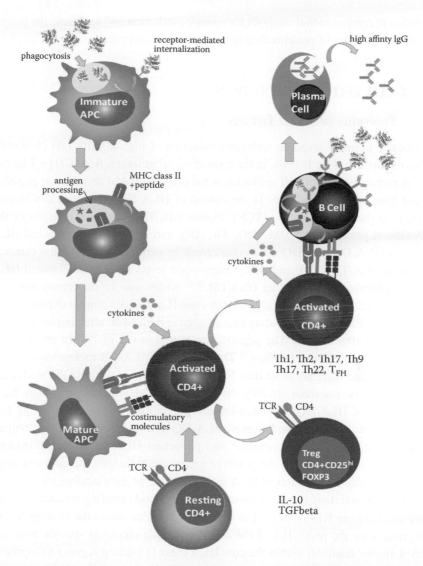

FIGURE 4.1 Summary of the adaptive humoral response for T-dependent antigens. A professional APC engulfs the therapeutic protein via non-specific phagocytosis or receptor-mediated internalization. Protein is processed into fragments by proteases in the endosomal pathway to produce peptide fragments, some of which may be bound by MHC class II, with the peptide–MHC II complex then being presented on the surface of the cell. Resting CD4[+] T-cells bind the peptide–MHC II complex via the T-cell receptor and, after engagement of co-stimulatory molecules and in the presence of cytokines, become activated. B-cells also act as APCs and may present the same peptide–MHCII complex on their surface, which can be engaged by the cognate-activated T-cell to provide growth support to B-cells via co-stimulatory factors. Thus, the B-cell matures to produce high-affinity, isotype-switched anti-therapeutic antibodies. Central to this process are the presentation of peptides in the context of MHC class II and the recognition of this complex by T-cells. (Modified from Baker, M. P., and Carr, F. J., *Current Drug Safety*, 5(4), 308–13, 2010.[23])

the different computational methods for epitope prediction and assessing the potential immunogenicity of proteins during preclinical development.

4.2   T-CELL EPITOPE PREDICTION

4.2.1   Properties of T-Cell Epitopes

An adaptive immune response with the production of high-affinity IgG class antibodies requires help for B-cells, in the form of co-stimulation, from CD4+ T helper cells. A prerequisite for T-cell activation is the presentation of short linear peptides derived from processed proteins in the context of HLA class II, which can be recognized via the T-cell receptor (TCR) (Figure 4.1). There are five isotypes of the HLA class II protein: HLA-DM, -DO, -DP, -DQ, and -DR. The non-classical HLA molecules HLA-DM and DQ act as chaperones in peptide loading of the classical HLA molecules -DP, -DQ, and -DR.[24] Expression levels among the classical HLA class II molecules are highest for HLA-DR,[25,26] which may be the reason why it is the best-studied HLA class II isotype. HLA class II molecules consist of two chains, α and β, that form a peptide binding groove open at both ends, allowing for peptides of variable lengths to bind. The peptides are bound in the groove in an extended conformation by hydrogen bonding.[27] The β chain of HLA-DR molecules is highly polymorphic, and there are more than 1700 alleles identified so far in the worldwide population,[28] whereas there are only three α chain alleles. In contrast, both the α and β chains of HLA-DQ and HLA-DP are polymorphic, potentially giving rise to thousands of combinations; however, HLA-DQ diversity is limited by incompatibility of certain α/β chain pairings,[29] and particular HLA-DP α/β combinations have been found to dominate the potential repertoire.[30] Despite the apparent polymorphic diversity of the classical HLA molecules, the peptide binding repertoire is not unlimited, and there is considerable overlap in peptide binding specificities both within and between HLA types.[31] The binding of peptides within the binding groove is dependent on the properties of the amino acid side chains at specific positions (termed anchor residues) within the core HLA class II binding register (comprising nine amino acids).[32] As a consequence, 9-mer binding motifs can be determined from the sequences of proteins that may bind to specific HLA class II molecules. However, the issue of peptide binding is more complicated than analysis of the core 9-mer since the HLA class II binding groove is open ended and can accommodate peptides of varying sizes. Therefore, the influence of residues (p1, p10, and p11) that are positioned outside of the core 9-mer are also considered to be of importance.[33–35] Peptide elongation in general has been reported to result in increased major histocompatibility complex (MHC) class II molecule affinity with an optimal peptide length of approximately 18–20 amino acids,[35] and in peptide elution experiments from dendritic cells (DCs), lengths between 18 and 25 amino acids were commonly observed for naturally presented peptides.[36,37]

4.2.2 *IN SILICO* T-CELL EPITOPE PREDICTION

4.2.2.1 HLA Class II Binding Prediction

Characterization of the peptide motifs that bind to different HLA class II molecules via *in vitro* peptide binding experiments has enabled the development of *in silico* algorithms or structural modeling tools for the prediction of HLA class II binding peptides (see Table 4.1 for a summary of freely available web-based tools). However, there are a number of limitations that must be considered when making use of these tools for immunogenicity prediction, not least of which is the fact that all tools base predictions on the binding of the core 9-mer within the HLA class II binding groove and do not consider additional interactions (as discussed in the previous section).

The fact that *in silico* algorithms are based on experimental peptide binding data causes some general limitations to the predictive ability of the algorithms. Since only a small fraction of the naturally occurring peptidome has been tested in these binding assays, the resulting peptide binding data set is limited and, ultimately, so are the *in silico* algorithms that use it as the basis for the scoring. Moreover, when comparing HLA class II binding data derived from experimental binding assays, the exact binding register of the peptide is often unclear, and this will impact the quality of the data set used to train *in silico* algorithms. Furthermore, the specificity and sensitivity of experimental binding assays are influenced by the assay conditions as well as the selection of test peptides. Peptides with poor solubility may not go into solution, giving false negative results, but could be presented on HLA class II molecules *in vivo* due to the binding of larger and more soluble fragments of the molecule and subsequent trimming to shorter peptides according to the "bind first, trim later model."[38] In addition, *in vivo* chaperones are involved in editing the peptide repertoire presented on antigen-presenting cells (APCs) in a way such that peptides with good overall affinity but fast off-rates are released by the chaperone HLA-DM[39] to promote the formation of HLA-DR–peptide complexes with long-term stability.

Another limitation of *in silico* HLA class II binding prediction algorithms lies in the way the probability of a peptide binding is calculated. As a result, the probabilities for the different binding pockets are considered without taking the overall structural peptide conformation into consideration. As *in silico* tools scan only the primary amino acid sequence, the impact of post-translational modifications or protein quality attributes such as aggregation cannot be assessed, and these factors have been shown to alter the number and type of presented peptides under certain circumstances.[37, 40] Finally, these tools only predict the interaction of the peptide with HLA class II, whereas the critical influence of antigen processing (by the APC) and recognition of the peptide–HLA class II complex by the TCR are not considered. Antigen processing within the professional antigen-presenting cell largely determines the size of the potential HLA class II peptide binding repertoire, and after presentation of this repertoire on the APC cell surface, only a limited number of peptides will engage with the T-cell receptor with sufficient affinity or frequency to stimulate cognate T-cell activation.[41] The remaining peptides are simply HLA class II binders that do not stimulate T-cell activation due to the non-responsiveness

TABLE 4.1

Summary of Freely Available Web-Based Tools for the Prediction of Peptide Binding to HLA Class II Molecules

Tool	Web Address	Alleles Interrogated			Comments
		HLA-DR	HLA-DQ	HLA-DP	
IEDB recommended	http://tools.immuneepitope.org/mhcii/	675	6	5	Collection of tools at the IEDB resource using different analysis methods, some of which provide large numbers of MHC class II alleles for analysis, including HLA-DR and some HLA-DQ and HLA-DP. Provides a useful interface for MHC class II binding prediction.
Consensus		55	6	5	
NetMHCIIpan		672	6	5	
NN-align		14	6	5	
SMM-align		15	6	5	
Combinatorial library		5	6	6	
Sturniolo		51	0	0	
MHC2PRED	http://www.imtech.res.in/raghava/mhc2pred/	25	10	0	Support vector machine approach providing the facility to set a user-defined cutoff with a useful number of alleles, including both HLA-DR and HLA-DQ.
MHCPRED 2.0	http://www.ddg-pharmfac.net/mhcpred/MHCPred/	3	0	0	A partial least-squares multivariate statistical approach. Method has limited utility for immunogenicity prediction due to limited number of alleles interrogated.

Name	URL			Description
PROPRED	http://www.imtech.res.in/raghava/propred/	51	0	0	Quantitative matrix approach interrogating up to 51 alleles with ability to set user-defined thresholds.
RANKPEP	http://imed.med.ucm.es/Tools/rankpep.html	37	9	4	Position-specific scoring matrix approach interrogating a total of 50 HLA alleles with ability to set user-defined thresholds.
SVRMHC	http://svrmhc.biolead.org/	6	0	0	Support vector machine method for quantitative modeling of the interaction between a peptide and MHC class II. Method has limited utility for immunogenicity prediction due to limited number of alleles interrogated.
NetMHCII 2.2	http://www.cbs.dtu.dk/services/NetMHCII/	14	6	6	Prediction method using artificial neural networks for predicting peptide binding affinity. Method has limited utility for immunogenicity prediction due to low number of alleles interrogated.
NetMHCIIpan 3.0	http://www.cbs.dtu.dk/services/NetMHCIIpan-3.0/	655	α chain: 28 β chain: 104	α chain: 16 β chain: 128	Prediction method using artificial neural networks trained on a large data set for predicting peptide binding affinity. Predictions can be made with any HLA allele of known sequence, although number of alleles for a single run is limited to 20. User-defined thresholds can be set.

(anergy) or absence of (through deletion during central tolerance) TCRs in the T-cell compartment.

Taken together, these limitations result in the fact that *in silico* methods consistently overpredict the number of actual T-cell epitopes present in protein sequences and can therefore, at best, provide a useful aid to identifying regions of a sequence that have the potential to bind to HLA class II. Subsequent *ex vivo* or *in vivo* T-cell activation analysis is normally required to identify genuine T-cell epitopes and rank peptides according to their potency to induce a T-cell response.

Despite these limitations, *in silico* HLA class II prediction tools have clear benefits for very specific applications during drug development, as outlined in Section 4.2.3. The latest developments in the computational evaluation of HLA class II peptide binding have been comprehensively reviewed recently,[42-45] and the most useful of these tools for predicting immunogenicity, such as RANKPEP,[46] PROPRED,[47] Tepitope,[48] and NetMHCIIpan,[49] are algorithms that interrogate a broad spectrum of human HLA class II alleles. The latter also has the ability to predict binding to HLA class II alleles where there is not yet any experimental binding data available. A collection of tools have been selected and curated at the Immune Epitope Database Analysis Resource (IEDB) (www.iedb.org[50]), providing a convenient web interface for analysis of protein sequences, and a method based upon a consensus of these tools is also provided.[51] Similarly, the iTope™ software[52] interrogates up to 34 HLA class II alleles and is analogous to both RANKPEP and PROPRED and combines data from *in vitro* HLA class II binding studies with alignments of large databases of peptides that have been demonstrated to be T-cell epitopes via *ex vivo* T-cell assays. When considering a broad spectrum of HLA class II alleles, the data output from these tools can be highly complex, especially when considering a large number of HLA class II alleles, and it is important to set data cutoffs based upon experimentally determined T-cell epitopes in order both to reduce the data sets to manageable proportions and to ensure that the predictions do not span extended tracts of sequence. This point is illustrated by Figure 4.2, where two algorithms (NetMHCIIpan and iTope) were compared for their ability to predict peptides binding to the same panel of 34 HLA class II alleles within the sequence of human interferon alpha 2a (IFNα2a), and the output was further compared to T-cell epitopes identified by *ex vivo* T-cell assays.[53] Figure 4.2a displays the results from NetMHCIIpan using the IEDB recommended cutoff values, whilst in Figure 4.2b, an additional criterion that a binding score must be generated for at least 50% of the alleles tested was applied. Figure 4.2c similarly displays the iTope data with high- and moderate-affinity peptides highlighted. There is a good correspondence between the two methods where either high- or moderate-affinity binding peptides are predicted that coincide with the T-cell epitopes; however, due to the limitations of the methods, as discussed above, HLA class II binding peptides and T-cell epitopes do not always align, and additional binding peptides are predicted that are not actual T-cell epitopes. Increasing the stringency of the cutoff for NetMHCIIpan (Figure 4.2b compared to Figure 4.2a) reduces the complexity of the output while not impacting upon the overall success of the prediction. The need to set appropriate cutoff values is illustrated by both methods, where, if all predicted binding

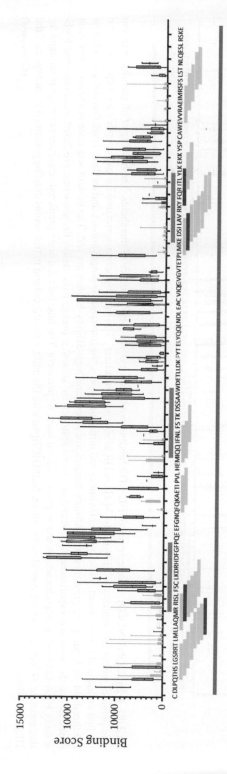

FIGURE 4.2a Comparison of HLA-DR binding prediction methods and influence of thresholds. The binding of all possible 9 amino acid peptides within the sequence of IFNα2a to 34 HLA-DR alleles was calculated. The binding scores generated for each peptide are depicted above the first amino acid of each 9-mer, showing the maximum and minimum scores, the interquartile range (boxed), and median (line). Overall scores below the threshold for binding are gray, moderate-affinity binders are green, and high-affinity binders are red. The protein sequence is shown below each chart. Above the sequence, experimentally determined T-cell epitopes are shown as orange bars. Below the sequence, predicted moderate- and high-affinity HLA-DR binding peptides are shown as green and red bars, respectively. The blue bars illustrate predicted sequence coverage if all peptides that generate a binding score are considered positive. NetMHCIIpan: median score ≤ 50 = high affinity, median score > 50 and ≤ 500 = moderate affinity.

(Continued)

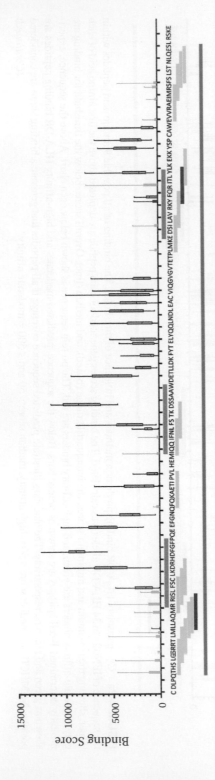

FIGURE 4.2b (CONTINUED)   Comparison of HLA-DR binding prediction methods and influence of thresholds. The binding of all possible 9 amino acid peptides within the sequence of IFNα2a to 34 HLA-DR alleles was calculated. The binding scores generated for each peptide are depicted above the first amino acid of each 9-mer, showing the maximum and minimum scores, the interquartile range (boxed), and median (line). Overall scores below the threshold for binding are gray, moderate-affinity binders are green, and high-affinity binders are red. The protein sequence is shown below each chart. Above the sequence, experimentally determined T-cell epitopes are shown as orange bars. Below the sequence, predicted moderate- and high-affinity HLA-DR binding peptides are shown as green and red bars, respectively. The blue bars illustrate predicted sequence coverage if all peptides that generate a binding score are considered positive. NetMHCIIpan with the addition to the criteria in that scores must be generated for at least half the HLA-DR alleles listed.

(Continued)

FIGURE 4.2c (CONTINUED) Comparison of HLA-DR binding prediction methods and influence of thresholds. The binding of all possible 9 amino acid peptides within the sequence of IFNα2a to 34 HLA-DR alleles was calculated. The binding scores generated for each peptide are depicted above the first amino acid of each 9-mer, showing the maximum and minimum scores, the interquartile range (boxed), and median (line). Overall scores below the threshold for binding are gray, moderate-affinity binders are green, and high-affinity binders are red. The protein sequence is shown below each chart. Above the sequence, experimentally determined T-cell epitopes are shown as orange bars. Below the sequence, predicted moderate- and high-affinity HLA-DR binding peptides are shown as green and red bars, respectively. The blue bars illustrate predicted sequence coverage if all peptides that generate a binding score are considered positive. iTope: High- (binding score ≥ 0.6) and moderate- (binding score ≥ 0.55 but < 0.6) affinity peptides highlighted where the median binding score to all alleles is >0.55.

peptides are considered, almost the entire sequence is predicted to contain overlapping HLA-DR binding peptides.

4.2.2.2 Databases of T-Cell Epitopes

The scientific literature provides the sequences of thousands of experimentally identified T-cell and B-cell epitopes. Furthermore, there are a number of manually curated databases that comprise searchable records of these previously identified epitopes.[54] One of the key advantages of these databases is that they contain very large data sets, although the disadvantage is that the inclusion criteria allow data to be included that have been generated using a variety of experimental methods and will therefore have a variable level of accuracy. By far the largest of these databases is the Immune Epitope Database (IEDB) project.[50,55] In spite of the size of the IEDB (currently contains 119,185 items on reported epitopes), it is still far from exhaustive. Indeed, databases such as the HLA binding and non-binding peptide database (MHCBN), which comprises a curated database of HLA binding and non-binding peptides, as well as known T-cell epitopes, contain sequences that are not present in the IEDB.[56] Thus, different databases offer value from the diversity of the data sets, as well as the unique searching features, such as the ability to identify non-HLA binding peptides, integrated BLAST search facilities,[57] and HLA–disease associations, all of which enable researchers to address specific research aims.

For protein therapeutics, specialist reference libraries of known T-cell epitopes have been developed that enable the screening of novel sequences by BLAST analysis to identify homologous (or partially homologous) peptides that have previously stimulated T-cells in *ex vivo* T-cell assays. The T-Cell Epitope Database™ (TCED)[58] comprises data from peptides that have been tested for T-cell reactivity via *ex vivo* T-cell epitope mapping analysis. More than 60% of the peptides in the TCED are derived from the T-cell epitope mapping of therapeutic antibody variable region sequences, and as a result, the TCED is particularly useful in homology searching for T-cell epitopes in novel antibody variable region sequences. When combined with conventional *in silico* HLA class II prediction software, additional BLAST search analysis of the TCED for sequence similarity between HLA class II binding peptides and previously identified T-cell epitopes from *ex vivo* T-cell assays can reduce the frequency of false positives when analyzing therapeutic antibody sequences. The TCED also contains information on peptides tested in *ex vivo* T-cell assays that have been shown not to stimulate T-cell activation, and this information has been utilized to enable the construction of novel antibody V region sequences in which T-cell epitopes are avoided during antibody humanization.[58]

The ability to combine HLA class II binding prediction and a searchable database of known T-cell epitopes for V region sequences has a demonstrable advantage over analysis of sequences by HLA class II binding prediction alone. Figure 4.3 shows the light chain of the murine chimeric anti-CD20 antibody (rituximab),[59] which was analyzed using *in silico* HLA class II binding software (iTope) for the presence of promiscuous moderate- and high-affinity HLA class II binding peptides. Four promiscuous high-affinity HLA class II binding peptides and two moderate-affinity HLA class II binding peptides were predicted in the light chain of rituximab.

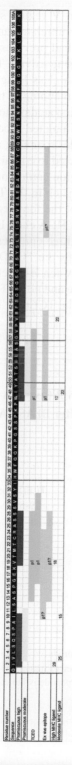

FIGURE 4.3 T-cell epitope analysis of the light-chain V region sequence of murine chimeric anti-CD20 MAb (rituximab). *In silico* tools to predict MHC class II binding peptides (iTope) and analysis of T-cell epitope databases (TCED) were used to identify T-cell epitopes by analyzing sequential overlapping 9-mer peptide fragments. The location of promiscuous high- and moderate-affinity peptides are indicated by red and yellow bars, respectively. The number of binding alleles out of 34 tested is shown for high-affinity binding MHC ligands (high MHC ligand) and moderate-affinity MHC ligands (moderate MHC ligand). The position of p1 anchor residues for promiscuous high- and moderate-affinity core 9-mers aligns with the number of binding alleles, for example, isoleucine at position 2 binds 29 MHC class II alleles with high affinity and is the p1 anchor residue for the core 9-mer IVLSQSPAI. TCED peptides that share homology to the rituximab sequence are highlighted as green bars, and 15-mer peptides that stimulated T-cell responses in an *ex vivo* T-cell epitope mapping assay of the rituximab light-chain V region sequence are shown as blue bars. The proposed position of p1 anchor residues for core 9-mers present in TCED and *ex vivo* T-cell epitope mapping peptides is indicated by p1 in the green or blue bars. Overlapping peptide analysis in *ex vivo* T-cell assays that reveal the location of two alternative p1s are indicated by p1. (From Bryson et al., *Biodrugs*, 24(1), 1–8, 2010.[44])

BLAST analysis of the TCED with the light-chain V region sequence revealed homology with three 15-mer peptide sequences from unrelated antibody sequences (two of which overlapped by 12 amino acids) that had previously been shown to contain T-cell epitopes (Table 4.2 and Figure 4.3) using *ex vivo* T-cell assays. Two of the four predicted promiscuous high-affinity HLA class II ligands were coincident with the TCED matches. *Ex vivo* T-cell epitope mapping of the rituximab light chain using overlapping 15-mer peptides (by 12 amino acids) showed that peptides spanning the two regions identified by the TCED stimulated T-cell responses measured by proliferation and cytokine secretion. Peptides spanning a third region also elicited an *ex vivo* T-cell response that was not predicted to contain high numbers of HLA-DR class II binding ligands (although this region could contain HLA-DQ or -DP restricted binding peptides) or identified as a match in the TCED. Thus, from the rituximab light-chain V region sequence, two of the four regions identified as containing promiscuous high- and moderate-affinity HLA class II binding peptides actually contained T-cell epitopes, and two of the three epitopes identified by the *ex vivo* T-cell epitope mapping were also identified by the TCED. These data show that while the TCED is useful in discriminating between HLA class II ligands and T-cell epitopes, the database is clearly not exhaustive and novel T-cell epitopes may be present in sequences that are not represented in the TCED.

TABLE 4.2

Distribution of MHC Class II Binding Peptides in Rituximab V Region Light Chain Compared to T-Cell Epitopes Identified after BLAST Analysis with TCED and *Ex Vivo* T-Cell Epitope Mapping

| Pocket 1 Anchor Sequence Location | Promiscuous MHC Class II Binding Ligands | | TCED Core 9-mer[a] | *Ex Vivo* T Cell Assay |
	Moderate Affinity	High Affinity		
I2		29	−	
V3	25		−	
I10	(15)		−	+
V19		18	+	+
W47		17	+	+
I48	23		+	+
V60		22		
T74-C88	−	−	−	+

Note: Moderate-affinity MHC class II ligands are peptides that produced a binding score of >0.55 in the iTope scoring function. High affinity refers to the number of ligands with a high binding score (>0.6). In parentheses, a moderate affinity MHC class II peptide is indicated, which binds a high but not promiscuous number of ligands.

[a] Identical core 9-mer match with TCED.

Source: Reproduced from Bryson et al. 2010.[44]

However, since the TCED peptides do not always share exactly identical sequence homology to the test sequence, changes at key HLA anchor and TCR contact positions will influence the ability of a peptide to stimulate a T-cell response. Hence, epitopes identified by TCED may not always be reflected in the partially homologous peptide's ability to stimulate a T-cell response in *ex vivo* T-cell assays. To minimize this effect, the physiochemical properties of the amino acid side chains are compared where residues differ in the core 9-mer between test sequence and a match in the TCED. Only core 9-mers that show >50% homology in the core 9-mer and have conserved residues at the non-homologous anchor positions are scored as hits.

Analysis of antibody V region sequences for T-cell epitopes using the combination of *in silico* HLA class II binding tools and T-cell epitope databases improves the accuracy of *in silico* immunogenicity prediction methods by narrowing down the number of potential T-cell epitopes that are identified by *in silico* HLA class II binding analysis alone. Nevertheless, the limitations of the discussed *in silico* tools highlight the importance of performing additional T-cell epitope analysis, such as whole protein *ex vivo* T-cell assays further downstream in the drug development pathway, in order to ensure that the optimal leads are selected before clinical studies are performed.

4.2.3 Use of HLA Binding Prediction Algorithms for Drug Development

In view of the issues with algorithms that predict peptide binding to HLA class II (as outlined above), the applications for these *in silico* tools need to be carefully considered. The limitation of *in silico* tools in their ability to predict the precise location and potency of T-cell epitopes in therapeutic proteins precludes their use in assessing the potential immunogenicity of lead proteins. However, such tools have found utility in screening large numbers of sequences, often generated during lead discovery, where T-cell epitope analysis by more accurate *ex vivo* methods is impractical. The frequency and promiscuity of HLA class II binding peptides in these novel sequences can be used as criteria in order to select sequences likely to contain fewer potential T-cell epitopes. Since the HLA class II loci are highly polymorphic in an outbred population, resulting in different individuals expressing different alleles to form a haplotype, it is useful for such tools to assess the binding of peptides to a large number of HLA class II alleles in order to be useful in the assessment or removal of immunogenicity from protein therapeutics, particularly since the different HLA class II molecules may have markedly different peptide binding specificities. Once the number of lead candidates has been reduced, for example, during early preclinical development, a more accurate assessment of the number of T-cell epitopes and their potency can be determined by using MHC-associated peptide proteomics (MAPPs)[37, 40] and *ex vivo* T-cell assays.

Another area of application is the analysis of peptides already identified as T-cell epitopes by *in vitro* or *ex vivo* methods in order to identify key residues that may be important for peptide–HLA binding and to select alternative residues that may reduce or eliminate this interaction (de-immunization).[53, 60–62] Furthermore, since it has been demonstrated that single amino acid changes within therapeutic proteins have the potential to modulate the risk of clinical immunogenicity,[63] *in silico* analysis of point mutations made within human therapeutic proteins (introduced to improve their properties) can be used as an initial high-throughput screen to select alternatives with the least immunogenic potential.

4.3 B-CELL EPITOPE PREDICTION

4.3.1 PROPERTIES OF B-CELL EPITOPES

The prediction of B-cell epitopes is complicated by the fact that most epitopes are conformational and discontinuous; that is, they are made up of protein residues that may be distantly separated in the protein sequence (or even on different protein subunits), but are brought into close proximity by the secondary, tertiary, and quaternary structures.[64] A minority of B-cell epitopes are linear, and prediction of these can be useful for designing peptides to be used in immunization strategies to raise antibodies against a particular protein. The structure of B-cell epitopes and binding by antibodies have been comprehensively reviewed.[65]

4.3.2 IN SILICO B-CELL EPITOPE PREDICTION

Early methods relied upon identifying regions within the linear protein sequence with high hydrophilicity, accessibility, and mobility to derive amino acid propensity scales that could be used to identify peptides that were highly likely to be solvent exposed on the surface of the protein and therefore visible to antibodies. A comprehensive analysis of these methods[66] demonstrated that their predictive value was scarcely better than random. As with the prediction of peptides binding to HLA class II, the use of sophisticated computational techniques that assess numerous features of B-cell epitopes has increased the reliability of prediction methods. For example, Chen et al.[67] claim a prediction accuracy of 72.5% when combining a support vector machine (SVM) classifier with a variety of amino acid propensity scales. An improved SVM approach was able to predict linear B-cell epitopes with an accuracy of 76%,[67] and a mathematical morphology approach was demonstrated to be effective, particularly for predicting epitopes with low to moderate antigenicity.[68] More recent approaches include a multiple linear regression tool[69] and a method based on large data sets of both known B-cell epitopes and known non-epitopes.[70] As with all sequence-based prediction tools, the quality and size of the training data set is critical for success.

The prediction of conformational epitopes is far more problematic, particularly if the structure of the antigen is unknown. Methods that use structural data to predict conformational epitopes for any protein include CEP,[71] DiscoTope 2.0,[72] ElliPro,[73] and CE-KEG.[74] Methods have also been developed that use experimental data in conjunction with computation in order to predict conformational epitopes. For example, Pep-3D-Search[75] uses sequence information from mimotopes (linear peptides that mimic conformational epitopes) generated experimentally from peptide phage display libraries and attempts to map the mimotope to the surface of the antigen. Current advances in the prediction of conformational epitopes have been recently reviewed and a summary of available web-based tools presented.[76] The most recent developments include a method that predicts B-cell epitopes with an accuracy of greater than 70%[77] and a combinatorial approach that integrates multiple prediction strategies.[78]

As with T-cell epitope prediction tools, a collection of B-cell epitope prediction tools have been selected and curated at the IEDB (www.iedb.org),[50] which provides a web interface for analysis of protein sequences using Discotope 2.0 and ElliPro.

4.3.3 Use of B-Cell Epitope Prediction Algorithms for Drug Development

The utility of B-cell epitope prediction lies mainly in the field of vaccine design, where it may be desirable to introduce or enhance B-cell epitopes; however, it is an oversimplification of the immune response to assume that successfully raising an antibody response against an antigen will lead to success in vaccination. Predicting the consequence of binding to a particular B-cell epitope, for example, virus neutralization, is likely to be unachievable for the foreseeable future.

From the perspective of engineering protein therapeutics to reduce the risk of immunogenicity, B-cell epitope prediction has been used for veneering, that is, removing identified B-cell epitopes by mutation in an attempt to make a protein invisible to the antibody and reduce/eliminate cross-reaction with antibodies already present in patient sera.[79, 80] Unfortunately, the antibody response is highly adaptive, and this type of engineering will probably result in other surface regions of the protein therapeutic being recognized instead via B-cell epitope spreading, a process that has been well described for autoimmune disorders.[81, 82]

4.4 CONCLUSIONS

There are now a wide variety of sophisticated web tools available for the prediction of both HLA class II binding peptides (potential T-cell epitopes) and B-cell epitopes within the sequences of proteins. HLA class II binding predictions can now be made with moderate accuracy; however, especially in terms of identifying T-cell epitopes, all methods are overpredictive due to their inability to model all the biological processes involved in the formation of a T-cell epitope, but may also fail to predict T-cell epitopes that are associated with peptides that bind weakly to HLA class II. This limitation impacts the utility of these tools for analyzing potential protein therapeutics for their potential for clinical immunogenicity; however, they have utility for high-throughput screening of large numbers of candidate molecules and for estimating the impact of specific mutations on HLA class II binding profiles.

B-cell epitope prediction is still challenging, primarily because most significant epitopes are conformational, thus requiring knowledge of the tertiary and quaternary structures. Furthermore, while T-cell epitope prediction may allow protein reengineering to remove a critical step in the development pathway of a robust ADA response, removal of B-cell epitopes may allow the engineered protein to escape binding by pre-existing antibodies, but may not prevent the adaptive immune response from generating new antibody specificities that will recognize either the engineered region or other surface regions of the protein.

REFERENCES

1. Vugmeyster, Y., Xu, X., Theil, F.-P., Khawli, L. A., and Leach, M. W. (2012). Pharmacokinetics and toxicology of therapeutic proteins: Advances and challenges. *World Journal of Biological Chemistry*, 3(4), 73–92. doi: 10.4331/wjbc.v3.i4.73.
2. Sethu, S., Govindappa, K., Alhaidari, M., Pirmohamed, M., Park, K., and Sathish, J. (2012). Immunogenicity to biologics: Mechanisms, prediction and reduction. *Archivum Immunologiae et Therapiae Experimentalis*, 60(5), 331–44. doi: 10.1007/s00005-012-0189-7.
3. Ordás, I., Feagan, B. G., and Sandborn, W. J. (2012). Therapeutic drug monitoring of tumor necrosis factor antagonists in inflammatory bowel disease. *Clinical Gastroenterology and Hepatology*, 10(10), 1079–87; quiz, e85–86. doi: 10.1016/j.cgh.2012.06.032.
4. Rini, B., Wadhwa, M., Bird, C., Small, E., Gaines-Das, R., and Thorpe, R. (2005). Kinetics of development and characteristics of antibodies induced in cancer patients against yeast expressed rDNA derived granulocyte macrophage colony stimulating factor (GM-CSF). *Cytokine*, 29(2), 56–66. doi: 10.1016/j.cyto.2004.09.009.
5. Hegen, H., Millonig, A., Bertolotto, A., Comabella, M., Giovanonni, G., Guger, M., et al. (2014). Early detection of neutralizing antibodies to interferon-beta in multiple sclerosis patients: Binding antibodies predict neutralizing antibody development. *Multiple Sclerosis*, 20(5), 577–87. doi: 10.1177/1352458513503597.
6. Casadevall, N., Eckardt, K.-U., and Rossert, J. (2005). Epoetin-induced autoimmune pure red cell aplasia. *Journal of the American Society of Nephrology*, 16(Suppl. 1), S67–69. Retrieved from http://www.ncbi.nlm.nih.gov/pubmed/15938038.
7. Stravitz, R. T., Chung, H., Sterling, R. K., Luketic, V. A., Sanyal, A. J., Price, A. S., et al. (2005). Antibody-mediated pure red cell aplasia due to epoetin alfa during antiviral therapy of chronic hepatitis C. *American Journal of Gastroenterology*, 100(6), 1415–19. doi: 10.1111/j.1572-0241.2005.41910.x.
8. Li, J., Yang, C., Xia, Y., Bertino, A., Glaspy, J., Roberts, M., and Kuter, D. J. (2001). Thrombocytopenia caused by the development of antibodies to thrombopoietin. *Blood*, 98(12), 3241–48. Retrieved from http://www.ncbi.nlm.nih.gov/pubmed/11719360.
9. Thistlethwaite, J. R., Stuart, J. K., Mayes, J. T., Gaber, A. O., Woodle, S., Buckingham, M. R., and Stuart, F. P. (1988). Complications and monitoring of OKT3 therapy. *American Journal of Kidney Diseases*, 11(2), 112–19. Retrieved from http://www.ncbi.nlm.nih.gov/pubmed/3277401.
10. Niaudet, P., Jean, G., Broyer, M., and Chatenoud, L. (1993). Anti-OKT3 response following prophylactic treatment in paediatric kidney transplant recipients. *Pediatric Nephrology*, 7(3), 263–67. Retrieved from http://www.ncbi.nlm.nih.gov/pubmed/8318095.
11. Baert, F., Noman, M., Vermeire, S., Van Assche, G., D'Haens, G., Carbonez, A., and Rutgeerts, P. (2003). Influence of immunogenicity on the long-term efficacy of infliximab in Crohn's disease. *New England Journal of Medicine*, 348(7), 601–8. doi: 10.1056/NEJMoa020888.
12. Cassinotti, A., and Travis, S. (2009). Incidence and clinical significance of immunogenicity to infliximab in Crohn's disease: A critical systematic review. *Inflammatory Bowel Diseases*, 15(8), 1264–75. doi: 10.1002/ibd.20899.

13. Tcheng, J. E., Kereiakes, D. J., Lincoff, A. M., George, B. S., Kleiman, N. S., Sane, D. C., et al. (2001). Abciximab readministration: Results of the ReoPro Readministration Registry. *Circulation*, 104(8), 870–75. Retrieved from http://www.ncbi.nlm.nih.gov/pubmed/11514371.

14. Almagro, J. C., and Fransson, J. (2012). Humanization of antibodies. *International Immunology*, 24(7), 1620–33. doi: 10.1093/intimm/dxs032.

15. Wang, S. (2011). Advances in the production of human monoclonal antibodies. *Antibody Technology Journal*, 1–4. Retrieved from http://search.ebscohost.com/login. aspx?direct=true&profile=ehost&scope=site&authtype=crawler&jrnl=22303170&AN =84959721&h=e3EblneNQmiS2ZD8GJ2sWS1wXhQavfOKJvjsIOJbQeg0gvEihr1fT X55vxdA/JZ0asfdyl0g0usWn+wOiwY6+w==&crl=c.

16. Getts, D. R., Getts, M. T., McCarthy, D. P., Chastain, E. M. L., and Miller, S. D. (2010). Have we overestimated the benefit of human(ized) antibodies? *mAbs*, 2(6), 682–94. Retrieved from http://www.pubmedcentral.nih.gov/articlerender.fcgi?artid=3011222&t ool=pmcentrez&rendertype=abstract.

17. Van Schouwenburg, P. A., Rispens, T., and Wolbink, G. J. (2013). Immunogenicity of anti-TNF biologic therapies for rheumatoid arthritis. *Nature Reviews: Rheumatology*, 9(3), 164–72. doi: 10.1038/nrrheum.2013.4.

18. Welt, S., Ritter, G., Williams, C., Cohen, L. S., John, M., Jungbluth, A., et al. (2003). Phase I study of anticolon cancer humanized antibody A33 phase I study of anticolon cancer humanized antibody A33. *Clinical Cancer Research*, 9, 1338–46. Retrieved from http://www.ncbi.nlm.nih.gov/pubmed/12684402.

19. Shankar, G., Shores, E., Wagner, C., and Mire-Sluis, A. (2006). Scientific and regulatory considerations on the immunogenicity of biologics. *Trends in Biotechnology*, 24(6), 274–80. doi: 10.1016/j.tibtech.2006.04.001.

20. Kirshner, S. 2009. Guidance for industry: Assay development for immunogenicity testing of therapeutic proteins. http://www.fda.gov/downloads/Drugs/.../Guidances/ UCM192750.pdf (accessed May 21, 2014).

21. Rosenberg, A. S. 2013. Guidance for industry: Immunogenicity assessment for therapeutic protein products. http://www.fda.gov/downloads/Drugs/GuidanceCompliance RegulatoryInformation/Guidances/UCM338856.pdf (accessed May 21, 2014).

22. Tovey, M. G., Legrand, J., and Lallemand, C. (2011). Overcoming immunogenicity associated with the use of biopharmaceuticals. *Expert Review of Clinical Pharmacology*, 4(5), 623–31. doi: 10.1586/ecp.11.39.

23. Baker, M. P., and Carr, F. J. (2010). Pre-clinical considerations in the assessment of immunogenicity for protein therapeutics. *Current Drug Safety*, 5(4), 308–13. Retrieved from http://www.ncbi.nlm.nih.gov/pubmed/20615174.

24. Kropshofer, H., Vogt, A. B., Thery, C., Armandola, E. A., Li, B. C., Moldenhauer, G., et al. (1998). A role for HLA-DO as a co-chaperone of HLA-DM in peptide loading of MHC class II molecules. *EMBO Journal*, 17(11), 2971–81. doi: 10.1093/ emboj/17.11.2971.

25. Laupéze, B., Fardel, O., Onno, M., Bertho, N., Drénou, B., Fauchet, R., and Amiot, L. (1999). Differential expression of major histocompatibility complex class Ia, Ib, and II molecules on monocytes-derived dendritic and macrophagic cells. *Human Immunology*, 60(7), 591–97. Retrieved from http://www.ncbi.nlm.nih.gov/pubmed/10426276.

26. Fernández-Viña, M. A., Klein, J. P., Haagenson, M., Spellman, S. R., Anasetti, C., Noreen, H., et al. (2013). Multiple mismatches at the low expression HLA loci DP, DQ, and DRB3/4/5 associate with adverse outcomes in hematopoietic stem cell transplantation. *Blood*, 121(22), 4603–10. doi: 10.1182/blood-2013-02-481945.

27. Stern, L. J., and Wiley, D. C. (1994). Antigenic peptide binding by class I and class II histocompatibility proteins. *Structure*, 2(4), 245–51. Retrieved from http://www.ncbi. nlm.nih.gov/pubmed/8087551.

28. Robinson, J., Halliwell, J. A., Hayhurst, J. D., Flicek, P., Parham, P., and Marsh, S. G. E. (2015). The IPD and IMGT/HLA database: Allele variant databases, 43(November 2014), 423–431. doi: 10.1093/nar/gku1161.

29. Kwok, W. W., Kovats, S., Thurtle, P., and Nepom, G. T. (1993). HLA-DQ allelic polymorphisms constrain patterns of class II heterodimer formation. *Journal of Immunology*, 150(6), 2263–72.

30. Castelli, F. A., Buhot, C., Sanson, A., Zarour, H., Pouvelle-Moratille, S., Nonn, C., et al. (2002). HLA-DP4, the most frequent HLA II molecule, defines a new supertype of peptide-binding specificity. *Journal of Immunology*, 169(12), 6928–34. doi: 10.4049/jimmunol.169.12.6928.

31. Greenbaum, J., Sidney, J., Chung, J., Brander, C., Peters, B., and Sette, A. (2011). Functional classification of class II human leukocyte antigen (HLA) molecules reveals seven different supertypes and a surprising degree of repertoire sharing across supertypes. *Immunogenetics*, 63(6), 325–35. doi: 10.1007/s00251-011-0513-0.

32. Sant'Angelo, D. B., Robinson, E., Janeway, C. A., and Denzin, L. K. (2002). Recognition of core and flanking amino acids of MHC class II-bound peptides by the T-cell receptor. *European Journal of Immunology*, 32(9), 2510–20. doi: 10.1002/1521 -4141(200209)32:9<2510::AID-IMMU2510>3.0.CO;2-Q.

33. Conant, S. B., and Swanborg, R. H. (2003). MHC class II peptide flanking residues of exogenous antigens influence recognition by autoreactive T-cells. *Autoimmunity Reviews*, 2(1), 8–12. Retrieved from http://www.ncbi.nlm.nih.gov/pubmed/12848969.

34. Zavala-Ruiz, Z., Strug, I., Anderson, M. W., Gorski, J., and Stern, L. J. (2004). A polymorphic pocket at the P10 position contributes to peptide binding specificity in class II MHC proteins. 11, 1395–1402. doi: 10.1016/j.

35. O'Brien, C., Flower, D. R., and Feighery, C. (2008). Peptide length significantly influences *in vitro* affinity for MHC class II molecules. *Immunome Research*, 4, 6. doi: 10.1186/1745-7580-4-6.

36. Max, H., Halder, T., Kropshofer, H., Kalbus, M., Müller, C. A., and Kalbacher, H. (1993). Characterization of peptides bound to extracellular and intracellular HLA-DR1 molecules. *Human Immunology*, 38(3), 193–200. Retrieved from http://www.ncbi.nlm. nih.gov/pubmed/8106277.

37. Karle, A. C., Oostingh, G. J., Mutschlechner, S., Ferreira, F., Lackner, P., Bohle, B., et al. (2012). Nitration of the pollen allergen bet v 1.0101 enhances the presentation of bet v 1-derived peptides by HLA-DR on human dendritic cells. *PloS One*, 7(2), e31483. doi: 10.1371/journal.pone.0031483.

38. Sercarz, E. E., and Maverakis, E. (2003). MHC-guided processing: Binding of large antigen fragments. *Nature Reviews: Immunology*, 3(8), 621–29. doi: 10.1038/nri1149.

39. Roche, P. A. (1996). Out damned CLIP! Out, I say! *Science*, 274(5287), 526–27. Retrieved from http://www.ncbi.nlm.nih.gov/pubmed/8928005.

40. Rombach-Riegraf, V., Karle, A. C., Wolf, B., Sordé, L., Koepke, S., Gottlieb, S., et al. (2014). Aggregation of human recombinant monoclonal antibodies influences the capacity of dendritic cells to stimulate adaptive T-cell responses *in vitro*. *PloS One*, 9(1), e86322. doi: 10.1371/journal.pone.0086322.

41. Cochran, J. R., Cameron, T. O., and Stern, L. J. (2000). The relationship of MHC-peptide binding and T-cell activation probed using chemically defined MHC class II oligomers. *Immunity*, 12(3), 241–50. doi: 10.1016/S1074-7613(00)80177-6.

42. Lin, H. H., Zhang, G. L., Tongchusak, S., Reinherz, E. L., and Brusic, V. (2008). Evaluation of MHC-II peptide binding prediction servers: Applications for vaccine research. *BMC Bioinformatics*, 9(Suppl. 12), S22. doi: 10.1186/1471-2105-9-S12-S22.

43. De Groot, A. S., McMurry, J., and Moise, L. (2008). Prediction of immunogenicity: *In silico* paradigms, *ex vivo* and *in vivo* correlates. *Current Opinion in Pharmacology*, 8(5), 620–26. doi: 10.1016/j.coph.2008.08.002.

44. Bryson, C. J., Jones, T. D., and Baker, M. P. (2010). Prediction of immunogenicity of therapeutic proteins: Validity of computational tools. *Biodrugs*, 24(1), 1–8. doi: 10.2165/11318560-000000000-00000.

45. Zhang, L., Udaka, K., Mamitsuka, H., and Zhu, S. (2012). Toward more accurate pan-specific MHC-peptide binding prediction: A review of current methods and tools. *Briefings in Bioinformatics*, 13(3), 350–64. doi: 10.1093/bib/bbr060.

46. Reche, P. A., Glutting, J.-P., Zhang, H., and Reinherz, E. L. (2004). Enhancement to the RANKPEP resource for the prediction of peptide binding to MHC molecules using profiles. *Immunogenetics*, 56(6), 405–19. doi: 10.1007/s00251-004-0709-7.

47. Singh, H., and Raghava, G. P. (2001). ProPred: Prediction of HLA-DR binding sites. *Bioinformatics*, 17(12), 1236–37. Retrieved from http://www.ncbi.nlm.nih.gov/pubmed/11751237.

48. Sturniolo, T., Bono, E., Ding, J., Raddrizzani, L., Tuereci, O., Sahin, U., et al. (1999). Generation of tissue-specific and promiscuous HLA ligand databases using DNA microarrays and virtual HLA class II matrices. *Nature Biotechnology*, 17(6), 555–61. doi: 10.1038/9858.

49. Karosiene, E., Rasmussen, M., Blicher, T., Lund, O., Buus, S., and Nielsen, M. (2013). NetMHCIIpan-3.0, a common pan-specific MHC class II prediction method including all three human MHC class II isotypes, HLA-DR, HLA-DP and HLA-DQ. *Immunogenetics*, 65(10), 711–24. doi: 10.1007/s00251-013-0720-y.

50. Kim, Y., Ponomarenko, J., Zhu, Z., Tamang, D., Wang, P., Greenbaum, J., et al. (2012). Immune epitope database analysis resource. *Nucleic Acids Research*, 40(web server issue), W525–30. doi: 10.1093/nar/gks438.

51. Wang, P., Sidney, J., Kim, Y., Sette, A., Lund, O., Nielsen, M., and Peters, B. (2010). Peptide binding predictions for HLA DR, DP and DQ molecules. *BMC Bioinformatics*, 11, 568. doi: 10.1186/1471-2105-11-568.

52. Perry, L. C. A., Jones, T. D., and Baker, M. P. (2008). New approaches to prediction of immune responses to therapeutic proteins during preclinical development. *Drugs in R&D*, 9(6), 385–96. Retrieved from http://www.ncbi.nlm.nih.gov/pubmed/18989990.

53. Jones, T. D., Hanlon, M., Smith, B. J., Heise, C. T., Nayee, P. D., Sanders, D. A., et al. (2004). The development of a modified human IFN-alpha2b linked to the Fc portion of human IgG1 as a novel potential therapeutic for the treatment of hepatitis C virus infection. *Journal of Interferon and Cytokine Research*, 24(9), 560–72. doi: 10.1089/jir.2004.24.560.

54. Peters, B., and Sette, A. (2007). Integrating epitope data into the emerging web of biomedical knowledge resources. *Nature Reviews: Immunology*, 7(6), 485–90. doi: 10.1038/nri2092.

55. Vita, R., Zarebski, L., Greenbaum, J. A., Emami, H., Hoof, I., Salimi, N., et al. (2010). The immune epitope database 2.0. *Nucleic Acids Research*, 38(database issue), D854–62. doi: 10.1093/nar/gkp1004.

56. Lata, S., Bhasin, M., and Raghava, G. P. S. (2009). MHCBN 4.0: A database of MHC/TAP binding peptides and T-cell epitopes. *BMC Research Notes*, 2, 61. doi: 10.1186/1756-0500-2-61.

57. Altschul, S. F., Gish, W., Miller, W., Myers, E. W., and Lipman, D. J. (1990). Basic local alignment search tool. *Journal of Molecular Biology*, 215(3), 403–10. doi: 10.1016/S0022-2836(05)80360-2.

58. Holgate, R. G. E., and Baker, M. P. (2009). Circumventing immunogenicity in the development of therapeutic antibodies. *IDrugs*, 12(4), 233–37. Retrieved from http://www.ncbi.nlm.nih.gov/pubmed/19350467.

59. Grillo-López, A. J., White, C. A., Dallaire, B. K., Varns, C. L., Shen, C. D., Wei, A., et al. (2000). Rituximab: The first monoclonal antibody approved for the treatment of lymphoma. *Current Pharmaceutical Biotechnology*, 1(1), 1–9. Retrieved from http://www.ncbi.nlm.nih.gov/pubmed/11467356.

60. Jones, T. D., Phillips, W. J., Smith, B. J., Bamford, C. A., Nayee, P. D., Baglin, T. P., et al. (2005). Identification and removal of a promiscuous CD4+ T-cell epitope from the C1 domain of factor VIII. *Journal of Thrombosis and Haemostasis*, 3(5), 991–1000. doi: 10.1111/j.1538-7836.2005.01309.x.

61. Cantor, J. R., Yoo, T. H., Dixit, A., Iverson, B. L., Forsthuber, T. G., and Georgiou, G. (2011). Therapeutic enzyme deimmunization by combinatorial T-cell epitope removal using neutral drift. *Proceedings of the National Academy of Sciences of the United States of America*, 108(4), 1272–77. doi: 10.1073/pnas.1014739108.

62. King, C., Garza, E. N., Mazor, R., Linehan, J. L., Pastan, I., Pepper, M., and Baker, D. (2014). Removing T-cell epitopes with computational protein design. *Proceedings of the National Academy of Sciences of the United States of America*, 111(23), 8577–82. doi: 10.1073/pnas.1321126111.

63. Mucha, J. M., Stickler, M. M., Poulose, A. J., Ganshaw, G., Saldajeno, M., Collier, K., et al. (2002). Enhanced immunogenicity of a functional enzyme by T-cell epitope modification. *BMC Immunology*, 3, 2. Retrieved from http://www.pubmedcentral.nih.gov/articlerender.fcgi?artid=65700&tool=pmcentrez&rendertype=abstract.

64. Sivalingam, G. N., and Shepherd, A. J. (2012). An analysis of B-cell epitope discontinuity. *Molecular Immunology*, 51(3–4), 304–9. doi: 10.1016/j.molimm.2012.03.030.

65. Van Regenmortel, M. H. V. (2009). What is a B-cell epitope? *Methods in Molecular Biology*, 524, 3–20. doi: 10.1007/978-1-59745-450-6_1.

66. Blythe, M. J., and Flower, D. R. (2005). Benchmarking B-cell epitope prediction: Underperformance of existing methods. *Protein Science*, 14(1), 246–48. doi: 10.1110/ps.041059505.

67. Chen, J., Liu, H., Yang, J., and Chou, K.-C. (2007). Prediction of linear B-cell epitopes using amino acid pair antigenicity scale. *Amino Acids*, 33(3), 423–28. doi: 10.1007/s00726-006-0485-9.

68. Chang, H.-T., Liu, C.-H., and Pai, T.-W. (2008). Estimation and extraction of B-cell linear epitopes predicted by mathematical morphology approaches. *Journal of Molecular Recognition*, 21(6), 431–41. doi: 10.1002/jmr.910.

69. Lian, Y., Ge, M., and Pan, X.-M. (2014). EPMLR: Sequence-based linear B-cell epitope prediction method using multiple linear regression. *BMC Bioinformatics*, 15(1), 414. doi: 10.1186/s12859-014-0414-y.

70. Singh, H., Ansari, H. R., and Raghava, G. P. S. (2013). Improved method for linear B-cell epitope prediction using antigen's primary sequence. *PLoS One*, 8(5), e62216. doi: 10.1371/journal.pone.0062216.

71. Kulkarni-Kale, U., Bhosle, S., and Kolaskar, A. S. (2005). CEP: A conformational epitope prediction server. *Nucleic Acids Research*, 33(web server issue), W168–71. doi: 10.1093/nar/gki460.

72. Kringelum, J. V., Lundegaard, C., Lund, O., and Nielsen, M. (2012). Reliable B-cell epitope predictions: Impacts of method development and improved benchmarking. *PLoS Computational Biology*, 8(12), e1002829. doi: 10.1371/journal.pcbi.1002829.

73. Ponomarenko, J., Bui, H.-H., Li, W., Fusseder, N., Bourne, P. E., Sette, A., and Peters, B. (2008). ElliPro: A new structure-based tool for the prediction of antibody epitopes. *BMC Bioinformatics*, 9, 514. doi: 10.1186/1471-2105-9-514.

74. Lo, Y.-T., Pai, T.-W., Wu, W.-K., and Chang, H.-T. (2013). Prediction of conformational epitopes with the use of a knowledge-based energy function and geometrically related neighboring residue characteristics. *BMC Bioinformatics*, 14(Suppl. 4), S3. doi: 10.1186/1471-2105-14-S4-S3.

75. Huang, Y. X., Bao, Y. L., Guo, S. Y., Wang, Y., Zhou, C. G., and Li, Y. X. (2008). Pep-3D-Search: A method for B-cell epitope prediction based on mimotope analysis. *BMC Bioinformatics*, 9, 538. doi: 10.1186/1471-2105-9-538.

76. Sun, P., Ju, H., Liu, Z., Ning, Q., Zhang, J., Zhao, X., et al. (2013). Bioinformatics resources and tools for conformational B-cell epitope prediction. *Computational and Mathematical Methods in Medicine*, 2013, 943636. doi: 10.1155/2013/943636.

77. Zhang, J., Zhao, X., Sun, P., Gao, B., and Ma, Z. (2014). Conformational B-cell epitopes prediction from sequences using cost-sensitive ensemble classifiers and spatial clustering. *Biomed Research International*, 2014, 689219. doi: 10.1155/2014/689219.

78. Hu, Y.-J., Lin, S.-C., Lin, Y.-L., Lin, K.-H., and You, S.-N. (2014). A meta-learning approach for B-cell conformational epitope prediction. *BMC Bioinformatics*, 15(1), 378. doi: 10.1186/s12859-014-0378-y.

79. Onda, M., Beers, R., Xiang, L., Lee, B., Weldon, J. E., Kreitman, R. J., and Pastan, I. (2011). Recombinant immunotoxin against B-cell malignancies with no immunogenicity in mice by removal of B-cell epitopes. *Proceedings of the National Academy of Sciences of the United States of America*, 108(14), 5742–47. doi: 10.1073/pnas.1102746108.

80. Weldon, J. E., Xiang, L., Zhang, J., Beers, R., Walker, D. A., Onda, M., et al. (2013). A recombinant immunotoxin against the tumor-associated antigen mesothelin reengineered for high activity, low off-target toxicity, and reduced antigenicity. *Molecular Cancer Therapeutics*, 12(1), 48–57. doi: 10.1158/1535-7163.MCT-12-0336.

81. Thrasyvoulides, A., and Lymberi, P. (2003). Evidence for intramolecular B-cell epitope spreading during experimental immunization with an immunogenic thyroglobulin peptide. *Clinical and Experimental Immunology*, 132(3), 401–7. Retrieved from http://www.pubmedcentral.nih.gov/articlerender.fcgi?artid=1808729&tool=pmcentrez&rendertype=abstract.

82. Routsias, J. G., Vlachoyiannopoulos, P. G., and Tzioufas, A. G. (2006). Autoantibodies to intracellular autoantigens and their B-cell epitopes: Molecular probes to study the autoimmune response. *Critical Reviews in Clinical Laboratory Sciences*, 43(3), 203–48. doi: 10.1080/10408360500523837.

5 Application of Mechanistic Pharmacokinetic– Pharmacodynamic Modeling toward the Development of Biologics

Pratap Singh, Abhinav Tiwari,*
Anson K. Abraham,† and Anup Zutshi*

CONTENTS

* Pharmacokinetics, Dynamics and Metabolism—New Biological Entities, Translational Modeling and Simulation (TMS), Worldwide Research and Development, Pfizer, Inc., Cambridge, Massachusetts
† Pharmacokinetics, Pharmacodynamics & Drug Metabolism (PPDM)—Quantitative Pharmacology & Pharmacometrics (QP2), Merck & Co., Inc., West Point, Pennsylvania

5.1   INTRODUCTION

Proteins and peptides have increasingly gained popularity as therapeutic agents since 1982, when human insulin was first approved by the U.S. Food and Drug Administration (FDA). Using proteins as therapeutics is an obvious treatment, as many diseases are caused by mutations in the proteins (e.g., β-glucocerebrosidase, lactase, pancreatic enzymes) or by an excess or lack of them (e.g., insulin, growth hormone, blood clotting factors). To date, more than 130 proteins or peptides have been approved by the FDA as therapeutics,[1] and more than 350 monoclonal antibodies, a class of protein therapeutics, are in various stages of preclinical and clinical development.[2]

This chapter highlights the utility of PK/PD and systems pharmacology models in addressing critical questions at key stages of biologic drug discovery and development (Figure 5.1). At the very early stage, when a target is being explored as a new intervention point, a systems pharmacology model can be used to examine the target's role in disease pathology and check the feasibility of the approach given the ubiquitous complexity of biological pathways. Following target validation, it becomes critical to identify lead compounds that are capable of inhibiting the target with high selectivity and potency. At this stage, mechanistic models that contain information about *in vivo* target expression and turnover rates can be applied to provide guidance on affinity requirements for lead biologics. We have provided general guidance on affinity requirements for targeting common types of receptors—soluble and membrane-bound. Once a lead compound with ideal affinity is selected for clinical development, empirical PK and PK/PD models can be applied to project human PK and target coverage in tissues of interest. In the last few sections, we introduced the basic concepts of TMDD models for biologics with non-linear PK, and SoA models for biologics against targets residing in specific tissues. Such models are instrumental for designing FIH trials and ensuring clinical testing of mechanisms.

The following sections focus on the disposition of a class of protein therapeutics called monoclonal antibodies (mAbs), 40 of which have been approved by the FDA.[2] These mAbs function by either delivering a payload to cells or sequestering a target of interest (receptor or ligand), thereby interfering with the mechanism that causes disease. Various sections of this chapter highlight key modeling and simulation (M&S) approaches that are being used to address specific questions and issues that arise during various stages of drug discovery (Figure 5.1).

5.1.1   MONOCLONAL ANTIBODY ISOTYPES

mAbs are a class of molecules that belong to the immunoglobulin Ig superfamily and contain two identical heavy and two identical light chains. Depending on the

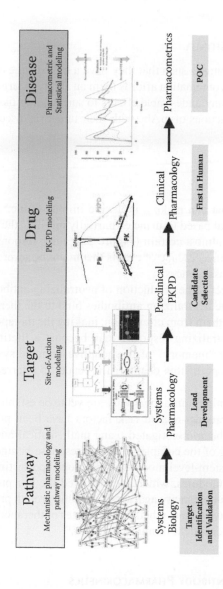

FIGURE 5.1 Modeling and simulation approaches are used throughout drug discovery.

characteristics of heavy chains present, mAbs can be classified further into 5 different isotypes (i.e., IgA, IgD, IgE, IgG, and IgM), as presented in Table 5.1. Two isotypes of light chains are present in the human body: κ and λ. Both heavy and light chains contain complementarity determining regions (CDRs) that recognize epitopes on their targets, such as soluble proteins or cell surface receptors.

5.1.2 SPECIFICITY AND AFFINITY

mAbs are designed with epitopes such that they have a high degree of specificity and affinity for the target. This combination affords them an ability to target antigens while at the same time reducing any undesirable effects to surrounding tissues. The impact of the affinity on the interactions of mAbs with their targets is discussed in detail in Section 5.3.

5.1.3 IMMUNOGENICITY

There are many factors that can make a molecule immunogenic.[4] If the Ig source for the therapeutic is animal based (murine or partially murine), the likelihood of an immunogenic response is almost certain (Figure 5.2). Such responses, caused by a foreign protein, can result in a cascade of events that might be severely anaphylactic, leading to catastrophic consequences.

A relatively benign effect is the production of neutralizing antibodies against the therapeutic antibody. These neutralizing antibodies bind to the therapeutic antibodies and effectively remove them from circulation, making the therapeutic ineffective. Unfortunately, such a neutralizing response may become a lifetime issue for the therapeutic because of the "memory" of the adaptive immune system. Additional details are beyond the scope of this chapter; the reader is referred to many good reviews on the subject.[5, 6]

The mechanisms governing immunogenicity are yet to be established, but several patient-, treatment-, and product-related risk factors have been proposed.[7, 8] Some of these risk factors have been investigated using mathematical modeling to gain a quantitative understanding of the complex immune response behind immunogenicity.[9–12] More recently, a system-level model was developed consisting of subcellular, cellular, and whole-body modules, which recapitulated the key processes underlying T-cell-dependent immunogenicity against therapeutic protein products.[13, 14] Later this model was also used to evaluate whether aggregates of therapeutic protein products could induce T-cell-dependent immune response.[15]

5.1.4 MONOCLONAL ANTIBODY PHARMACOKINETICS

The molecular size of a mAb determines its disposition from the body, and since most mAbs are IgG molecules, their disposition behavior is very similar to that of endogenous immunoglobulins. As shown in Table 5.1, IgG has the longest half-life (~21 days) and is the most popular construct of choice for therapeutic development.

TABLE 5.1

Human Ig Subtypes and Their Properties

Ig Isotype	Valence	Molecular Weight (kDa)	Steady-State Concentration in Serum (mg/ml)	Self-Association	Proportion of Total Ig Concentration	Serum Half-Life (days)	Function
IgA1	2, 4	160	3	Monomer, dimer	15	6	Mucus immunity
IgA2	2, 4	160	0.5		2	6	
IgD	2	184	0.03	Monomer	<1	3	B-Cell antigen receptor
IgE	2	188	0.00005		<1	2	Mammalian hypersensitivity
IgG1	2	146	9		50	21	Activation, antibody-dependent cell cytotoxicity, neonatal immunity, feedback
IgG2	2	146	3		15	20	
IgG3	2	170	1		5	7	
IgG4	2	146	0.5		3	21	
IgM	10, 12	970	1.5	Pentamer, hexamer	10	10	B-Cell antigen receptor, complement activation

Source: Lobo, E.D., et al., *J Pharm Sci*, 93(11): 2645–68, 2004.

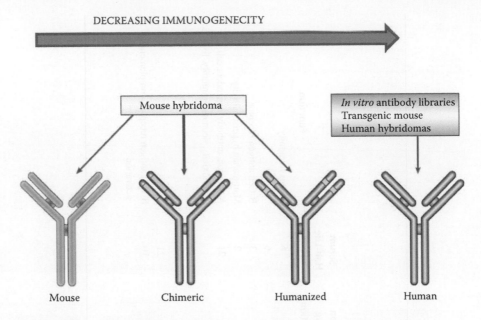

DECREASING IMMUNOGENECITY

Mouse hybridoma

In vitro antibody libraries
Transgenic mouse
Human hybridomas

Mouse   Chimeric   Humanized   Human

FIGURE 5.2   Engineered mAbs. (Adapted from Brekke, O.H., and Sandlie, I., *Nat Rev Drug Discov*, 2(1): 52–62, 2003.[16])

Intravenous (IV) administration of IgG therapeutic antibodies exhibits a classic biexponential decay profile if the doses result in plasma levels that are at least tenfold higher than the target concentrations (the reasons for this condition will become clear in Section 5.3). Usually the distribution-dominated phase (alpha-phase) lasts about 3–7 days, followed by a 20- to 27-day elimination-dominated phase (beta-phase). The pharmacokinetic (PK) parameters for such a typical profile are assumed to be known in models that are developed during the early phases (prelead development) of drug discovery. It is important to note that these parameters are useful only when the PK of a mAb is linear; that is, the expected concentrations of mAb are significantly greater ($\geq$ tenfold) than the target concentrations, particularly if the target is membrane-bound. The typical PK parameters for an IV or subcutaneous (SC) dose are summarized in Table 5.2,[17] and in the absence of additional information, these are used as a first approximation in simulations.

5.1.5   ADME Properties of mAbs

This section focuses on essential ADME properties of mAbs; for a more general understanding, the reader is referred to the several good reviews on this topic.[3, 17–20] The differences in the ADME properties of large-molecule biologics and small-molecule drugs are presented in Table 5.3.

TABLE 5.2

Typical Linear PK Parameters for an IgG-Based mAb

Parameter	Value
Clearance (CL)	0.26 L/day
Distributional clearance (Q)	0.56 L/day
Volume of distribution of central compartment (Vc)	3.06 L
Volume of distribution of peripheral compartment (Vp)	3.1 L
Volume of distribution at steady state (Vss)	6.2 L
SC bioavailability	60%
Absorption rate constant (k_a)	0.26 L/day

5.1.5.1 Absorption

Most mAbs are administered parenterally, as absorption via the enteral route is limited for the reasons described earlier. Although IV dosing is readily feasible, it requires a medical professional to administer it. On the other hand, SC and intramuscular (IM) administrations are easier and favored commercially by pharmaceutical companies. However, both SC and IM administrations are limited by the volume that can be delivered without significant pain or discomfort, which in the case of the former, is approximately a 1–2 ml/injection. Nonetheless, novel techniques have been developed that expand the volume of SC space by using hyaluronidase enzymes that hydrolyze connective tissues (Halozyme® Therapeutics). The volume limitations, in combination with the solubility limits of mAbs in aqueous solutions (typically 100 mg/ml,[21] although some advanced formulations have achieved 150 mg/ml), restrict the dose of mAb that can be delivered via a single SC administration to about 100–150 mg. A higher-dose requirement necessitates an increase in the number of SC injections. Another aspect that needs to be considered during development is that concentrations approaching 100 mg/ml may result in unacceptably high viscosity (>20 cP), which makes the formulation difficult to deliver using a syringe. In comparison with small-molecule drugs, the absorption of mAbs from the SC site into systemic circulation is a slow process, as it involves transfer through lymph ducts. This slow absorption ($t_{1/2} \sim 2.7$ days) results in a T_{max} around 36–72 h. Since lymphatic fluid is sampled constantly by the immune cells in lymph nodes, the initial interactions between mAb and immune cells may result in an immunogenic response at or close to the site of administration. The bioavailability of several mAbs, evaluated in the clinic and preclinically in non-human primates,[22] ranges from 30% to 90%. The bioavailability observed in non-human primates is generally higher than that in humans; nonetheless, it is a good surrogate,[22] and typically a value of 60% is used as a first approximation for modeling.

5.1.5.2 Distribution

Just like absorption, distribution of a mAb into the systemic circulation and tissue interstitium is also restricted by its size. The distribution of mAbs from the systemic

TABLE 5.3
Differences in the ADME Properties of Large-Molecule Biologics and Small-Molecule Drugs

Property	Biologics	Small Molecules
Size	Large Molecules, MWt ~ 150 kDa	Small Molecules, MWt ~ < 1 kDa
Physicochemical Properties	Limited knowledge, Complex properties, Limited Solubility, Viscosity, Charge, Isoelectric point, Stability	Extensive Knowledge, Well understood properties, Solubility, Stability, Ionization
Synthesis/Manufacture	Biological process, Heterogenous	Chemical Process; Homogenous
ADME Properties	Knowledge still evolving, Relatively simple properties	Advanced knowledge, Properties can be quite complex
Administration	Only parenteral methods (IV, SC, IM); No Oral Administration	Oral is most common route followed by parenteral
Distribution	Restricted distribution due to size, Volume of distribution to blood and extracellular fluid space (interstitium)	Extensive distribution to all body spaces including intracellular space. Volumes of distribution can exceed body water
Metabolism	Catabolism by proteolytic enzymes, Clearance by FcRn recycling	Metabolism by hepatic/renal Cytochrome P450 enzymes
Excretion	Excretion of small fragments by renal excretion	Biliary, Renal, and Pulmonary are major routes of excretion
Pharmacokinetics	Can be non-linear at low doses because of Target Mediated Disposition. Usually linear at higher doses. Pharmacokinetics are relatively similar between mAb constructs	Is linear at low doses and non-linear at doses that saturate a clearance mechanism. Pharacokinetics can be multivariate and very different between similar analogues
Systemic Half-Life	Long $t_{1/2}$, days and weeks	Relatively shorter $t_{1/2}$, minutes to days
Selectivity	Designed to be highly selective to its target, imparts some targeting ability	Selectivity can be a problem. Coupled with the extensive distribution, can result in cross-talk to other targets and potential toxicity
Special Binding	Physiological property of binding to the FcRn (Neonatal Fc receptor) allows the body to extend the residence time of the mAb	No physiological related binding behavior; Plasma and tissue protein binding is possible due to certain physiochemical properties of the molecule
Drug–Drug Interactions	Very few known; related more to the pharmacodynamic response	Depending on the Cytochrome P450 isomer involved in the Clearance, potential drug interactions affected by the same clearance pathway can be significant

(Continued)

TABLE 5.3 (CONTINUED)
Differences in the ADME Properties of Large-Molecule Biologics and Small-Molecule Drugs

Property	Biologics	Small Molecules
Immunogenecity	The biological source can generate an immunogenic response	Very rare; only possible if covalent interactions between drug and some systemic protein
Analytics	Ligand Binding Assays, Radioimmunoassays, Limited Mass spectroscopic techniques	Very high end chemical analytics using mass spectral methods and sample concentration techniques

Source: Shi, S., *Curr Drug Metab*, 15(3): 271–90, 2014.

circulation to tissues is a slow process, as evident by the relatively (compared with small molecules) slow appearance of the terminal kinetic phase in the biphasic plasma profile of mAb. Using radiolabeled ^{111}In, Baxter et al.[23] showed that about 10% of plasma levels of mAb distributed into tissue spaces after a pseudoequilibrium is reached. This ratio was observed to be similar in mice.[24] Moreover, since the total interstitial volume is about one-third of the total body water (0.6 L/kg),[25] one can assume that the interstitial space in every tissue is also about a one-third of the tissue volume.[26] Consequently, 30% of the plasma levels of mAb are expected to distribute into the tissue interstitial space. So if the mAb concentration in plasma is known, then it can be used to approximate the mAb concentration in tissue interstitium. An elegant experiment conducted at Novartis confirmed that about 30% of the plasma levels of mAb are found in the skin interstitium. In this experiment, they studied Secukinumab, an anti-IL-17 mAb for psoriasis, by perfusing the dermal interstitium and measuring the plasma, as well as the interstitial perfusate for IL17 and Secukinumab. They observed 23% and 35% of the plasma levels of mAb in the dermal interstitial perfusate for healthy and diseased individuals, respectively.[27]

Note that intracellular distribution of a molecule the size of an antibody is not possible by diffusion. Typically, receptors on the cell surfaces interact with the antibody and internalize it into the cell. This process is known as target-mediated drug disposition (TMDD)[28] and is responsible for affecting mAb kinetics in the blood (see Section 5.4 for details).

5.1.5.3 Metabolism

Biologics with well-defined protein structures are not subjected to the cytochrome P450-based metabolism, which is common for small molecules. In general, mAbs are cleared by three pathways: (1) catabolism by peptidases, (2) recycling by the FcRn receptor (also called the Brambell receptor, and (3) TMDD. Since these biologics are protein molecules, they are expected to be degraded by proteolytic enzymes such as peptidases.

FcRn recycling is an endogenous process that regulates the concentrations of IgG-based molecules in the body. Specifically, FcRn receptors on different cells (e.g., vascular endothelial, macrophages, dendritic, hepatocytes, epithelial, and kidney) bind IgG molecules at low pH upon internalization, thereby protecting them from catabolism. During and after internalization, a fraction of IgG molecules are broken down in the lysosomal vacuoles and cleared, while the remaining fraction is returned to the cell surface and secreted back into the interstitium. This extends the molecule's residence time in the body. Many pharmaceutical and biotechnology companies have exploited this mechanism to design molecules with half-lives much longer than the usual 21 days observed for a standard IgG-based molecule.[29]

The third clearance mechanism is TMDD. When a mAb binds to its target present on circulating immune cells, endothelial cells or in tissue spaces, the bound antibody can be internalized and degraded. This depletion of a mAb causes the plasma levels to drop precipitously, which is characterized by a classic convex shape of a non-linear kinetic process (see Figure 5.8 and Section 5.4.1.2 for details). TMDD is observed at lower concentrations of mAb, and as the concentrations of mAb in plasma and interstitium increase, the process gets saturated, after which the antibody kinetics become linear. This is in stark contrast to small molecules where saturable kinetic behavior is observed at higher drug concentrations, at which the clearance mechanism has saturated (usually cytochrome P450-based enzymes).

5.1.5.4 Excretion

Most of the peptides and proteins that are produced by proteolytic degradation of mAbs are assimilated by the body and utilized. In some instances (MW < 60 kDa), these catabolized fragments can be excreted by kidneys. A more detailed treatise on ADME properties is presented in the review by Shi.[20]

5.2 TARGET SELECTION AND VALIDATION USING SYSTEMS PHARMACOLOGY MODELS

Quantitative systems pharmacology (QSP) models are increasingly being used in the pharmaceutical industry at all stages of drug discovery, as they provide a quantitative and integrated perspective for decision making instead of empirical observations and qualitative reasoning. Specifically, QSP models utilize systems biology methods to identify and validate novel targets for therapeutic intervention, elucidate mechanisms of action of new and existing drugs, and create a repository of knowledge that guides modulation of complex networks.[30,31] As an example, Figure 5.3 depicts the coagulation pathway that governs the human blood clotting process. Such models have been developed for several diseases, such as breast cancer,[32,33] arrhythmias,[34] insulin resistance,[35] and depression,[36] and have provided value at various stages of drug discovery and development.[37,33,35] A good example of QSP modeling's impact on biologic drug discovery is demonstrated by the ErbB network model that led to the discovery of MM-121, a monoclonal antibody showing a potent effect in the mice tumor xenograft model.[37,38] A similar modeling approach also led to the development of MM-141, a tetravalent bispecific antibody targeting IGF-IR and ErbB3.[39] QSP models can also help in the design of clinical trials with a robust biomarker

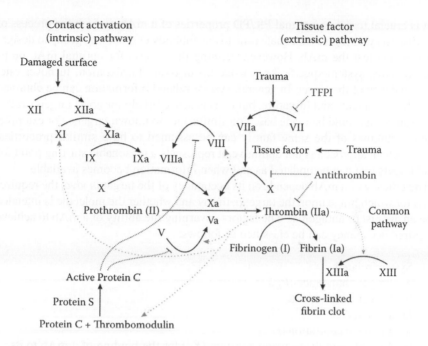

FIGURE 5.3 Systems model of blood coagulation pathway in humans. (From Coagulation cascade, 2007, http://commons.wikimedia.org/wiki/File%3ACoagulation_full.svg.[41])

strategy. Recently, a QSP model was developed for this pathway to better understand the effects of *in vitro* modulation of various network components on coagulation.[40]

Despite its promise, QSP modeling faces many challenges: uncertain network topology, limited quantitative data, largely unknown parameter space, and experimentally unobservable state variables. Often, available experimental data allow the estimation of only a few parameters, rendering many critical parameters unidentifiable. Even when experimental assays are in place for data generation, it is challenging to determine optimal sampling and dosing strategies. Nevertheless, it is imperative that one continues to strive for a quantitative systems-level understanding of biological processes by developing mathematical models. In the next section, a simple binding model is discussed that can be used to guide the process of mAb lead optimization.

5.3 MONOCLONAL ANTIBODY LEAD OPTIMIZATION

Successful development of a mAb is primarily governed by its PK half-life, selectivity and affinity for the target, physiochemical stability, degradation mechanisms, and immunogenicity risk.[42] Owing to the technological advances in antibody engineering, many of these mAb properties can be tailored to meet specific requirements.[43–46] This section primarily focuses on the optimization of PK/PD properties of mAbs, that is, PK half-life and target affinity. Details on the mathematical analysis discussed below will be submitted elsewhere (manuscript in preparation).

It is crucial to define optimal PK/PD properties of a mAb early in the process of drug discovery, as it gives adequate time to the antibody engineering group to design, produce, and test the mAb. However, defining the criteria for optimal mAb properties requires system-specific knowledge about tissue localization, turnover rate, and abundance of the target. In general, system-related information can be obtained through experiments and literature, but in its absence (likely for novel targets), rational assumptions could be made based on similarities to known targets. For example, soluble cytokines of the same family can be assumed to have similar properties. While such an approach is not definitive, it represents a reasonable starting point for novel targets that can be updated as and when information becomes available.

The efficacy of a mAb depends on its occupancy of the target *in vivo*, the requirements for which hinge upon the target pathway and whether the molecule is intended to be an agonist or antagonist. The factors governing the ability of a mAb to achieve a required occupancy can be classified as follows:

1. System related
   a. Target abundance (R_{tot})
   b. Target turnover rate
2. Drug-related
   a. mAb concentration (D_{tot})
   b. Equilibrium dissociation constant (K_D) for the binding of a mAb to its target

System-related factors are primarily driven by the disease state and cannot be modified, but their knowledge allows the optimization of PK half-life and K_D of a mAb as per the requirements. In general, clinical mAbs have well-characterized PK properties,[17] which in turn determine the dosing concentrations. Below, a relationship is derived between three of the four factors listed above: D_{tot}, R_{tot}, and K_D. For simplicity, the synthesis and degradation of the target and elimination of the mAb and mAb–target complex is ignored, as these processes tend to occur on a much slower timescale. As a result, the system reflects the conditions found *in vitro*. Furthermore, the binding of a mAb to the target is assumed to be in rapid equilibrium. Hence,

$$K_D = \frac{[D][R]}{[DR]}$$ (5.1)

Also, due to conservation of mass,

$$D_{tot} = [D] + [DR] \text{ and } R_{tot} = [R] + [DR]$$ (5.2)

Substituting the expressions for [D] and [R] from Equation 5.2 into Equation 5.1 and rearranging,

$$D_{tot} = \frac{[DR]}{R_{tot} - [DR]} K_D + [DR]$$ (5.3)

Dividing the above equation by R_{tot} and replacing $[DR]/R_{tot}$ by occupancy (occ),

$$\frac{D_{tot}}{R_{tot}} = \frac{occ}{(1-occ)}\frac{K_D}{R_{tot}} + occ \qquad (5.4)$$

Plots characterizing the biphasic relationship between D_{tot} and K_D for various R_{tot} concentrations have been reported elsewhere.[47] Here, the above relationship is plotted after normalization by R_{tot} (see Equation 5.4 and Figure 5.4), as it allows a more general characterization of the drug-related parameters vis-à-vis the target. It becomes readily evident that the inflection point (1) occurs when $K_D \sim 0.1\ R_{tot}$ and (2) is independent of the requirements of target occupancy. When $K_D > 0.1\ R_{tot}$ (shaded region in Figure 5.4), the total mAb concentrations (D_{tot}) required to achieve desired target occupancy increase exponentially. As dose is correlated with D_{tot}; a mAb that lies in this region will require a significantly higher dose and thus may be termed as suboptimal. A suboptimal mAb with $K_D = R_{tot}$ would require doses of $\sim 10\ R_{tot}$ to achieve 90% target occupancy. On the other hand, for a mAb whose target affinity has been enhanced such that $K_D = 0.1\ R_{tot}$, the dose required to achieve similar target occupancy is $\sim R_{tot}$. Further improvements in affinity, that is, $K_D < 0.1\ R_{tot}$, will not result in proportionate dose reductions, and hence the mAb may be considered optimal.

Given some basic understanding of the target occupancy requirements, antibody engineering teams are generally faced with the choice of optimizing either the K_D or the PK half-life. For a suboptimal mAb (shaded region in Figure 5.4), the dose required to achieve 90% target occupancy ($dose_{90}$) is not affected by the PK half-life, but is significantly reduced by increasing the affinity (see arrow from point A to B in Figure 5.5). By contrast, for an optimal mAb (unshaded region in Figure 5.5) $dose_{90}$ can be only reduced, albeit marginally, by increasing the PK half-life (see arrow from point B to C). Although improvements in PK half-life may not reduce the dose considerably, they do confer the advantage of less frequent dosing. Note that the above conclusions are independent of the target occupancy requirements.

Next, the relationship between K_D and $dose_{90}$ is evaluated for the impact of target properties such as baseline concentration ($R_{tot,0}$), elimination rate (k_{elR}), and soluble

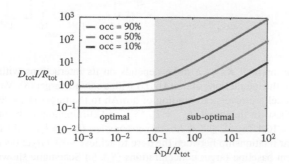

FIGURE 5.4 Biphasic relationship between D_{tot}/R_{tot} and K_D/R_{tot} (Equation 5.4) for 10%, 50%, and 90% target occupancy (occ). Shaded and unshaded regions represent the linear and constant phase, respectively, of the biphasic relationship.

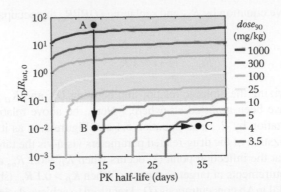

FIGURE 5.5 Optimizing PK/PD properties of a mAb. Contour plots show the impact of modifying mAb properties, namely, PK half-life and affinity relative to target levels ($K_D/R_{tot,0}$), on the dose required to achieve 90% target occupancy ($dose_{90}$) for a monthly dosing regimen. Shaded and unshaded regions have the same meaning as in Figure 5.4. Model parameters: $k_{elR} = 16.4$ day^{-1}, $k_{elDR} = k_{elD}$, $k_{on} = 86.4$ nM^{-1} day^{-1}, $k_{off} = k_{on}*K_D$, $k_{synR} = k_{elR}*R_{tot,0}$. Note: This model reflects *in vivo* conditions, as it accounts for target synthesis and elimination of mAb, target, and mAb–target complex.

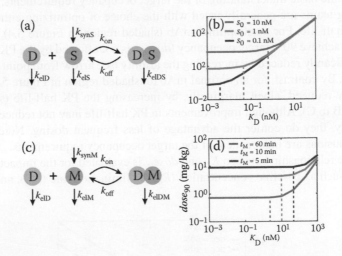

FIGURE 5.6 The optimal K_D of a mAb depends on its interactions with the target. (a) Schematic showing the interactions of a mAb (D) with a soluble target (S). Various processes included in the model are reversible binding of a mAb to the target ($k_{on} = 86.4$ nM^{-1} day^{-1}, $k_{off} = k_{on}*K_D$), synthesis of the target ($k_{synS} = k_{elS}*S_0$), and elimination of the mAb ($k_{elD} = 0.03$ day^{-1}), target ($k_{elS} = 16.4$ day^{-1}), and mAb–target complex ($k_{elDS} = k_{elD}$). (b) Plots characterizing the non-linear relationship between the dose to achieve 90% target occupancy ($dose_{90}$) and K_D for different baseline target concentrations (S_0). (c) Schematic showing the interactions of a mAb (D) with a membrane-bound target (M). The model processes are the same as in (a), with differences in the following parameters: $k_{synM} = k_{elM}*M_0$, $M_0 = 1$ nM, $k_{elM} = 0.693/t_M$, and $k_{elDM} = k_{elM}$. (d) Plots characterizing the non-linear relationship between $dose_{90}$ and K_D for different half-lives (t_M) of elimination for membrane-bound target.

versus membrane-bound characteristics. For clarity, $R_{tot,0}$ and k_{elR} for soluble and membrane-bound targets are defined as S_0 and M_0, and k_{elS} and k_{elM}, respectively. Simulations using a soluble target model (Figure 5.6a) show that decreasing K_D results in lower $dose_{90}$ until a threshold (dashed lines in Figure 5.6b), below which there is no dose reduction. Moreover, this threshold increases with the baseline concentration of the soluble target (Figure 5.6b), a relationship that also holds true for the membrane-bound target (not shown). Another factor that contributes to the location of threshold is the turnover of the mAb–target complex. For a membrane-bound target (Figure 5.6c), the turnover of the mAb–target complex is assumed to be equal to that of the free target, and consequently, the threshold is also governed by the half-life of target elimination (Figure 5.6d).

5.4   PROJECTION OF HUMAN PK, TARGET OCCUPANCY, AND STARTING AND EFFICACIOUS DOSES FOR FIH TRIALS

Paracelsus, the founder of toxicology, wrote in the sixteenth century: *dosis facit venenum* ("it's the dose that makes poison"). This centuries-old quote truly resonates with the modern drug discovery and development process, where a fine balance is required between safety and efficacy as the lead candidate advances toward clinical testing. To ensure a successful clinical testing, the drug candidate should be tested in a dose range that maximizes the probability of observing therapeutic effects without any safety concerns. Thus, a rational selection of the starting dose and subsequent dose escalation scheme to the top dose is key for the developability of a new biological entity (NBE). In the following section, the current approaches for the prediction of human PK of a NBE are presented. These approaches allow for robust predictions of exposure parameters, such as maximum drug concentration (C_{max}), area under the curve (AUC), and minimum drug concentration (C_{min}) at a given dose. Subsequently, quantitative models for predicting human pharmacodynamics (PD), target occupancy, and efficacious concentration are discussed in detail. Finally, a rational approach to select the FIH starting dose and the maximum top dose is presented based on the minimum anticipated biological effect level (MABEL) and target occupancy concept.

5.4.1   Prediction of Human PK

5.4.1.1   Linear PK Models

Most mAbs and protein biotherapeutics exhibit linear PK as a function of dose. PK data obtained in preclinical species such as rat, mouse, or monkeys are typically modeled by simple mathematical models describing monoexponential, biexponential, or multiexponential profiles seen in plasma or serum exposures. Depending on the number of phases observed, PK data can be fitted by appropriate compartmental models, as shown in Figure 5.7a and b. These models are empirical in nature and contain parameters related to *in vivo* drug distribution and clearance processes, which are estimated from exposure data from PK studies in preclinical studies. Eventually, these parameters are scaled using allometric scaling techniques to predict human

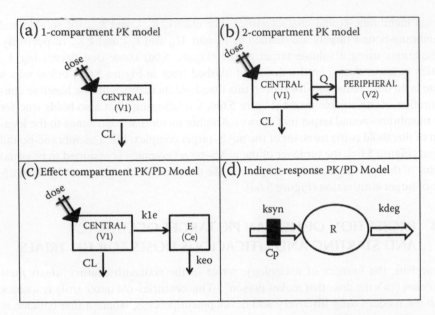

FIGURE 5.7 Representative examples of PK and PK/PD models typically applied within drug discovery and development. (a) One-compartment PK model. (b) Two-compartment PK model. (c) PK/PD model linked to describe drug distributional delays. (d) An indirect-response PK/PD model incorporating the effect of a drug on the receptor synthesis rate.

exposures at intended doses. Mordenti and colleagues[48] were the first to explore the scaling of clearance and volume of distribution, and showed that for each protein, the data were well described by an allometric equation ($Y = aW^b$), where a is allometric coefficient, b is allometric exponent, and W is body weight. Based on this data set, $0.65 < b < 0.84$ for the clearance parameter and $0.83 < b < 1.05$ for the volume of distribution at steady state. Subsequently, others have performed similar analyses on larger data sets of 15 protein drugs,[49,50] 34 therapeutic proteins,[51] and 14 mAbs.[52] More recent work includes those from Deng et al.[53] and Oitate et al.[54,55] Most of the work described above utilized noncompartmental analysis for estimating clearance and other parameters. While this is a general approach that works for any PK profile, it is well recognized that mAbs and many biologics exhibit a biphasic response indicative of a rapid distribution phase, followed by a long terminal phase. More scholarship is needed in this area to explore scaling of 2-compartmental PK parameters.

5.4.1.2 TMDD Models

While the majority of mAbs and protein biotherapeutics display linear PK in the therapeutic dose range, many biologics currently in the market exhibit non-linear PK due to TMDD.[17] Figure 5.8 presents a typical non-linear PK profile seen for antibodies with TMDD characteristics. As shown, faster clearance is seen at lower doses since a significant portion of dose is cleared through target engagement and subsequent internalization and degradation of the drug–target complex via trafficking to lysosomes. At higher doses, target-mediated clearance gets saturated, resulting

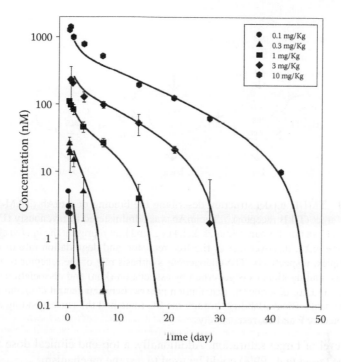

FIGURE 5.8   Characteristic non-linear PK profile observed for antibodies exhibiting target-mediated drug disposition (TMDD) behavior. Symbols represent observed data, and solid lines represent fitting of the data using a TMDD model similar to the one shown in Figure 5.9. Doses were administered intravenously. (From doi:10.1208/s 12248-014-9690-8. 2015 Mar, 19(2):389–99. Supplemental info.)

in a longer terminal half-life. The magnitude of TMDD depends greatly on target expression and the turnover rate of the drug–target complex.[28] Naturally, for biologics with TMDD, projection of human PK can be challenging, especially at the starting doses, where serum exposures are not high enough to fully saturate the target. A recent report[56] addressed this by proposing a set of translational rules that can be applied to monkey data to project human PK of mAbs. These rules provide specific guidance on how drug- or target-related parameters should be scaled based on the modeling of monkey data, using models that describe drug–target binding and various turnover processes. As an illustration of this approach, Figure 5.9 presents a TMDD model (assuming quasi-steady-state approximation) built to project human PK of the anti-TrkB mAb (called TAM163) based on the fitting of cynomolgus and rhesus monkey PK data.[57] Available preclinical data allowed the estimation of drug-related (e.g., drug elimination and distributional rate constants) and target-related (e.g., receptor expression) parameters. Furthermore, *in vitro* assays were developed to estimate the binding rate constants, kon and koff, and receptor turnover parameters, kdeg and kint. Such experimental measurements allowed complete characterization of the TMDD model, which in turn could be used for human projections. A well-characterized TMDD model is especially beneficial, as it provides information about the *in vivo* receptor occupancy and the clinical doses that will result in

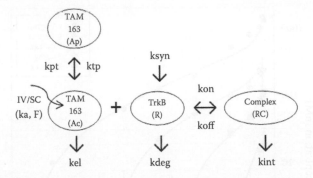

FIGURE 5.9 TMDD model structure describing the binding of a mAb (TAM-163) to its endogenous target (TrkB receptor).[57] The mAb was administered intravenously (IV) or subcutaneously (SC) in non-human primates. kel, kdeg, and kint represent the *in vivo* elimination rate of the free mAb, turnover rate of the free receptor, and degradation rate of the mAb–receptor complex, respectively. The endogenous synthesis rate of the receptor is denoted by ksyn, whereas binding kinetics is governed by association (kon) and dissociation (koff) rate constants. kpt and ktp describe the distribution process between central (Ac) and peripheral (Ap) compartments. Bioavailability through the SC route and the corresponding absorption rate are denoted by F and ka, respectively.

a specific level of target saturation. Additionally, a top-end clinical dose that fully saturates the target (e.g., 99%) could be used to test the mechanism.

5.4.1.3   Physiology-Based Pharmacokinetic Models

Classic compartmental modeling reduces complex processes of drug distribution and elimination in various organs into a limited number of rate constants that generally do not have a clear physiological meaning, unless they are decoupled into readily interpretable quantities such as clearance and volume of distribution. Furthermore, while simple compartmental models are amenable to fitting and provide a descriptive framework for the preclinical or human data, they generally lack the mechanistic framework for scaling *in vitro* data to *in vivo* settings toward predicting serum or tissue exposure profiles. Physiology-based pharmacokinetic (PBPK) models (Figure 5.10a) describe the major physiological organs or tissues in the human body and the blood flow between them in a quantitative fashion.[58] From their origins in environmental toxicology research to small-molecule drug distribution,[59–61] the PBPK modeling approach is now being applied to biologics, and several recent publications have developed a platform PBPK model to describe the serum and tissue disposition of mAbs and other biotherapeutics.[62–65] Over the years, the antibody PBPK models have evolved from modeling each organ empirically to incorporating IgG-FcRn interactions in all tissues and dividing individual organs into vascular, interstitial, and cellular subcompartments with endosomal trafficking,[24] as shown in Figure 5.10b.

5.4.2   Starting Dose Projections for FIH Trials

The traditional approach to select the starting dose for FIH trials is based on the guidance issued by FDA's Center for Drug Evaluation and Research (CDER) in

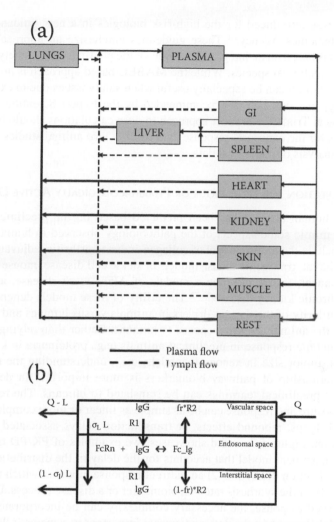

FIGURE 5.10 A PBPK model describing IgG disposition. (a) The most important organs connected by plasma (solid lines) and lymphatic flows (dashed lines). (b) The model structure within each organ compartment comprising vascular, endosomal, and interstitial space. (Adapted from Garg, A., and Balthasar, J.P., *J Pharmacokinet Pharmacodyn*, 34(5): 687–709, 2007.[24])

2005[66] that lists the following key steps: (1) establish a no observable adverse effect level (NOAEL) in the preclinical species and convert it to a human equivalent dose (HED) based on the body surface area or weight normalization, (2) select the HED from the most appropriate species (typically, the most sensitive species) and apply a safety factor (at least 10-fold) to estimate a maximum recommended starting dose (MRSD), and (3) adjust the MRSD based on the pharmacologically active dose. While these guidelines are easy to follow, they are solely based on the NOAEL from toxicology studies, are focused on dose rather than exposures, and do not account for the drug effect on the target. After the shocking adverse events seen in 2006 during the FIH administration of TGN1412 (a CD28 superagonist mAb),[67] the concept of MABEL was introduced for the high-risk biologics in a new guidance by the European Medicines Agency.[68] These guidelines emphasize an approach to dose selection that is based on an integrated analysis of the pharmacology, safety, and efficacy data in preclinical species. While the MABEL-based approach is not required for every program, it can be especially useful when safety issues due to exaggerated pharmacology are expected. As recommended by the Expert Scientific Group on Phase I Clinical Trials,[69] a broader approach to dose calculations should be utilized that accounts for the exposure–response data from *in vivo* animal studies as well as the PK/PD analysis of preclinical data.

5.4.3 PROJECTION FOR EFFICACIOUS OR PHARMACOLOGICALLY ACTIVE DOSE

In the past, human efficacy has been predicted based on the preclinical animal models that mimic some aspects of the pathobiology observed in human disease. Examples include the mouse model of collagen-induced arthritis, adjuvant-induced arthritis in the rat, transgenic mouse models of sickle cell disease, mouse xenograft models of cancer, transgenic mouse models of Alzheimer's disease, and rodent models of chronic kidney disease.[70-72] The utility of these models depends greatly upon the similarity of the disease drivers in animals versus humans and the translatability of the animal disease response to humans. Rather than relying solely on the gross outcome response in preclinical animals (e.g., proteinuria in kidney disease models, tumor size in xenograft models, etc.), understanding the exposure–response relationship of pathway biomarkers is more important in determining how well the preclinical learnings can be translated to humans. The relationship between exposure and response can be as simple as linear or more complex, involving feedback loops, rebound effects, or transduction delays associated with protein expression. Figure 5.7c and d showcases two examples of PK/PD models: (1) a linked compartment model that accounts for the delay in the distribution of drug into the site of action (SoA) and (2) an indirect-response model in which the drug is expected to affect the synthesis rate of a biomarker or a disease process. Depending on the observed response, the necessary complexity can be incorporated into the PK/PD models to explain the data and extract important parameters related to the drug (i.e., *in vivo* potency or IC50) or system (i.e., biomarker synthesis rate, target turnover, etc.). These models, in spite of their empirical nature, can be valuable in developing a quantitative framework for translating preclinical data to clinics, and guiding the clinical dosing regimen.

Exposure–response data from animal models can be inherently limiting due to fundamental differences between animal and human disease. However, objective measures such as target occupancy can be extremely useful in projecting a pharmacologically active dose, as the binding of biologics to their target on a cell surface or in circulation triggers the complex downstream effects. Thus, determining a target occupancy profile upon drug administration, building an occupancy–response profile in preclinical species, and ultimately translating the learnings to humans are immensely useful while predicting pharmacologically active doses in humans. Emerging ideas such as SoA models[73,74] or minimal PBPK models[75] seek to predict the level of target occupancy in disease-relevant tissues (i.e., the site of action) as a

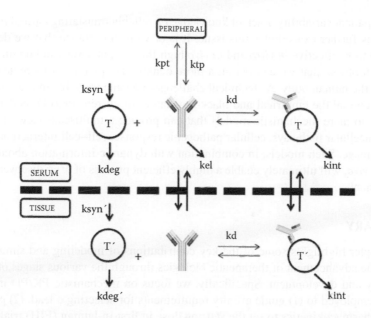

FIGURE 5.11   A site-of-action model to predict target coverage in tissue. kdeg and kdeg' are target turnover rates in serum (T) and tissue (T'), respectively. ksyn and ksyn' are zero-order target synthesis rates in serum and tissue, respectively. kel and kint are elimination rates of free antibody and complex, respectively. kpt and ktp are antibody distribution rates from serum to peripheral, and peripheral to serum, respectively. kd is the equilibrium binding constant. (Adapted from Brodfuehrer, J., et al., *Pharm Res*, 31(3): 635–48, 2014.[73])

function of dosing regimen. Figure 5.11 shows the structure of an SoA model that was applied to predict target coverage in the spleen. Such models account for the distribution of biologics into tissues, the binding of drug to the target at the site, and the *in vivo* turnover of the target, drug, and drug–target complex. In the case of soluble targets, where an increase in the total target (sum of free and bound targets) levels is observed due to the slower turnover of the complex compared with the free target, bioanalytical assays can be developed to characterize their time course. Subsequently, these data can be fed to SoA models to estimate the target occupancy, the apparent binding constant, and the *in vivo* turnover rate of the free target. When adequate data are available, the model can be easily translated to humans to simulate various dosing scenarios and corresponding occupancy profiles.

5.5   CONCLUSIONS AND FUTURE DIRECTIONS

While the PK/PD modeling approaches described here are increasingly being utilized by the pharmaceutical industry, regulatory agencies, and academicians, several challenges limit their impact. One of the key challenges is the scarcity of data available on human diseases, such as quantitative information on disease initiation and progression, physiologically relevant biomeasures (e.g., tissue expression and target turnover rates), and biomarkers that can predict patient response to therapy and interpatient variability. Lack of knowledge on reliably translating animal models to humans further exacerbates this issue. It is of vital importance that we develop cost- and time-effective *in vitro* and *in vivo* high-throughput assays and noninvasive imaging tools, so that we can obtain a more quantitative picture of biological networks in the human body. As technical challenges become tractable in the future, it is expected that the empirical and black-box nature of current PK/PD models will give way to more mechanistic models that can provide a multiscale view of a target's intracellular pathways, cellular pathogenic response, cell–cell interactions, and organ damage. Such models, in combination with dynamic information about drug action *in vivo*, will ultimately enable a highly efficient process of drug discovery and development.

SUMMARY

This chapter highlights some of the key contributions of modeling and simulation toward the advancement of therapeutic biologics through the various stages of drug discovery and development. Specifically, we focus on mechanistic PK/PD models that are employed to (1) guide affinity requirements for selecting a lead, (2) predict human pharmacokinetics to set the starting dose in first-in-human (FIH) trials, and (3) project pharmacologically active or efficacious doses to help establish proof of mechanism in humans. This chapter starts with a brief introduction on absorption, distribution, metabolism, and excretion (ADME) properties of monoclonal antibodies. Subsequently, we discuss the mathematical models that can be used to study the antibody-dependent activation or suppression of pharmacological targets within signal transduction pathways. Such interactions may lead to changes in pathway biomarkers, which can also be included in the model and can thereby help in the design of clinical protocols that efficiently demonstrate desirable therapeutic outcomes. At appropriate places, we have provided details of specific modeling approaches along with select examples.

REFERENCES

1. Leader, B., Q.J. Baca, and D.E. Golan. Protein therapeutics: A summary and pharmacological classification. *Nat Rev Drug Discov*, 2008; 7(1): 21–39.
2. Smith, A.J. New horizons in therapeutic antibody discovery: Opportunities and challenges versus small-molecule therapeutics. *J Biomol Screen*, 2014.
3. Lobo, E.D., R.J. Hansen, and J.P. Balthasar. Antibody pharmacokinetics and pharmacodynamics. *J Pharm Sci*, 2004; 93(11): 2645–68.
4. Schellekens, H, Factors influencing the immunogenicity of therapeutic proteins. *Nephrol Dial Transplant*, 2005; 20(Suppl 6): vi3–9.
5. Gorovits, B., et al. Recommendations for the characterization of immunogenicity response to multiple domain biotherapeutics. *J Immunol Methods*, 2014; 408: 1–12.
6. Wullner, D., et al. Considerations for optimization and validation of an *in vitro* PBMC derived T-cell assay for immunogenicity prediction of biotherapeutics. *Clin Immunol*, 2010; 137(1): 5–14.
7. Singh, S.K. Impact of product-related factors on immunogenicity of biotherapeutics. *J Pharm Sci*, 2011; 100(2): 354–87.

8. Moss, A.C., V. Brinks, and J.F. Carpenter. Review article: Immunogenicity of anti-TNF biologics in IBD—the role of patient, product and prescriber factors. *Aliment Pharmacol Ther*, 2013; 38(10): 1188–97.

9. Bell, G.I. Mathematical model of clonal selection and antibody production. *J Theor Biol*, 1970; 29(2): 191–232.

10. Blyuss, K.B., and L.B. Nicholson. Understanding the roles of activation threshold and infections in the dynamics of autoimmune disease. *J Theor Biol*, 2015 June 21; 375:13–20.

11. Ciupe, S.M., R.M. Ribeiro, and A.S. Perelson. Antibody responses during hepatitis B viral infection. *PLoS Comput Biol*, 2014; 10(7): e1003730.

12. Chen, X., et al. A mathematical model of the effect of immunogenicity on therapeutic protein pharmacokinetics. *AAPS J*, 2013; 15(4): 1141–54.

13. Chen, X., T. Hickling, and P. Vicini. A mechanistic, multi-scale mathematical model of immunogenicity for therapeutic proteins. Part 2: Model applications. *CPT Pharmacometrics Syst Pharmacol*, 2014; 3:e134.

14. Chen, X., T.P. Hickling, and P. Vicini. A mechanistic, multiscale mathematical model of immunogenicity for therapeutic proteins. Part 1: Theoretical model. *CPT Pharmacometrics Syst Pharmacol*, 2014; 3: e133.

15. Yin, L., et al. The role of aggregates of therapeutic protein products in immunogenicity: An evaluation by mathematical modeling (in press)

16. Brekke, O.H., and I. Sandlie. Therapeutic antibodies for human diseases at the dawn of the twenty-first century. *Nat Rev Drug Discov*, 2003; 2(1): 52–62.

17. Dirks, N.L., and B. Meibohm. Population pharmacokinetics of therapeutic monoclonal antibodies. *Clin Pharmacokinet*, 2010; 49(10): 633–59.

18. Dostalek, M., et al. Pharmacokinetics, pharmacodynamics and physiologically-based pharmacokinetic modelling of monoclonal antibodies. *Clin Pharmacokinet*, 2013; 52(2): 83–124.

19. Grimm, H.P. Gaining insights into the consequences of target-mediated drug disposition of monoclonal antibodies using quasi-steady-state approximations. *J Pharmacokinet Pharmacodyn*, 2009; 36(5): 407–20.

20. Shi, S. Biologics: An update and challenge of their pharmacokinetics. *Curr Drug Metab*, 2014; 15(3): 271–90.

21. Wang, W., E.Q. Wang, and J.P. Balthasar. Monoclonal antibody pharmacokinetics and pharmacodynamics. *Clin Pharmacol Ther*, 2008; 84(5): 548–58.

22. Richter, W.F., S.G. Bhansali, and M.E. Morris. Mechanistic determinants of biotherapeutics absorption following SC administration. *AAPS J*, 2012; 14(3): 559–70.

23. Baxter, L.T., et al. Biodistribution of monoclonal antibodies: Scale-up from mouse to human using a physiologically based pharmacokinetic model. *Cancer Res*, 1995; 55(20): 4611–22.

24. Garg, A., and J.P. Balthasar. Physiologically-based pharmacokinetic (PBPK) model to predict IgG tissue kinetics in wild-type and FcRn-knockout mice. *J Pharmacokinet Pharmacodyn*, 2007; 34(5): 687–709.

25. Davies, B., and T. Morris. Physiological parameters in laboratory animals and humans. *Pharm Res*, 1993; 10(7): 1093–95.

26. Elaine, M. *Essentials of Human Anatomy and Physiology*. Benjamin Cummings, San Francisco, 7th ed. 2003.

27. Bruin, G.E.A. Secukinumab exploratory study to investigate distribution in dermal interstitial fluid using dermal open flow microperfusion after single 300 mg subcutaneous administration: Pharmacokinetic assessment in healthy volunteers and patients with moderate-to-severe plaque psoriasis. Presented at International Investigative Dermatology Congress, Edinburgh, Scotland, 2013.

28. Mager, D.E., and W.J. Jusko. General pharmacokinetic model for drugs exhibiting target-mediated drug disposition. *J Pharmacokinet Pharmacodyn*, 2001; 28(6): 507–32.

29. Giragossian, C., et al. Neonatal Fc receptor and its role in the absorption, distribution, metabolism and excretion of immunoglobulin G-based biotherapeutics. *Curr Drug Metab*, 2013; 14(7): 764–90.

30. Sorger, P.K. Quantitative and systems pharmacology in the post-genomic era: New approaches to discovering drugs and understanding therapeutic mechanisms. 2011. NIH white paper. http://www.nigms.nih.gov/News/reports/Documents/SystemsPharmaWP Sorger2011.pdf

31. Bordbar, A., et al. A multi-tissue type genome-scale metabolic network for analysis of whole-body systems physiology. *BMC Syst Biol*, 2011; 5: 180.

32. Kirouac, D.C., et al. Computational modeling of ERBB2-amplified breast cancer identifies combined ErbB2/3 blockade as superior to the combination of MEK and AKT inhibitors. *Sci Signal*, 2013; 6(288): ra68.

33. Lee, M.J., et al. Sequential application of anticancer drugs enhances cell death by rewiring apoptotic signaling networks. *Cell*, 2012; 149(4): 780–94.

34. Berger, S.I., A. Ma'ayan, and R. Iyengar. Systems pharmacology of arrhythmias. *Sci Signal*, 2010; 3(118): ra30.

35. DeSouza, C., and V. Fonseca. Therapeutic targets to reduce cardiovascular disease in type 2 diabetes. *Nat Rev Drug Discov*, 2009; 8(5): 361–67.

36. Tanaka, K., and G.J. Augustine. A positive feedback signal transduction loop determines timing of cerebellar long-term depression. *Neuron*, 2008; 59(4): 608–20.

37. Schoeberl, B., et al. An ErbB3 antibody, MM-121, is active in cancers with ligand-dependent activation. *Cancer Res*, 2010; 70(6): 2485–94.

38. Schoeberl, B., et al. Therapeutically targeting ErbB3: A key node in ligand-induced activation of the ErbB receptor-PI3K axis. *Sci Signal*, 2009; 2(77): ra31.

39. Fitzgerald, J.B., et al. MM-141, an IGF-IR- and ErbB3-directed bispecific antibody, overcomes network adaptations that limit activity of IGF-IR inhibitors. *Mol Cancer Ther*, 2014; 13(2): 410–25.

40. Nayak, S., et al. Using a systems pharmacology model of blood coagulation network to predict the effects of various therapies on biomarkers. *CPT Pharmacometrics Sys Pharmacol*, 2015; 4(7): 396–405.

41. D., J. Coagulation cascade. 2007. http://commons.wikimedia.org/wiki/File%3A Coagulation_full.svg (accessed February 6, 2015).

42. Mahmood, I., and M.D. Green. Pharmacokinetic and pharmacodynamic considerations in the development of therapeutic proteins. *Clin Pharmacokinet*, 2005; 44(4): 331–47.

43. Boder, E.T., M. Raeeszadeh-Sarmazdeh, and J.V. Price. Engineering antibodies by yeast display. *Arch Biochem Biophys*, 2012; 526(2): 99–106.

44. Doerner, A., et al. Therapeutic antibody engineering by high efficiency cell screening. *FEBS Lett*, 2014; 588(2): 278–87.

45. Ducancel, F., and B.H. Muller. Molecular engineering of antibodies for therapeutic and diagnostic purposes. *MAbs*, 2012; 4(4): 445–57.

46. Igawa, T., et al. Engineering the variable region of therapeutic IgG antibodies. *MAbs*, 2011; 3(3): 243–52.

47. Dubel, S. Molecular engineering II: Affinity maturation. In *Handbook of Therapeutic Antibodies: Technologies, Emerging Developments and Approved Therapeutics*. Wiley, Hoboken, NJ, 2010.

48. Mordenti, J., et al. Interspecies scaling of clearance and volume of distribution data for five therapeutic proteins. *Pharm Res*, 1991; 8(11): 1351–59.

49. Mahmood, I. Interspecies scaling of protein drugs: Prediction of clearance from animals to humans. *J Pharm Sci*, 2004; 93(1): 177–85.

50. Mahmood, I. Pharmacokinetic allometric scaling of antibodies: Application to the first-in-human dose estimation. *J Pharm Sci*, 2009; 98(10): 3850–61.

51. Wang, W., and T. Prueksaritanont. Prediction of human clearance of therapeutic proteins: Simple allometric scaling method revisited. *Biopharm Drug Dispos*, 2010; 31(4): 253–63.

52. Ling, J., et al. Interspecies scaling of therapeutic monoclonal antibodies: Initial look. *J Clin Pharmacol*, 2009; 49(12): 1382–402.

53. Deng, R., et al. Projecting human pharmacokinetics of therapeutic antibodies from nonclinical data: What have we learned? *MAbs*, 2011; 3(1): 61–66.

54. Oitate, M., et al. Prediction of human pharmacokinetics of therapeutic monoclonal antibodies from simple allometry of monkey data. *Drug Metab Pharmacokinet* 2011; 26(4): 423–30.

55. Oitate, M., et al. Prediction of human plasma concentration-time profiles of monoclonal antibodies from monkey data by a species-invariant time method. *Drug Metab Pharmacokinet* 2012; 27(3): 354–59.

56. Singh, A.P., et al. Quantitative prediction of human pharmacokinetics for mAbs exhibiting target-mediated disposition. *AAPS J*, 2015; 17(2):389–99.

57. Vugmeyster, Y., et al. Agonistic TAM-163 antibody targeting tyrosine kinase receptor-B: Applying mechanistic modeling to enable preclinical to clinical translation and guide clinical trial design. *MAbs*, 2013; 5(3): 373–83.

58. Jones, H.M., I.B. Gardner, and K.J. Watson. Modelling and PBPK simulation in drug discovery. *AAPS J*, 2009; 11(1): 155–66.

59. Jones, H., and K. Rowland-Yeo. Basic concepts in physiologically based pharmacokinetic modeling in drug discovery and development. *CPT Pharmacometrics Syst Pharmacol*, 2013, 2. e63.

60. Jones, H.M., K. Mayawala, and Poulin, P. Dose selection based on physiologically based pharmacokinetic (PBPK) approaches. *AAPS J*, 2013; 15(2): 377–87.

61. Rowland, M., C. Peck, and G. Tucker. Physiologically-based pharmacokinetics in drug development and regulatory science. *Annu Rev Pharmacol Toxicol*, 2011; 51: 45–73.

62. Shah, D.K., and A.M. Betts. Towards a platform PBPK model to characterize the plasma and tissue disposition of monoclonal antibodies in preclinical species and human. *J Pharmacokinet Pharmacodyn*, 2012; 39(1): 67–86.

63. Abuqayyas, L., and J.P. Balthasar. Application of PBPK modeling to predict monoclonal antibody disposition in plasma and tissues in mouse models of human colorectal cancer. *J Pharmacokinet Pharmacodyn*, 2012; 39(6): 683–710.

64. Davda, J.P., et al. A physiologically based pharmacokinetic (PBPK) model to characterize and predict the disposition of monoclonal antibody CC49 and its single chain Fv constructs. *Int Immunopharmacol*, 2008; 8(3): 401–13.

65. Heiskanen, T., T. Heiskanen, and K. Kairemo. Development of a PBPK model for monoclonal antibodies and simulation of human and mice PBPK of a radiolabelled monoclonal antibody. *Curr Pharm Des*, 2009; 15(9): 988–1007.

66. FDA, ed. Guidance for industry: Estimating the maximum safe starting dose in initial clinical trials for therapeutics in adult healthy volunteers. Rockville, MD: FDA, 2005.

67. Farzaneh, L., N. Kasahara, and F. Farzaneh. The strange case of TGN1412. *Cancer Immunol Immunother*, 2007; 56(2): 129–34.

68. Agoram, B.M. Use of pharmacokinetic/pharmacodynamic modelling for starting dose selection in first-in-human trials of high-risk biologics. *Br J Clin Pharmacol*, 2009; 67(2): 153–60.

69. Expert Scientific Group on Phase I Clinical Trials (Duff, G., Chair). Final report. Norwich, UK: Stationary Office, November 30, 2006.

70. Gotz, J., and L.M. Ittner. Animal models of Alzheimer's disease and frontotemporal dementia. *Nat Rev Neurosci*, 2008; 9(7): 532–44.

71. Kobezda, T., et al. Of mice and men: How animal models advance our understanding of T-cell function in RA. *Nat Rev Rheumatol*, 2014; 10(3): 160–70.
72. Marchesi, V. Breast cancer: Stable breast cancer xenograft models. *Nat Rev Clin Oncol*, 2013; 10(8): 426.
73. Brodfuehrer, J., et al. Quantitative analysis of target coverage and germinal center response by a CXCL13 neutralizing antibody in a T-dependent mouse immunization model. *Pharm Res*, 2014; 31(3): 635–48.
74. Chudasama, V.L., et al. Simulations of site-specific target-mediated pharmacokinetic models for guiding the development of bispecific antibodies. *J Pharmacokinet Pharmacodyn*, 2015; 42(1):1–18.
75. Li, L., et al. Incorporating target shedding into a minimal PBPK-TMDD model for monoclonal antibodies. *CPT Pharmacometrics Syst Pharmacol*, 2014; 3: e96.

6 Challenges in High-Concentration Biopharmaceutical Drug Delivery
A Modeling Perspective

*Anuj Chaudhri**

CONTENTS

6.1 INTRODUCTION

Passive immunotherapy via biologics such as therapeutic monoclonal antibodies (mAbs) and antibody-based biotherapeutics is a multi-billion-dollar industry with more than 40 U.S. Food and Drug Administration (FDA) approved mAbs currently in the market[1] and many more in development.[2] These monoclonals need to be administered to the patient in very high doses (>1 mg/kg) for potency issues. This

---

* Computational Research Division, Lawrence Berkeley National Lab, Berkeley, California

is usually done intravenously, which requires highly skilled workers and high costs of patient care. There is a growing need to develop a more convenient route of administration, such as subcutaneous (SC) delivery. However, subcutaneous delivery poses an upper limit on the dosage volume that can be administered, typically <1.5 ml. This necessitates the development of SC formulations with concentrations of >100 mg/ml. At such high concentrations, these protein solutions pose several formulation challenges, such as high viscosity during manufacturing,[3,4] protein stability issues leading to association-based aggregation,[5,6] solution degradation,[7,8] and undesired immunogenic responses in the body.[9] Hence, developing a more fundamental understanding of issues such as self-association and aggregation in high-concentration protein formulations can lead to development of more stable therapeutic drugs that can be administered more easily and in a risk-free environment.

6.1.1 HIGH-CONCENTRATION PROTEIN SOLUTIONS

At high concentrations, protein solutions are affected greatly by protein–protein interactions (PPIs) that lead to nonideal, viscoelastic behavior.[10] The intermolecular interactions arise due to the presence of a large number of protein molecules in solution and shorter separation distance between them. These intermolecular interactions affect the solution osmotic pressure via the extra virial terms in the pressure expansion, thus causing deviations of solution behavior from ideality.[11] In solution, protein molecules interact via numerous forces, such as electrostatic, hydrophobic, excluded volume (steric), hydrogen bonding, and van der Waals dispersion forces.[12] In dilute solutions, hydrogen bonding, excluded volume, and van der Waals dispersion forces are not very significant compared to electrostatic and hydrophobic forces. Solution behavior is often characterized by the potential of mean force (PMF), which is the average force of interaction between two protein molecules when they are treated as rigid spherical molecules.[13] The PMF is related to the second virial coefficient (B_{22}), which can help characterize PPI in dilute solutions. In the case of dilute solutions, treating the entire protein as a rigid colloidal sphere is a good model to explain solution behavior in most cases. However, for high-concentration proteins, these assumptions cannot explain the nonideal, viscoelastic behavior completely.[10] The irregularities in the shape of the protein molecules, surface roughness, charge heterogeneity, and geometric complementarity play a vital role for protein solutions at high concentrations.

6.1.2 SELF-ASSOCIATION AND HIGH SOLUTION VISCOSITY

One of the consequences of strong intermolecular interactions at high concentrations is the reversible oligomerization of native protein structures in solution known as self-association. It is hypothesized that the strong intermolecular interactions lead to transient networks of proteins that also cause the solution viscosity to increase dramatically. Liu et al.[14] studied the behavior of three monoclonal antibodies and found that one of them (mAb1) is 60-fold more viscous than a solution without them. Adding salt reduced the viscosity, thus confirming the important role of charge–charge electrostatics in the self-association of mAbs. Another mAb (mAb2) did not show the sharp viscosity increase that mAb1 did. Its viscosity behavior could be explained by the extended Mooney equation that takes into account only excluded volume effects. Kanai et al.[15] further studied the behavior of mAb1 solutions and concluded that Fab–Fab interactions contribute significantly to the self-association of these mAbs.

Yadav et al.[16–18] performed the most elaborate experimental studies on these mAbs as a function of pH, ionic strength, and concentration using dynamic light scattering to probe the interactions between them. Interactions were characterized by the second virial coefficient, interaction parameter, and solution storage modulus. The interactions were found to be strong, attractive, and domain specific at pH 6.0, suggesting formation of self-associating transient multivalent networks that lead to viscosity changes in mAb1. They also concluded that the viscosity profiles for mAb1 solutions cannot be explained completely by electroviscous effects or using an effective molecular volume. The viscosity behavior was attributed to short-range electrostatic attractive potentials such as charge–charge and charge–dipole that exist between specific regions of the mAbs and lead to self-associating transient networks. Further studies by Yadav et al.[19, 20] suggest the sensitive role of the amino acid sequence in the self-associating behavior of mAbs. Changing a few residues in the complementarity determining region (CDR) of the mAbs changes the self-associating behavior quite dramatically. It was found that merely treating the mAb as a charged colloidal sphere is not enough to explain all the effects. Charge heterogeneity is important in understanding how specific domains of the antibodies interact with one another and self-associate.

6.1.3 MULTISCALE MODELING

In order to test the hypothesis that charge heterogeneity is an important factor in the self-association of proteins compared to the net charge of the protein, it becomes important to perform very systematic controlled experiments similar to the ones performed by Yadav et al.[19, 20] These experiments were, however, confined to specific residues in the CDR and in general can involve a large number of permutations and combinations. Theoretical and computational tools can help fill the gap where experiments can become difficult to perform.[21]

Molecular modeling and simulation techniques have contributed significantly to our current understanding of chemical and biological processes, which continues to evolve as computational power grows multifold every year.[22] One of the most popular techniques is molecular dynamics (MD),[23] which treats systems atomistically and essentially solves Newton's equations of motion for all the atoms in the system. The current capabilities of MD simulations allow one to model millions of atoms for a few nanoseconds. However, for large biomolecules, most of the interesting processes (e.g., protein folding, self-assembly, self-association, aggregation) that span over a wide range of time and length scales remain inaccessible via traditional MD.[24] As an example, consider modeling a mAb in an electrolyte solution. This simulation involves solving the equations for well over 300,000 atoms (mAb + water + ions). Now consider doing the same for a solution of mAbs containing hundreds of thousands of mAbs in order to study self-association or aggregation. The number of atoms involved in the simulation is exponentially large and intractable by MD. Coarse-grained (CG) structurally reduced models of biomolecules can help bridge this gap of disparate time and length scales and make biological processes more accessible.[24–26] In CG modeling, the complex atomistic system is reduced by grouping a large number of atoms/molecules into relatively fewer CG particles whose motion is followed by solving the relevant equations of motion (Figure 6.1). The CG particles are often modeled as CG sites (point particles) that interact via effective interactions. The key to CG modeling is of course determining the correct form of the effective interactions that retain the relevant physics pertaining to the system of interest. A greatly simplified version of the relevant physical system via a low-resolution model can provide more insight into its behavior for processes that are not accessible using MD. The CG methodology has been used for a large number of condensed-phase and small biomolecular systems[27] and is being extended to include large biomolecules as well.[28]

In this chapter, the current understanding of this highly complex problem of self-association and challenges with regard to modeling protein–protein interactions in high-concentration solutions is reviewed from a computational perspective. In Section 6.2, the problem of self-association is studied using coarse-grained modeling. Section 6.3 describes some of the challenges that are still present in coming up with good theoretical representations of protein–protein interactions in high-concentration solutions, and Section 6.4 offers some suggestions to overcome them. This is followed by a brief summary in Section 6.5.

6.2 COARSE-GRAINED MODELING OF SELF-ASSOCIATION OF mAbs

Computational studies were performed on a variety of mAbs and their variants using coarse-grained modeling to help understand the phenomena of self-association.[29,30] In these studies, coarse-grained models of two mAbs—mAb1 and mAb2—and their mutants were developed from the underlying atomistic structure. These two mAbs were picked since they lie at two extremes of a spectrum of viscosity behavior with concentration; mAb1 shows a dramatic increase in viscosity with concentration, whereas mAb2 does not. The amino acid sequence of the two mAbs is the same except in the CDRs. Hence, it is pertinent to understand why mAb1 shows such dramatic changes in viscosity, whereas mAb2 does not. A CG representation requires identification of CG sites and a model describing the site–site interactions. The CG sites were picked based on the long-wavelength dynamics of the mAb[29] by constructing an elastic network model of the mAb and performing normal-mode analysis.[31] Two low-resolution models were developed: a 12-site and a 26-site model, as shown in Figure 6.1. The 12-site model contains 12 sites that represent the 12 domains of the mAb. Long-wavelength dynamics of the mAb showed that each domain of

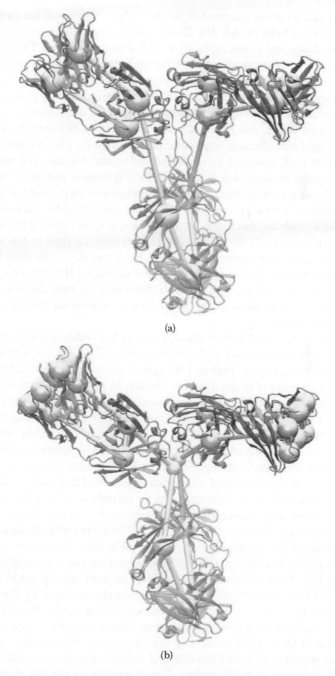

(a)

(b)

FIGURE 6.1 Coarse-grained representations of a mAb overlaid on its all atom model: (a) 12 sites and (b) 26 sites. In the 12-site model, each site represents the underlying domain. In the 26-site model, additional sites are used for the CDR and hinge regions. Note that the sizes of the sites in the figure have not been drawn to proportion; in general, their radius of interaction extends beyond the edges of the domain they represent.

the antibody tends to move independent of each other, which was the basis for the construction of the 12-site model. The 26-site model was built on top of the 12-site model with the hinge and CDRs explicitly modeled using CG sites. The masses and charges of each site were calculated from the underlying atomic structure. In this way, each CG site represents a structurally reduced representation of the fully atomistic domain, with the mass and charge of the domain lumped at its center of mass.

The CG variables, positions, and momenta represent the collective motion of a large number of atoms, and hence effective interactions between the sites must represent averaged and effective large-scale protein motion. The interprotein interactions included both screened electrostatics using the Yukawa potential (more about this potential in Section 6.3) and near-field repulsion and dispersion using the Lennard–Jones potential. The intraprotein interactions were based on harmonic spring interactions, whose spring constants were calculated by performing MD simulations on a mAb and calculating effective interactions between the domains[29] ("Supporting Information"). Simulations were also done on rigid mAbs where the intraprotein springlike interactions gave way to hard, rodlike constraints. Langevin CGMD simulations were then performed on 1000 mAbs at pH 6.0 in a periodic box for 5 μs at a time step of 1 ps, with concentrations ranging from 20 to 120 mg/ml. Property data were collected over the last 4 μs to calculate all the time-averaged properties and behavior.

The results of the simulations[29] showed that mAb1—the antibody that shows a dramatic increase in viscosity with concentration—forms dense self-associated structures at concentrations as high as 120 mg/ml, compared to mAb2. Simulations predicted the formation of dense dynamic clusters in mAb1 solutions at high concentrations, whereas no such clusters were found in mAb2 solutions at any time. On average, 15 dynamic clusters were found in mAb1 at 120 mg/ml, with average sizes ranging from 5 to 45 mAbs.[30] High-shear rheology studies on mAb1 and mAb2[32] have shown considerable shear-thinning behavior in mAb1 solutions at high-shear rates, confirming the reverse arrangement of the deformable clusters predicted by CG simulations. On the other hand, experiments indicated that mAb2 solutions did not exhibit any shear-thinning behavior, and no clusters were found in these solutions at any concentration, confirming the predictions by CG simulations.

The potential of mean force calculated from the intersite radial distribution functions (RDFs), using CG simulations, showed considerable short-range structure in mAb1, which points to a large number of nearest and next-nearest neighbors. Both Fab–Fab and Fab–Fc interactions were found to be dominant in mAb1 solutions. Fab–Fab interactions have already been experimentally explored by Kanai et al.[15] and were found to be responsible for self-association in mAb1. Recent light and small-angle x-ray scattering (SAXS) scattering experiments by Lilyestrom et al.[33] (on a different mAb that self-associates via short-range electrostatic attractions) indicated the presence of both Fab–Fab and Fab–Fc interactions that led to the formation of oligomers of stoichiometry 2–9 depending on the salt concentration. Hence, the experimental evidence confirms the predictions from simulations that in mAbs with strong short-range attractions, both Fab–Fab and Fab–Fc interactions will be important to form dense clusters that can lead to high viscosities at high concentrations.

The CG simulations also predicted that mAb2 solutions formed uniform structures with no clustering (as opposed to mAb1 forming randomly distributed clusters) based on Fab–Fc interactions. In order to understand this, the charge distribution for the mAb1 and mAb2 12 site models can be compared in Figure 6.2. mAb1 (Figure 6.2a) has a strong charge quadrupole on the Fab arm (sites 1, 2, 5, and 6), which makes short-range attractions very dominant. In the case of mAb2 (Figure 6.2b), sites 1 and 5 are highly positively charged, and hence Fab–Fab repulsions are more dominant. On the other hand, Fc sites 4 and 10 are negatively charged, thus making Fab–Fc interactions more dominating in mAb2. The importance of repulsions in mAb2 solutions has recently been confirmed by small-angle neutron scattering (SANS) experiments by Yearley et al.[34] Their experiments indicate that the attractions between mAb1 are highly anisotropic consistent with the nonuniform surface charge distribution. In mAb2, the overall repulsive interactions dominate over the entire concentration range, which is the reason for these solutions to be dominated by monomeric units, as predicted by the CG simulations. Further evidence of clustering in mAb1 comes from extensive SAXS/SANS and neutron–spin echo experiments by Yearley

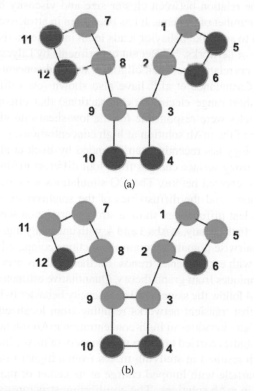

(a)

(b)

FIGURE 6.2 Charge distribution in (a) mAb1 and (b) mAb2 for the 12-site model. Blue represents positive charge, whereas red indicates negative charge; note that mAb1 has strong dipoles in the Fab and Fc region, whereas strongly positive Fab regions are found in mAb2. This makes short-range Fab–Fab attractions more attractive in mAb1 versus short-range repulsions in mAb2.

et al.[35] Clusters were formed in mAb1 solutions at concentrations starting from 10 mg/ml, with the average cluster size being uniform at high concentrations.

CG simulations were also able to pick out the differences in behavior of the charge-swap mutants based on equilibrium structure formation at high concentrations.[30] One of the outcomes of the study was the prediction of self-associating behavior in the charge-swap mutants. This was used to compare the CG model with existing experimental observations[19] on these mutants. Overall, the CG simulations predicted behaviors that corroborated well with experiments. Experimentally, the only oddly behaved mAb was found to be mutant M6. It was observed that even though M6 has considerable short-range attractions (similar to mAb1), it did not show high-viscosity dependence at high concentration (unlike mAb1). CG simulations on M6 indicated the presence of short-range attractions that corroborated with experimental observations. Further, simulations predicted that a large number of small dynamic clusters (average size of 5–25 mAbs) were found in equilibrium simulations of M6, compared to mAb1, which formed a small number of large clusters (average size of 5–45 mAbs). The smaller-than-expected viscosity is hence the result of the smaller clusters that can slide over each other easily during rheological experiments. The relation between cluster size and viscosity has been verified experimentally in a number of systems. It has been seen in attractive colloids that the transition from fluid to solidlike behavior leads to an increase in viscosity by formation of large clusters or networks.[36,37] Recent experiments by Lilyestrom et al.[33] have also shown a strong correlation between oligomer size and concentration-dependent viscosity behavior. Castellanos et al.[38] have also shown (on a different mAb that self-associates via short-range electrostatic attractions) that oligomers and fractal submicrometer particles were responsible for the low-shear-rate viscosity and non-Newtonian behavior of the mAb solution at high concentrations.

The CG methodology has recently been extended by Buck et al.[39] by parameterizing the CG model using surface charges from four different mAbs (mAbs 1–4, different from the ones reported before). The CG simulations were performed at many different concentrations, and the diffusivities of the solutions were calculated. The concentration-dependent diffusivities show a strong correlation with experimentally observed viscosities. In this study, mAbs 3 and 4, with unusually high viscosities, show a large number of pairwise domain–domain interactions compared to the other two, which corroborates with experimental trends. Further analysis was done on network formation using techniques from graph theory. Quantitative estimates of network density for mAbs 3 and 4 follow the same trend as viscosity behavior (with concentration), validating the idea that transient networks resulting from localized interactions can potentially lead to high viscosities in high-concentration mAb solutions.

Coarse-grained studies carried out on a wide variety of mAbs indicate that charge heterogeneity, which resulted in studying mAbs from a higher resolution than simply as a colloidal particle with lumped charge at its center of mass, is responsible for self-association in mAb solutions. The equilibrium structures suggest that mAb molecules interact via effective short-range attractions that result in the formation of dynamic self-associating networks. It is inferred that the formation of these networks results in high-viscosity changes with concentration. This hypothesis has been confirmed by recent experiments on various mAb solutions that show self-associating behavior.[32–35, 38] The CG models and simulations have proven to be extremely useful in rationalizing completely different high-concentration viscosity behaviors of mAbs that only differ by a few amino acids. This suggests that CG modeling can be effectively used as an investigation probe that can qualitatively check the effect of sequence changes in proteins and draw conclusions about the self-associating and viscosity behavior based on equilibrium structure formation.

6.3   MODELING CHALLENGES

In order to design better models for proteins at high concentrations, the nature of protein–protein interactions has to be understood in more detail. The CG models described in the previous section require two things: identification of CG sites that is based on the long-wavelength dynamics of an elastic network model of the protein (slow-moving modes) and the set of effective interactions between the sites that are picked based on intuitive reasoning and experimental data. Picking the CG sites is relatively simpler compared to elucidating the role of protein–protein interactions in these models. Under dilute conditions, proteins can be treated as charged colloids and effective potentials can be written for the interactions. However, for proteins at high concentrations, interactions with the solvent, ions, and other proteins have to be taken into account, keeping in mind the charge heterogeneity of the protein molecule itself. Measurement of dynamic properties in solution such as viscosity requires one to model the solvent explicitly, which can be quite a challenge in itself. Implicit solution models also lead to formation of networks that exhibit glassy behavior at high concentrations,[39] which elucidates the need for good solvent models that can help alleviate that. It is thus important to take a closer look at the role of electrostatics, dispersion forces, and hydrodynamics in order to build a more robust model of high-concentration proteins.

6.3.1   Electrostatic and Dispersion Effects

To model electrostatic effects in biomolecules, a large number of excellent reviews are available.[40–44] In dilute solutions, the dispersion and electrostatic interactions between the proteins can be modeled quite well using the DLVO (Derjaguin–Landau–Verwey–Overbeek) theory.[45, 46] The DLVO theory is based on the Poisson–Boltzmann approximation, where the ions are treated as point charges and no ion–ion correlations are taken into account. This approximation holds for very low salt concentrations (<0.05 M) only and is not very helpful in biological systems of interest.[47] The DLVO potential is a balance between the attractive van der Waals dispersion forces and the repulsive electric double-layer forces that exist on all colloidal particles in solution. At higher concentrations, the ion–ion effects begin to play an important role and the DLVO theory becomes no longer applicable. It is then instructive to see how one could model the interactions between protein molecules surrounded by solvent and ions from a multiscale perspective.

6.3.1.1   Atomistic Modeling

In an all-atom (AA) model, the protein, solvent, and ions are modeled as point charges that interact via Coulombic forces,

$$\psi_i^C(r) = \frac{q_i}{4\pi\varepsilon_0 r} \tag{6.1}$$

where $\psi_i^C(r)$ is the electrostatic potential from a simple ion with charge q_i located at position r, and ε_0 is the dielectric permittivity of free space. The electrostatic potential energy $\phi(r)$ for a system of charges is then given by simple superposition as

$$\phi(r) = \sum_{\substack{i,j \\ i \neq j}} \frac{q_i q_j}{4\pi\varepsilon_0 r_{ij}} \tag{6.2}$$

where $r_{ij} = r_i - r_j$. If polarization effects via induced dipoles are included in the AA model, the charge–dipole and dipole–dipole interactions also have to be accounted for, apart from the simple charge–charge interactions. The AA models are mainly used in MD simulations of small biomolecular systems where the solvent and ions need to be modeled explicitly. Though the AA models are very robust, care needs to be taken when considering infinite systems using periodic boundary conditions.[48]

6.3.1.2   Continuum Modeling

At the other end of the spectrum are continuum models that treat the system as a dielectric continuum. These models solve the Poisson–Boltzmann (PB) equation to give the electrostatic potential in and around the protein. The Poisson–Boltzmann theory was first introduced by Guoy[49] and Chapman[50] and generalized later by Debye and Hückel.[51] The PB model mainly assumes extremely dilute solutions and treats the protein as a low dielectric medium and the electrolyte (solvent and ions) as a high dielectric medium.[52] The ions are also treated as point charges with no ion–ion correlations. The discrete solvent effects are not accounted for in PB theory. In spite of these assumptions, PB theory has been found to be very useful to look at electrostatic potentials of isolated small and large biomolecules.[53] There are many software packages available that solve the non-linear and linear PB equations, the most popular being APBS,[54] which uses some advanced solution techniques.

The electrostatic potential from a point charge in an electrolyte arises from the charge itself and from the ions surrounding the charge. The ions surrounding the charge shield the charge from other charges in its vicinity and screen the electrostatic potential. The shape and size of the ion clouds, along with the nature of the electrolyte solution, determine the interaction energy between the charges in the electrolyte solutions.[55,56] In the linear PB approximation (Debye–Hückel theory), the effects of the electrolyte are implicitly incorporated by using the screened Coulomb potential (Yukawa function). In this limit, the electrostatic potential $\psi_i^{PB}(r)$ from a point charge in an electrolyte is given by

$$\psi_i^{PB}(r) = q_i \frac{e^{-\kappa_D r}}{4\pi\varepsilon_0\varepsilon_r r} \tag{6.3}$$

where the Yukawa function

$$\frac{e^{-\kappa_D r}}{r}$$

models the electrostatic screening due to attraction of counter-ions and repulsion of co-ions, ε_r is the dielectric constant of the solvent, $1/\kappa_D$ and is the Debye length with the Debye parameter κ_D given by

$$\kappa_D^2 = \sum_i \frac{n_i q_i^2}{\varepsilon_0\varepsilon_r k_B T} \tag{6.4}$$

where n_i is the number density of the ionic species, q_i is the charge on the ion, k_B is Boltzmann's constant, and T is the absolute temperature. In the full non-linear PB limit, the bare charge q_i can no longer be used since the ion atmosphere around the charge affects the response of the electrolyte solution. The dressed ion theory (DIT) and dressed molecule theory (DMT) developed by Kjellander and coworkers[57–60] address the issues in linear PB by incorporating ion–ion correlations, finite ion sizes, and discrete solvent effects in a rigorous statistical mechanical manner. The response of the electrolyte in the non-linear PB limit remains exponential at large distances, but the potential from a simple charge is now approximately given by

$$\psi_i^{NPB}(r) \sim q_i^r \frac{e^{-\kappa_D r}}{4\pi\varepsilon_0\varepsilon_r r} \tag{6.5}$$

where q_i^r is now the renormalized charge (effective charge) around the particle that arises from the bare charge itself and the ion atmosphere surrounding the charged particle. In the linear limit, this renormalized charge q_i^r equals the bare charge q_i due to the response being weak, and the linear approximation becomes adequate.

6.3.1.3   Coarse-Grained (Mesoscale) Modeling

Coarse-grained models lie in between the AA and continuum models where the protein, solvent, and ions are partially or fully resolved, depending on the problem at hand. For example, at one level, the protein could be fully resolved, whereas the solvent and ions are modeled as continuum entities. At another level, the protein itself could be modeled as a collection of CG sites in a continuum electrolyte. In both cases, the electrolyte is still modeled implicitly, but the effect of the solvent and ions includes ion–ion correlations and finite ion size effects that were previously neglected in the PB approximation. In the first case, where the protein is modeled as a collection of point charges in a continuum electrolyte, the electrostatic potential is still the screened Yukawa function. It was shown by Kjellander and coworkers that the potential $\psi_i^r(r)$ from a point charge in an electrolyte where ion–ion correlations and finite ion effects are included is given by

$$\psi_i^r(r) \sim q_i^r \frac{e^{-\kappa r}}{4\pi\varepsilon_0 E_r r} \tag{6.6}$$

where κ is the screening parameter of the electrolyte and E_r is the dielectric constant of the electrolyte. In the infinite dilution limit, the screening parameter κ equals the Debye screening parameter κ_D and E_r equals the dielectric constant of the solvent ε_r.

In the case where the protein itself is partially resolved as in a CG model, the CG sites are treated as charge densities $\rho(r)$ interacting with a continuum electrolyte. The protein is then modeled as a collection of charges, dipoles, and higher-order multipoles. In such a case, the electrostatic potential is given by a screened multipole expansion,[61]

$$\psi_i^r(r) = \left[q_i^r - \mu_i^r \cdot \nabla + \frac{1}{3}\Theta_i^r : \nabla\nabla + \quad \right] \frac{e^{-\kappa r}}{4\pi\varepsilon_0 E_r r} \tag{6.7}$$

where q^r is the renormalized charge (effective Yukawa charge),

$$q^r = \int ds\, \rho^r(s) \frac{\sinh(\kappa s)}{\kappa s} \tag{6.8}$$

μ^r is the renormalized dipole moment,

$$\mu^r = 3 \int ds\, \rho^r(s)s \left[\frac{\cosh(\kappa s)}{(\kappa s)^2} - \frac{\sinh(\kappa s)}{(\kappa s)^3} \right] \tag{6.9}$$

and Θ^r is the renormalized quadrupole moment,

$$\Theta^r = \frac{15}{2} \int ds\, \rho^r(s)(3ss - s^2 I) \times \left[\frac{\sinh(\kappa s)}{(\kappa s)^3} - 3\frac{\cosh(\kappa s)}{(\kappa s)^4} + 3\frac{\sinh(\kappa s)}{(\kappa s)^5} \right] \tag{6.10}$$

In the above equations, ρ^r is the renormalized charge density that is the sum of the bare charge density of the CG site itself and a dress charge due to the ion atmosphere surrounding the site.[56] It is interesting to note that for pure solvent ($\kappa = 0$, $E_r = \varepsilon_r$), Coulomb's law is recovered and Equation 6.7 becomes the multipolar expansion for the Coulomb potential. In this case, the atmosphere surrounding the CG site is simply from the solvent molecules that do not contribute to the net charge. Hence, the renormalized charge q^r equals the bare charge q of the site, and Coulomb's law follows.

The previous cases dealt with examples that modeled the solvent implicitly. A number of additional CG models can be created where the electrolyte is included explicitly via a CG model or a continuum one. Development of such models is extremely challenging and still in progress (see next section). If, in addition to electrostatics, dispersion forces are also considered in the model, the analysis becomes complicated, with the dispersion forces dominating at very close distances. For more details on the effects of dispersion forces, refer to the work by Kjellander and Ramirez.[62]

6.3.2 Solvent Hydrodynamics

The CG models described in the previous section incorporate the effect of the electrolyte implicitly while modeling the electrostatic interactions. However, these models are limited by the fact that dynamic properties such as viscosity cannot be calculated accurately without some form of explicit solvent. Implicit solvent systems also exhibit glassy behavior at concentrations higher than 100–120 mg/ml, which limits the applicability of the method. In order to overcome these limitations, some form of hydrodynamic interactions (solvent-mediated effects) needs to be incorporated into the CG models.

Atomistic solvent involves a large number of solvent degrees of freedom to track, which makes CG modeling an attractive alternative. One way to incorporate hydrodynamic effects would be to use a CG model for solvent (electrolyte) that would replicate these effects. However, good CG models for solvent are still under active research.[63–65] An alternative is to add hydrodynamic interactions in an ad hoc manner, as in Stokesian dynamics,[66] or use a mesoscale solvent using dissipative particle dynamics.[67] But these models do not account for the electrostatic interactions that come from the solvent and dissolved ions. Hence, there is a growing need to develop CG models of electrolyte solutions that can model both the hydrodynamic interactions and the electrostatic effects in a consistent manner.

Another way to model the solvent is to represent it as a fluctuating continuous medium and incorporate hydrodynamics via the continuum Landau–Lifshitz fluctuating hydrodynamic equations.[68] In this way, a hybrid methodology[69] can be constructed where the protein is modeled as an AA or CG model in a fluctuating continuum solvent. Deterministic continuum solvents have been shown to give large errors in the fluctuation spectrum in coupled particle simulations.[70] Fluctuating hydrodynamic models for electrolytes are nonexistent as of now and will be a subject of future research.

6.4 FUTURE WORK

The main parameters of the electrostatic problem involve the renormalized charge q^r, screening parameter κ, and dielectric constant E_r. In the CG models, the selection of the sites can be made in such a way that the net theoretical lumped charge on a site is small, and hence as a first approximation, the renormalized charge can be replaced by the bare charge. The screening parameter and dielectric constant have to be computed systematically in a self-consistent manner.

The dielectric constant is perhaps the most difficult parameter to calculate in biological systems. It arises due to the decrease in the net electric field inside a dielectric medium due to induced polarization. Though continuum studies have defined dielectric constants in terms of induced polarization effects, there has been a lot of debate in literature surrounding the real meaning of dielectric constants.[71] Statistical mechanics defines dielectric constants in terms of fluctuations of the average dipole moment of a system.[72, 73] A number of computational studies have focused on calculating the dielectric constant using the Kirkwood–Frohlich theory for simple physical systems,[74, 75] as well as proteins.[76, 77] However, no study has focused on calculating the dielectric constant of the medium between two charged distributions (in our case, CG sites) in a systematic fashion. For our purposes, we can assume that the CG sites contain lumped charges at the center of mass of the site. The CG sites then interact via the potential $\psi_i^r(r)$ in Equation 6.6 or 6.7. The two parameters can be computed by minimizing the quantity,

$$\left\| \psi^C - \psi^r \right\|^2 \tag{6.11}$$

in a least-squares sense.[78] In this way, the CG model tries to reproduce the same potential that one would find between the two sites as if they were interacting via an AA model with an explicit electrolyte (solvent and ions) between them. However, this has to be done for each CG site interaction since the dielectric constant is a local property in proteins[76] due to a heterogeneous surface charge distribution. Work in this area is still in progress and will be the subject of future investigations, along with development of CG/fluctuating hydrodynamic models of electrolytes.

6.5 SUMMARY

Computational modeling of biomolecular systems has come a long way[79, 80] since the first molecular dynamics simulations by Alder and Wainwright[81] a few decades ago. Computational experiments now frequently complement physical experiments and stand alone in their own right. The problem of aggregation and self-association in high-concentration proteins is highly complex and expensive to understand by physical experiments alone. Molecular dynamics simulations of coarse-grained mAbs have revealed that short-range electrostatic interactions are responsible for forming transient networks in solution that can possibly lead to unusually high viscosities at high concentrations. These models corroborate experimental viscosity data qualitatively at present since the analysis depends exclusively on equilibrium structure. Quantitative comparison of viscosity measurements can be achieved by building a computational model that incorporates explicit solvent and higher-order electrostatic effects. Several challenges and suggestions for calculating electrostatic interactions of high-concentration protein solutions were listed in this chapter. Incorporating these will help build better mAb models that can help understand issues in more complex situations such as phase separation, centrifugation, and chromatography.

ACKNOWLEDGMENTS

The author would like to thank Dr. Gregory Voth at the University of Chicago, under whose guidance and supervision the coarse-grained modeling was carried out. The author would also like to thank Dr. Isidro Zarraga, Dr. Steve Shire, Dr. Tom Patapoff, Dr. Sandeep Yadav, and Paul Brandt at Genentech, and Dr. Tim Kamerzell at the University of Kansas Medical Center for their support and encouragement during the course of this work. The author would like to acknowledge the support of Dr. John Bell at the Computational Research Division, Lawrence Berkeley National Lab, while writing this chapter.

REFERENCES

1. Aggarwal, S. 2014. What's fueling the biotech engine—2012 to 2013. *Nature* 32: 32–39.
2. Reichert, J. M. 2013. Antibodies to watch in 2014. *mAbs* 6: 5–14.
3. Shire, S. J., Shahrokh, Z., and Liu, J. 2004. Challenges in the development of high protein concentration formulations. *J. Pharm. Sci.* 93: 1390–1402.
4. Shire, S. J. 2009. Formulation and manufacturability of biologics. *Curr. Opin. Biotechnol.* 20: 708–714.
5. Cromwell, M. E., Hilario, E., and Jacobson, F. 2006. Protein aggregation and bioprocessing. *AAPS J.* 8: E572–E579.
6. Shire, S. J., Cromwell, M. E., and Liu J. 2006. Concluding summary: Proceedings of the AAPS Biotec Open Forum on "Aggregation of Protein Therapeutics." *AAPS J.* 8: E729–E730.
7. Frokjaer, S., and Otzen, D. E. 2005. Protein drug stability: A formulation challenge. *Nat. Rev. Drug Discov.* 4: 298–306.
8. Daugherty, A. L., and Mrsny, R. J. 2006. Formulation and delivery issues for monoclonal antibody therapeutics. *Adv. Drug Deliv. Rev.* 58: 686–706.
9. Rosenberg, A. S. 2006. Effects of protein aggregates: An immunologic perspective. *AAPS J.* 8: E501–E507.
10. Saluja, A., and Kalonia, D. S. 2008. Nature and consequences of protein-protein interactions in high protein concentration solutions. *Int. J. Pharm.* 358: 1–15.
11. Ross, P. D., and Minton, A. P. 1977. Analysis of non-ideal behavior in concentrated hemoglobin solutions. *J. Mol. Biol.* 112: 437–452.
12. Larson, R. G. 1999. *The Structure and Rheology of Complex Fluids.* Oxford University Press, New York.
13. Elcock, A. H., and McCammon, J. A. 2001. Calculation of weak protein-protein interactions: The pH dependence of the second virial coefficient. *Biophys. J.* 80: 613–625.
14. Liu, J., Nguyen, M. D. H., Andya, J. D., and Shire, S. J. 2005. Reversible self-association increases the viscosity of a concentrated monoclonal antibody in aqueous solution. *J. Pharm. Sci.* 94: 1928–1940.
15. Kanai, S., Liu, J., Patapoff, T. W., and Shire, S. J. 2008. Reversible self-association of a concentrated monoclonal antibody solution mediated by Fab-Fab interaction that impacts solution viscosity. *J. Pharm. Sci.* 97: 4219–4227.
16. Yadav, S., Liu, J., Shire, S. J., and Kalonia, D. S. 2010a. Specific interactions in high concentration antibody solutions resulting in high viscosity. *J. Pharm. Sci.* 99: 1152–1168.
17. Yadav, S., Shire, S. J., and Kalonia, D. S. 2010b. Factors affecting the viscosity in high concentration solutions of different monoclonal antibodies. *J. Pharm. Sci.* 99: 4812–4829.

18. Yadav, S., Shire, S. J., and Kalonia, D. S. 2012a. Viscosity behavior of high-concentration monoclonal antibody solutions: Correlation with interaction parameter and electroviscous effects. *J. Pharm. Sci.* 101: 998–1011.

19. Yadav, S., Alavattam, S., Kanai, S., Liu, J., Lien, S., Lowman, H., Kalonia, D. S., and Shire, S. J. 2011. Establishing a link between amino acid sequences and self-associating and viscoelastic behavior of two closely related monoclonal antibodies. *Pharm. Res.* 28: 1750–1764.

20. Yadav, S., Laue, T. M., Kalonia, D. S., Singh, S. N., and Shire, S. J. 2012b. The influence of charge distribution on self-association and viscosity behavior of monoclonal antibody solutions. *Mol. Pharmaceutics* 9: 791–802.

21. Laue, T., and Demeler, B. 2011. A postreductionist framework for protein biochemistry. *Nat. Chem. Biol.* 7: 331–334.

22. Stone, J. E., Hardy, D. J., Ufimtsev, I. S., and Schulten, K. 2010. GPU-accelerated molecular modeling coming of age. *J. Mol. Graph. Model.* 29: 116–125.

23. Allen, M. P., and Tildesley, D. J. 1987. *Computer Simulation of Liquids.* Oxford University Press, Oxford.

24. Saunders, M. G., and Voth, G. A. 2013. Coarse-graining methods for computational biology. *Annu. Rev. Biophys.* 42: 73–93.

25. Saunders, M. G., and Voth, G. A. 2012. Coarse-graining of multiprotein assemblies. *Curr. Opin. Struct Biol.* 22: 144–150.

26. Baaden, M., and Marrink, S. J. 2013. Coarse-grain modeling of protein-protein interactions. *Curr. Opin. Struct. Biol.* 23: 878–886.

27. Voth, G. A. 2008. *Coarse-Graining of Condensed Phase and Biomolecular Systems.* CRC Press, Taylor & Francis Group, Boca Raton, FL.

28. Dama, J. F., Sinitskiy, A. V., McCullagh, M., Weare, J., Roux, B., Dinner, A. R., and Voth, G. A. 2013. The theory of ultra-coarse-graining. 1. General principles. *J. Chem. Theory Comput.* 9: 2466–2480.

29. Chaudhri, A., Zarraga, I. E., Kamerzell, T. J., Brandt, J. P., Patapoff T. W., Shire, S. J., and Voth, G. A. 2012. Coarse-grained modeling of the self-association of therapeutic monoclonal antibodies. *J. Phys. Chem. B* 116: 8045–8057.

30. Chaudhri, A., Zarraga, I. E., Yadav, S., Patapoff, T. W., Shire, S. J., and Voth, G. A. 2013. The role of amino acid sequence in the self-association of therapeutic monoclonal antibodies: Insights from coarse-grained modeling. *J. Phys. Chem. B* 117: 1269–1279.

31. Zhang, Z., Pfaendtner, J., Grafmüller, A., and Voth, G. A. 2009. Defining coarse-grained representations of large biomolecules and biomolecular complexes from elastic network models. *Biophys. J.* 97: 2327–2337.

32. Zarraga, I. E., Taing, R., Zarzar, J., Luoma, J., Hsiung, J., Patel, A., and Lim, F. J. 2013. High shear rheology and anisotropy in concentrated solutions of monoclonal antibodies. *J. Pharm. Sci.* 102: 2538–2549.

33. Lilyestrom, W. G., Yadav, S., Shire, S. J., and Scherer, T. M. 2013. Monoclonal antibody self-association, cluster formation and rheology at high concentrations. *J. Phys. Chem. B* 117: 6373–6384.

34. Yearley, E. J., Zarraga, I. E., Shire, S. J., Scherer, T. M., Gokarn, Y., Wagner, N. J., and Liu, Y., 2013. Small angle neutron scattering characterization of monoclonal antibody conformations and interactions at high concentrations. *Biophys. J.* 105: 720–731

35. Yearley, E. J., Godfrin, P. D., Perevozchikova, T., Zhang, H., Falus, P., Porcar, L., Nagao, M., Curtis, J. E., Gawande, P., Taing, R., Zarraga I. E., Wagner, N. J., and Liu, Y. 2014. Observation of small cluster formation in concentrated monoclonal antibody solutions and its implications to solution viscosity. *Biophys. J.* 106: 1763–1770.

36. Trappe, V., Prasad, V., Cipelletti, L., Segre, P. N., & Weltz, D. A. (2001). Jamming phase diagram for attractive particles. *Nature,* 411(6839), 772–775.

37. Lu, P. J., Zaccarelli, E., Ciulla, F., Schofield, A. B., Sciortino, F., & Weitz, D. A. (2008). Gelation of particles with short-range attraction. *Nature,* 453(7194), 499–503.
38. Castellanos, M. M., Pathak, J. A., Leach, W., Bishop, S. M., and Colby, R. H. 2014. Explaining the non-Newtonian character of aggregating monoclonal antibody solutions using small-angle neutron scattering. *Biophys. J.* 107: 469–476.
39. Buck, P. M., Chaudhri, A., Kumar, S., and Singh, S. K. 2015. Highly Viscous Antibody Solutions are a Consequence of Network Formation Caused by Domain–Domain E;ectrostatic Complementarities: Insights from Coarse-Grained Simulations. *Mol. Pharmaceutics,* 12: 127–139.
40. Warshel, A., and Russell, S. T. 1984. Calculations of electrostatic interactions in biological systems and in solutions. *Q. Rev. Biophys.* 17: 283–422.
41. Matthew, J. B. 1985. Electrostatic effects in proteins. *Ann. Rev. Biophys. Biophys. Chem.* 14: 387–417.
42. Sharp, K. A., and Honig, B. 1990. Electrostatic interactions in macromolecules: Theory and applications. *Ann. Rev. Biophys. Biophys. Chem.* 19: 301–332.
43. Nakamura, H. 1996. Roles of electrostatic interactions in proteins. *Q. Rev. Biophys.* 29: 1–90.
44. Simonson, T. 2003. Electrostatics and dynamics of proteins. *Rep. Prog. Phys.* 66: 737–787.
45. Derjaguin, B. V., and L. D. Landau. 1941. The theory of stability of highly charged lyophobic sols and coalescence of highly charged particles in electrolyte solutions. *Acta Physicochim. URSS* 14: 633–652.
46. Verwey, E. J. W., and Overbeek, J. Th. G. 1948. *Theory of the Stability of Lyophobic Colloids.* Elsevier, Amsterdam.
47. Boström, M., Williams, D. R. M., and Ninham, B. W. 2001. Specific ion effects: Why DLVO theory fails for biology and colloid systems. *Phys. Rev. Lett.* 87: 168103 (1–4).
48. Warshel, A., Sharma, P. K., Kato, M., and Parson, W. W. 2006. Modeling electrostatic effects in proteins. *Biochim. Biophys. Acta* 1764: 1647–1676.
49. Guoy, M. 1910. Sur la constitution de la charge électrique a la surface d´un electrolyte. *J. Phys.* 9: 457–468.
50. Chapman, D. L. 1913. A contribution to the theory of electro-capillarity. *Phil. Mag.* 25: 475–481.
51. Debye, P., and Hückel, E. 1923. Zur theorie der electrolyte. *Phys. Zeitschr.* 24: 185–206.
52. Tanford, C., and Kirkwood, J. G. 1957. Theory of protein titration curves. I. General equations for impenetrable spheres. *J. Am. Chem. Soc.* 79: 5333–5339.
53. Fogolari, F., Brigo, A., and Molinari, H. 2002. The Poisson-Boltzmann equation for biomolecular electrostatics: A tool for structural biology. *J. Mol. Recognit.* 15: 377–392.
54. Baker, N. A., Sept, D. Joseph, S., Holst, M. J., and McCammon, J. A. 2001. Electrostatics of nanosystems: Application to microtubules and the ribosome. *Proc. Natl. Acad. Sci.* 98: 10037–10041.
55. Kjellander, R. 2001. Distribution function theory of electrolytes and electrical double layers. In *Electrostatic Effects in Soft Matter and Biophysics,* ed. C. Holm, P. Kékicheff, and R. Podgornik. NATO Science Series. Kluwer Academic, Dordrecht, 317–366.
56. Kjellander, R. 2007. Fundamental aspects of electrostatic interactions and charge renormalization in electrolyte systems. *Colloid J.* 69: 20–28.
57. Kjellander, R., and Mitchell, D. J. 1994. Dressed ion theory for electrolyte solutions: A Debye Hückel-like reformulation of the exact theory for the primitive model. *J. Chem. Phys.* 101: 603–626.
58. Kjellander, R., and Mitchell, D. J. 1995. Dressed ion theory for electric double layer structure and interactions; an exact analysis. *Mol. Phys.* 91: 173–188.
59. Kjellander, R., and Ramirez, R. 2005. Screened Coulomb potential and the renormalized charges of ions and molecules in electrolyte solutions. *J. Phys. Condens. Matter* 17: S3409–S3421.

60. Ramirez, R., and Kjellander, R. 2003. Dressed molecule theory for liquids and solutions: An exact charge renormalization formalism for molecules with arbitrary charge distributions. *J. Chem. Phys.* 119: 11380–11395.

61. Ramirez, R., and Kjellander, R. 2006. Effective multipoles and Yukawa electrostatics in dressed molecule theory. *J. Chem. Phys.* 125: 144110 (1–13).

62. Kjellander, R., and Ramirez, R. 2008. Yukawa multipole electrostatics and nontrivial coupling between electrostatic and dispersion interactions in electrolytes. *J. Phys. Condens. Matter* 20: 494209 (1–16).

63. Wang, H., Junghans, C., and Kremer, K. 2009. Comparative atomistic and coarse-grained study of water: What do we lose by coarse-graining? *Eur. Phys. J. E* 28: 221–229.

64. Yesylevskyy, S. O., Schäfer, L. V., Sengupta, D., and Marrink, S. J. 2010. Polarizable water model for the coarse-grained MARTINI force field. *PLoS Comput. Biol.* 6: e1000810 (1–17).

65. Riniker, S., and van Gunsteren, W. F. 2011. A simple, efficient polarizable coarse-grained water model for molecular dynamics simulations. *J. Chem. Phys.* 134: 084110 (1–13).

66. Brady, J. F., and Bossis, G. 1988. Stokesian dynamics. *Annu. Rev. Fluid Mech.* 20: 111–157.

67. Hoogerbrugge, P. J., and Koelman, J. M. V. A. 1992. Simulating microscopic hydrodynamic phenomena with dissipative particle dynamics. *Europhys. Lett.* 19: 155–160.

68. Landau, L. D., and Lifshitz, E. M. 2013. *Fluid Mechanics: Landau and Lifshitz: Course of Theoretical Physics*. Vol. 6. Elsevier, Amsterdam.

69. Mohamed, K. M., and Mohamad, A. A. 2010. A review of the development of hybrid atomistic–continuum methods for dense fluids. *Microfluidics Nanofluidics* 8: 283–302.

70. Donev, A., Bell, J. B., Garica, A. L., and Alder, B. J. 2010. A hybrid particle-continuum method for hydrodynamics of complex fluids. *Multiscale Model. Simul.* 8: 871–911.

71. Schutz, C. N., and Warshel, A. 2001. What are the dielectric "constants" of proteins and how to validate electrostatic models? *Proteins Struct. Funct. Bioinformatics* 44: 400–417.

72. Kirkwood, J. G. 1939. The dielectric polarization of polar liquids. *J. Chem. Phys.* 7: 911–919.

73. Fröhlich, H. 1948. General theory of the static dielectric constant. *Trans. Faraday Soc.* 44: 238–243.

74. Neumann, M. 1983. Dipole moment fluctuation formulas in computer simulations of polar systems. *Mol. Phys.* 50: 841–858.

75. Neumann, M., Steinhauser, O., and Pawley, G. S. 1984. Consistent calculation of the static and frequency-dependent dielectric constant in computer simulations. *Mol. Phys.* 52: 97–113.

76. Nakamura, H., Sakamoto, T., and Wada, A. 1988. A theoretical study of the dielectric constant of protein. *Protein Eng.* 2: 177–183.

77. Simonson, T., and Brooks III, C. L. 1996. Charge screening and the dielectric constant of proteins: Insights from molecular dynamics. *J. Am. Chem. Soc.* 118: 8452–8458.

78. Press, W. H., Teukolsky, S. A., Vetterling, W. T., and Flannery, B. P. 2007. *Numerical Recipes: The Art of Scientific Computing*. 3rd ed. Cambridge University Press, Cambridge.

79. Levitt, M. 2013. Birth and future of multi-scale modeling of biological macromolecules [Nobel lecture]. Nobel Media AB. http://www.nobelprize.org/nobel_prizes/chemistry/laureates/2013/levitt-lecture.html (accessed June 29, 2014).

80. Warshel, A. 2013. Computer simulations of biological functions: From enzymes to molecular machines [Nobel lecture]. Nobel Media AB. http://www.nobelprize.org/nobel_prizes/chemistry/laureates/2013/warshel-lecture.html (accessed June 29, 2014).

81. Alder, B. J., and Wainwright, T. E. 1957. Phase transition for a hard sphere system. *J. Chem. Phys.* 27: 1208–1209.

Section II

Developability Practices in the Biopharmaceutical Industry

7 Best Practices in Assessment of Developability of Biopharmaceutical Candidates

Steffen Hartmann* and Hans P. Kocher*

CONTENTS

---

* Novartis Pharma AG, Integrated Biologics Profiling, Biologics Technical Development and Manufacturing, Basel, Switzerland.

7.1 INTRODUCTION

Researchers in discovery organizations tend to select drug candidates based on biological properties, for example, in the case of antibodies, based on binding affinity, potency, efficacy, cross-reactivity with target of species for toxicological studies, and epitope. However, to develop a candidate into a drug, additional properties, such as expression and purification yields, aggregation propensity, stability, viscosity, physicochemical profile, compatibility with *in vivo* environment, and immunogenicity, need to be assessed. Departments or contract research organizations responsible for developing a process to manufacture a safe biologic drug of high quality in a cost-efficient, reproducible way are interested in these additional properties of the drug. Consequently, selection of the best molecule, taking both biology and developability into consideration, is a crucial step for successful development of biologic drugs (Figure 7.1). Developability addresses risk assessment and risk mitigation to improve the likelihood that selected biologic drug candidates can be successfully developed into medicines available to patients. Since major process factors are investigated during the developability assessment, an additional output of the assessment is a judgment on whether or not a drug candidate can be manufactured with a platform process. In the context of this chapter, the term *developability assessment* of a biologic drug candidate is understood as experimental, not computational, investigation to assess manufacturing feasibility (productivity, stability, process), ease of formulation for a specific route of administration, and importantly, compatibility with *in vivo* conditions (termed *in vivo* fitness in this chapter), such as cross-reactivity, half-life, stability in the *in vivo* environment, and immunogenicity. In short, during developability assessment, drug candidates are selected or, if necessary, engineered to fulfill manufacturing, formulation, and safety characteristics before expensive development efforts are initiated. The approach defines best candidates with respect to manufacturing, formulation, and safety, in contrast to adapting manufacturing processes to enable preparation of a difficult-to-produce drug. Adaptation of processes to prepare high-quality biologic drugs at later development stages becomes increasingly costly. *In silico* identification of potential post-translational modification (PTM) sites

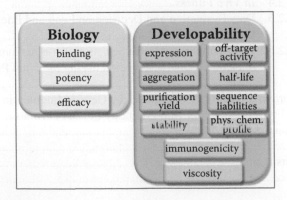

FIGURE 7.1 A biologic candidate needs to excel in both biological and developability characteristics to eventually become a successful drug.

known to possibly impact protein integrity, and determination of the isoelectric point (pI) are the only computational components of this chapter.

7.2 WHY DEVELOPABILITY MATTERS

Innovative biologic medicines are costly to develop and produce on a commercial scale. A typical development program for the production of clinical material consists of cell line development, process development, production of material for toxicology studies, and finally, production of material for clinical trials. In the course of several years, volumetric productivity and product quality are optimized, and the manufacturing process is scaled up to meet expected market demand. With the present-day understanding of protein expression and purification, it is possible to produce the large majority of potential therapeutic biologic drugs. Amazing progress in setting up a multitude of efficient pro- and eukaryotic expression systems to enable preparation of therapeutic proteins with increasing complexity has been made.[1,2] Nevertheless, costs and timelines to develop a suitable process for proteins that are difficult to express and purify and formulate may become exorbitant, to the extent that development of such therapeutic agents has to be abandoned. In contrast, development of protein entities that can be expressed, purified, and formulated in platform processes, that is, established, well-optimized processes suitable to manufacture proteins with minor modifications to the process, is commercially attractive. It thus makes sense to not only select a drug candidate on biological considerations and put forth whatever effort it takes to prepare the material into cell line and process development, but also select a candidate with a good developability profile before initiation of cell line and process development activities. Assessment of developability, conceived as comprehensive characterization of drug candidates in relation to anticipated efforts necessary to develop the drug, enables informed decision taking and risk mitigation. Candidates may be found to fit an established platform process, and have low risk and contained costs for technical development. Developability assessment may also lead to the conclusion that the cost to develop a candidate is medium or even high. Hence, developability assessment enables a project team, and company management, to make conscious, fact-based decisions with respect to development costs. Developability assessment de-risks the development process and reduces efforts and costs.[3–5]

7.3 ANTIBODY CANDIDATE ASSESSMENT

The generation of antibodies by phage display, hybridoma, or immortalization technologies typically yields a considerable number of potential drug candidates. First, research teams identify candidates, taking biological properties like binding, epitope, potency, and cross-reactivity into consideration. Subsequently, up to 50 candidates identified by the research team may enter developability assessment. At Novartis, developability assessment of candidates is divided into an early selection phase and a final profiling phase. During the selection phase, the number of

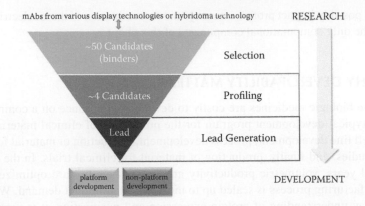

FIGURE 7.2 Many candidates are characterized in high-throughput assays during the selection phase. In this phase, the focus is on feasibility for manufacturing. A small number of candidates identified in the selection phase are subsequently subjected to in-depth characterization in the candidate profiling phase. During the profiling phase, feasibility for manufacturing is evaluated in more detail, and in addition, characteristics of the candidates with respect to formulation and *in vivo* fitness are assessed. At the end of the process, a lead and a backup candidate are identified. Based on the data generated during selection and profiling, it is possible to assess whether a candidate can likely be prepared in a low-cost platform process or needs to be processed in a resource-intense, non-platform process.

candidates is typically reduced to about four, which are ultimately characterized in detail during the profiling phase (Figure 7.2).

In order to assess a wide spectrum of candidates in the profiling phase, sequence and epitope diversity are taken into consideration, in addition to the developability characteristics of the candidates, because structurally closely related candidates may all show comparable developability issues.

The developability assessment process ends with the identification of a lead and a backup candidate. Based on the comprehensive results generated during candidate selection and profiling, it is possible to appraise whether a candidate can most likely be developed in a low-cost platform process or needs to be prepared in a resource-intense nonplatform process.

7.3.1 Early Selection Phase

In the early selection phase, potential PTM sites in the complementarity determining regions (CDRs) and the theoretical Isoelectric point (pI) of the candidates are determined *in silico*. Expression titer, aggregation, conformational stability, and hydrophobicity are assessed in *in vitro* assays.

7.3.1.1 Post-Translational Modifications

As a first step, the amino acid sequences of CDRs of the candidates are scanned for potential PTM sites *in silico*. Post-translational protein modifications may affect product stability, impact manufacturing process and formulation development, reduce potency, and increase risk of immunogenicity. Asparagine deamidation, aspartate isomerization, oxidation of methionine and tryptophan residues, unpaired cysteine residues, and potential sites for glycosylation are labeled.[6] The presence of a N-glycosylation site in the CDRs, for example, results in elimination of the candidate or, should no alternative candidate be available, initiation of efforts to modify the site by genetic engineering. Less critical modifications are highlighted and assessed along with the *in vitro* data generated for each candidate.

7.3.1.2 Critical Risk Factors

The feasibility to express a candidate and the aggregation propensity of candidates are considered critical risk factors. Critical risk factors affect candidate ranking directly and result in de-selection of a candidate.

7.3.1.2.1 Expression

Poor or low expression will cause difficulties during preparation of material for profiling and functional assessment in the early selection phase. Expressability of candidates is determined by transient expression of the candidates in human embryonic kidney cells (HEK293), as described by Geisse and Fux.[7] Polyethylenimine (PEI) is used as a transfection agent for cost reasons. Transfections are performed in 1 L roller bottles containing 100 ml of medium. After cultivation for 10 days, cells are removed by centrifugation and the supernatant is clarified by filtration. Protein content is determined on a high-performance liquid chromatography (HPLC) system equipped with an analytical protein A column. Transient expression in HEK293 cells at the 100 ml scale typically yields tens of milligrams of candidate protein, which is sufficient to perform the early selection phase assays and identify candidates for later detailed characterization. To date, transient expression at the 100 ml scale in Chinese hamster ovary (CHO) cells did not steadily yield enough material for characterization. Consequently, transient expression in HEK cells is used in early selection, although CHO cells will eventually be used for manufacturing. Chromatography on protein A is used to purify the antibody candidates. The identity and integrity of each purified candidate is checked by mass spectrometry. Expression levels in a transient expression system are not predictive for a stable expression system. Therefore, de-selection of candidates based on poor expression is due to the noncompatibility of these candidates with the high-throughput early-phase process. The short timelines do not allow us to individually upscale poor-expression candidates. Losing poor-expression candidates is, however, not a problem, because total candidate numbers at this stage are typically high.

7.3.1.2.2 Aggregation

Aggregation can occur during expression, processing, formulation, and use of protein drugs. It may lead to lower *in vivo* efficacy, increased variability among batches in manufacturing, and importantly, immunogenicity in patients.[8] Therefore, aggregation propensity is a key parameter that is determined and used to de-select candidates in the early selection phase, as well as in the profiling phase. The amount of antibody aggregates is assessed by size exclusion chromatography–multiangle light scattering.[9]

7.3.1.3 Cumulative Risk Factors

Cumulative risk factors affect rating in combination and enable comprehensive data driven ranking of candidates. In the early selection phase, the isoelectric point, conformational stability, surface hydrophobicity, and interface stability are assessed as cumulative risk factors.

7.3.1.3.1 Isoelectric Point

It is of interest to select candidates that fulfill requirements for purification in an established downstream processing platform. pI is a predictor in this respect and is determined *in silico* based on the amino acid sequence. As solubility in the context of drug formulation is assessed experimentally during the profiling phase, pI is, for example, not used to estimate the optimal pH for the best solubility of a candidate.

7.3.1.3.2 Conformational Stability

Conformational stability is assessed by differential scanning fluorimetry using the Thermofluor assay in a high-throughput mode.[10] Rising temperature eventually causes a folded, globular protein to unfold or melt. Determination of the melting point (Tm) provides a measure of protein stability, with more stable candidates requiring higher temperatures to unfold.

7.3.1.3.3 Hydrophobicity

Candidates with increased hydrophobicity tend to show liabilities with respect to aggregation, viscosity, and solubility. For this reason, hydrophobicity of candidates is assessed in the early selection phase. Surface hydrophobicity is experimentally measured by determination of the ammonium sulfate concentration at which the candidates elute from a hydrophobic interaction chromatography (HIC) column.

7.3.1.3.4 Interface Stability

Random association of heavy and light chain regions in libraries may generate candidates with good antigen binding properties. However, since there is no *in vivo* selection and maturation, as with candidates obtained by hybridoma technology or human B-cell immortalization, unstable heavy and light chain pairs may form. Interface stability between heavy and light chain regions in Fab fragments generated by phage display technology is tested by adsorption of heavy chain histidine tagged, non-cysteine bonded Fab fragments to a metal chelate affinity chromatography column. Following a wash with 30% isopropanol, the quantity of the eluted light chain is assessed.

7.3.1.4 Early Developability Assessment

For each candidate evaluated, a "traffic light" classification is used, with red for a high-risk, yellow for a medium-risk, and green for a low-risk assessment. This cumulative risk label for an individual candidate is generated from two critical factors, namely, aggregation and titer, and four risk factors (pI, Tm, hydrophobicity, and interface stability), respectively (Figure 7.3). Low-, medium-, and high-risk

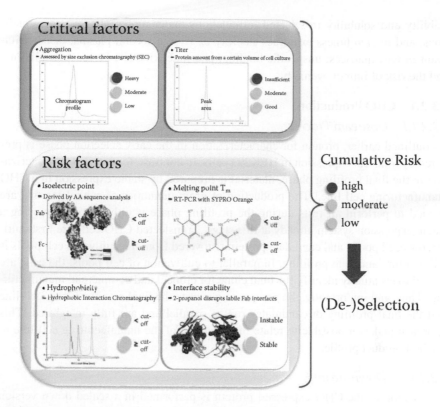

FIGURE 7.3 Candidates with considerable liabilities with respect to expression titer and aggregation propensity are de-selected. Cumulative risk factors are taken into consideration for further candidate ranking.

classification builds on thresholds for each risk factor. Thresholds are set based on scientific knowledge and historical data. The two critical factors are also classified using a traffic light label for each of them. In contrast, the four risk factors are only classified in two categories: green and yellow. Only the combination of unfavorable risk factors will lead to a negative impact on the overall risk label. High-risk candidates are de-selected and the remaining candidates ranked, taking technical developability, sequence, and epitope diversity, as well as biological activity, into consideration. Best candidates are characterized in more detail during the subsequent final profiling phase.

7.3.2 FINAL PROFILING PHASE

The final profiling phase covers more detailed evaluation of the candidates addressing CHO production, molecular profiling, preformulation, and *in vivo* fitness. CHO production encompasses assessments with respect to up and downstream process requirements relative to clinical manufacturing. Molecular profiling addresses physicochemical characterization with a focus on integrity, purity, heterogeneity, modification, viscosity, and conformational stability. Preformulation studies evaluate stability and solubility in selected formulations under representative stress conditions, and *in vivo* fitness examines stability *ex vivo* in serum, plasma, or other relevant *in vivo* matrices; unspecific binding; pharmacokinetics; Fc receptor binding; and the risk of immunogenicity.

7.3.2.1   CHO Production

7.3.2.1.1   Upstream Process

As outlined earlier, protein for characterization in the early selection phase is prepared by transient expression in HEK293 cells. In contrast, for detailed characterization in the final profiling phase, material is prepared by stable expression in a CHO manufacturing cell line.[1] The production of gram quantities of candidates that are needed to perform all tests during the profiling phase is only feasible by using a stable expression system already at this stage. Transfected CHO cells are first cultivated as cell pools, and candidate protein recovered from the cultivated cell pools is utilized for candidate profiling. In parallel to cultivation of cell pools, the cells are cloned to eventually identify the final clone for clinical and commercial manufacturing. Thus, and importantly, the process to generate the final manufacturing cell line and the final profiling of candidates run in parallel and are fully integrated. This approach makes it possible to relate the selection of the manufacturing cell line to the best product profile.

7.3.2.1.2   Downstream Process

Purification of the CHO-expressed protein is performed in a scaled down version of the standard antibody purification platform process. The cell harvest is clarified by centrifugation and filtration, and the antibody candidate captured on a protein A chromatography column. Subsequently, the pH of the eluate is lowered to mimic the virus inactivation step, and different ion-exchange chromatography resins and requirements are tested to identify optimal conditions to polish the protein. The purification evaluation allows us to assess whether a candidate can most likely be prepared in a low-resource platform process or needs to be handled in a more resource-intense non-platform approach.[11]

7.3.2.2   Molecular Profiling

Integrity, modification, purity, heterogeneity, viscosity, and conformational stability are assessed in molecular profiling. Integrity, purity, heterogeneity, and modification are process dependent, whereas viscosity and stability are largely associated with the amino acid sequence of the candidate.

7.3.2.2.1   Testing for Integrity and Modifications

The identity, integrity, and presence of modifications in the candidate protein after expression and purification are assessed by electrospray ionization (ESI) mass spectrometry (MS), sodium dodecyl sulfate capillary electrophoresis (SDS-CE), and electrophoresis on a microfluidics gel electrophoresis system. Peptide maps generated to enable detailed analysis and localization of modifications like deamidation, isomerization, oxidation, clipping, and glycosylation pattern are characterized by liquid chromatography–mass spectrometry (LC-MS).

7.3.2.2.2   Aggregation and Degradation Products

Quantification of aggregates, degradation products, and contaminations is based on size exclusion chromatography (SEC) and SDS-CE data.

7.3.2.2.3   Heterogeneity of Molecule

Charge variants as a result of deamidation, for example, are identified by capillary zone electrophoresis (CZE) and cation-exchange chromatography. Detailed characterization of the variants is based on peptide mapping and LC-MS.

7.3.2.2.4   Viscosity

Viscosity is determined in a capillary rheometer at high concentrations (e.g., 150 mg/ml) and by measuring colloidal stability by dynamic light scattering (DLS) as long as only restricted quantities of the candidate at low concentrations (1–10 mg/ml) are available. DLS measurements at low concentration and rheometry at high concentrations were found to be reasonably well correlated.[12, 13]

7.3.2.2.5   Conformational Stability

Conformational stability is assessed by differential scanning calorimetry during the profiling phase. Data obtained by differential scanning calorimetry during profiling were found to correlate well with data generated by high-throughput differential scanning fluorimetry during the early selection phase.

7.3.2.2.6   Isoelectric Point

pI can be determined by capillary electrophoresis–isoelectric focusing (CE-IEF). The experimentally determined pI was found to correlate well with the pI determined in silico during the early selection phase. Therefore, experimental pI measurements are only performed in exceptional cases during the final profiling phase.

7.3.2.3   Preformulation

The route of administration and formulation should be considered during developability assessment as well. These aspects have traditionally been addressed in later development phases, that is, at a point in time when necessary adaptations get expensive. The candidate profiling assay package includes determination of solubility and accelerated stability testing so that these important aspects can be considered for candidate selection.

7.3.2.3.1   Solubility

Solubility of a candidate is determined at a candidate protein concentration of typically 100 mg/ml in a pH range between 4.0 and 8.0. The solution is incubated for 1 and 7 days, and solubility assessed by measuring turbidity at 405 nm and, after centrifugation, UV absorption at 280 and 330 nm.

7.3.2.3.2   Accelerated Stability

As previously described, conformational stability of candidates is assessed under molecular profiling. In preformulation, the stability of candidates is tested after application of thermal and mechanical stress, and for long time periods. Candidate protein solutions at concentrations of approximately 100 mg/ml in a pH range between 5.0 and 6.5, with and without excipients like sucrose and arginine, are prepared. To assess thermal stability, the solutions are incubated at 25°C and 40°C for several weeks. Mechanical stress is applied by freeze–thawing the solutions and by shaking, respectively. Long-term stability is investigated following storage of the protein solution at 5°C and 25°C for up to 1 year. Stability data generated from material stored for several weeks are taken into account for developability assessment and selection of the lead and backup candidate. Longer-term stability data only become available after a lead decision has been made, but are useful in the context of later formulation activities of the lead candidate. Assessment of aggregation, heterogeneity, and modifications after stress uses methods described under molecular profiling and encompasses size exclusion chromatography, capillary zone electrophoresis, dynamic light scattering, and turbidity measurements. Detailed characterization of degradation products by mass spectrometry is performed as needed.

7.3.2.4   *In Vivo* Fitness

It is important to understand that properties of an antibody that allow successful production, purification, and formulation do not necessarily correlate with an optimal behavior *in vivo* (and vice versa). To investigate these risks, several assays are included during profiling and are termed *in vivo* fitness assessment.

7.3.2.4.1   FcRn Binding

IgGs are protected from lysosomal degradation by a recycling mechanism mediated by the neonatal Fc receptor (FcRn).[14, 15] The ability of IgG to bind to FcRn in an acidic environment and to dissociate at physiological pH conditions is a prerequisite for the long half-life of an antibody. Therefore, during the *in vivo* fitness assessment, the binding affinities of IgGs to human, cynomolgus monkey (cyno), rat, and mouse FcRn are determined at acidic (6.0) and physiological (7.4) pH values with a surface plasmon-based assay.

7.3.2.4.2   FcγR Binding

Measurement of binding against a selection of cyno and human Fcγ receptors allows predicting the presence or absence of antibody-dependent cellular cytotoxicity (ADCC) functionality. These measurements are performed on all IgG formats that are designed to either enhance or silence the ADCC functionality.[16]

7.3.2.4.3   Unspecific Binding

Off target or unspecific binding to other proteins is investigated using protein microarrays.[17] In the case of the protein microarray sold by Protagen, 384 human proteins are spotted on a nitrocellulose-coated glass surface, together with additionally selected proteins like the target antigen. Ideal candidates will only bind to their target antigen. Problematic candidates might possess excellent binding affinities to their target antigen, but also exhibit off target binding to many of the spotted proteins. The intention of the assay is to provide a multitude of protein epitopes for the antibody candidates and assess if the binding is restricted to the specifically designed target or not. The underlying assumption is that unspecific or off target binding activity of a candidate observed on a protein microarray will likely cause problems *in vivo* where the candidate will be exposed to an even more complex situation with an enormous variety of protein epitopes. Binding to less than 10 of the proteins spotted on the microchip is considered to be of no concern with respect to developability of a candidate as long as the signal intensity of these unspecific interactions stays below 10% of the signal intensity of the target antigen.

7.3.2.4.4   Serum Stability

Proteolytic stability and solubility are assessed *ex vivo* in cyno and human serum and plasma or other relevant *in vivo* matrices to judge compatibility with *in vivo* conditions. The candidates are incubated at 37°C for 2 weeks at a concentration equivalent to a 10 mg/kg dose. Sampling is performed in analogy to a pharmacokinetic (PK) study. Samples are typically analyzed via a surface plasmon resonance assay monitoring, in a parallel setup, the binding to the target, as well as the integrity of the Fc portion of the antibody.

7.3.2.4.5   Rat Triage Pharmacokinetics

A simple one-dose rat triage PK study with a single intravenous injection is used to determine if the distribution and half-life of the candidate matches the expected behavior of a human antibody in a rat. Antibody concentration is monitored for 28 days using a generic enzyme-linked immunosorbent assay (ELISA)-like assay employing a Gyrolab instrument. In rare cases of a suspected rat anti-human antibody response, a surface plasmon resonance-based immunogenicity assay is used to evaluate the respective samples.

The study is not intended to replace a full PK study, but rather aims to judge the compatibility of the candidate with the complex *in vivo* situation in an animal. Ideally, the antibody does not cross-react with the rat target to avoid any target-mediated disposition effects. Candidates with an abnormal PK profiles are de-selected.

7.3.2.4.6   Immunogenicity Risk Assessment

Assessment of the risk of immunogenicity during the early stages of preclinical drug development became increasingly important in recent years. The consequences of developing anti-drug antibodies (ADAs) can vary a lot. In most cases, anti-drug antibodies against a human antibody drug are relatively rare and do not pose a safety risk—in contrast to ADAs against a human protein analog, which can be a severe safety risk.

In silico methods for immunogenicity risk assessment are discussed in Chapter 4, but due to their various limitations, we are using major histocompatibility complex (MHC) class II–associated peptide proteomics. These *in vitro* experiments characterize binding patterns of a drug to MHC class II receptors on human dendritic cells.[8] Combined with T-cell response assay data, the risk of immunogenicity is assessed.

7.3.2.5 Data Evaluation and Decision on Lead and Backup Candidate

A decision on the lead and backup candidate based on data generated during the final profiling phase uses the traffic light system previously described, with green indicating a low, yellow a medium, and red a high risk for development. In this example of developability assessment in the profiling phase, monoclonal antibody candidates directed against a soluble human serum protein were characterized. Following identification of candidates based on biological properties, and reduction of the number of candidates by methodologies described under Section 7.3.1, four candidates were assessed in the final profiling phase. Two of the candidates, labeled mAb1 and mAb2, were generated by phage display, and two, labeled mAb3 and mAb4, by hybridoma technology. mAb1, mAb2, mAb3, and mAb4 had similar binding affinities in the low-picomolar range. The four candidates cross-reacted with the target in the animal species anticipated to be used for toxicological studies. The treatment regimen is expected to be by subcutaneous injection, and it is expected that high doses will be needed in a clinical setting. The result of the developability assessment of the four candidates is summarized in Figure 7.4. Note that immunogenicity risk and FcγR binding are not routinely determined for standard antibody formats.

Based on this information, the following conclusions can be drawn:

mAb1: Developability assessment of mAb1 uncovered a number of weaknesses: (1) a high tendency to aggregate under stress conditions in accelerated stability evaluations, (2) significant proteolysis (clipping) in a complementarity determining region, resulting in candidate heterogeneity, and (3) high viscosity at high protein concentrations and a tendency to aggregate. The sum of the identified weaknesses results in a cumulative risk that is considered to be high. Major efforts and resources would be needed to produce and formulate such an antibody.

mAb2: The aggregation propensity of mAb2 was found to be elevated after chromatography on protein A and stayed elevated during subsequent platform downstream processing procedures. As elaborated earlier, aggregation is a major concern affecting drug efficacy, downstream processing, formulation, and immunogenicity. Thus, downstream processing of candidate mAb2 is expected to require special efforts and resources. In addition, accelerated stability testing revealed the formation of basic variants, a property that would entail special efforts for drug formulation. The cumulative risk of the identified weaknesses was considered to be medium—not a no-go decision, but a flag that technical development efforts will exceed standard resource needs; a candidate to be taken into consideration should no alternative with improved properties exist.

mAb3: mAb3 displays an unusual developability profile. Risks for manufacturing and formulation were found to be low based on assessed developability properties, that is, a well-behaving antibody from a technical developability perspective at first sight. However, mAb3 showed an atypical

Property assessed	mAb1	mAb2	mAb3	mAb4
CHO production				
Upstream process	low	low	low	low
Downstream process	medium	medium	low	low
Molecular profiling				
Aggregation	high	medium	low	low
Post-Translational Modifications	medium	medium	low	low
Clipping	high	low	low	low
Heterogeneity	high	low	low	low
Viscosity	high	low	medium	medium
Conformational stability	low	low	low	low
Hydrophobicity	medium	low	medium	medium
Isoelectric point (calculated)	low	low	low	low
Pre-formulation				
Solubility	low	low	low	low
Accelerated stability	high	medium	low	low
In vivo fitness				
FcRn binding	medium	low	low	low
FcγR binding	n.d.	n.d.	n.d.	n.d.
Unspecific binding	low	low	low	low
Serum stability	low	low	low	low
Triage pharmacokinetics	low	low	high	low
Immunogenicity risk assessment	n.d.	n.d.	n.d.	n.d.
Cumulative risk	high	medium	high	low

FIGURE 7.4 Results of developability risk assessment of antibodies mAb1, mAb2, mAb3, and mAb4 directed against a human serum protein, respectively. n.d., not determined.

pharmacokinetic (PK) profile with an uncommonly fast clearance in a PK study. Fast clearance may be the result of unspecific binding to proteins and tissues, or a target-dependent or non-target-dependent sink or internalization, for example. The underlying reasons in the particular case of mAb3 were not elaborated. Based on the findings in the *in vivo* fitness developability category, mAb3 was rated as a high-risk candidate, resulting in its de-selection, since there is no fix for such a clearance issue.

mAb4: Assessment of mAb4 revealed a medium developability risk with respect to viscosity and hydrophobicity, and an overall low cumulative risk for technical development. A quick evaluation of excipients commonly used in formulations resulted in reduced viscosity. This finding indicated that the viscosity issue can most likely be overcome with minor efforts. As a consequence, mAb4 was chosen as a lead candidate to start development activities.

In summary, as a result of developability assessments, the cumulative risk was considered high for antibody mAb1, medium for mAb2, high for mAb3, and low for mAb4. mAb4 was selected as the lead candidate, and the development organization started development activities with this candidate. Risks with respect to manufacturing were found to be low, and risks for formulation were considered to be manageable. mAb2 was selected as a backup candidate. Aggregation propensity during downstream processing and charge heterogeneity following accelerated stability assessments were identified as risks. Development of a process for manufacturing and optimization of a formulation is considered possible, albeit with resources that exceed resources commonly allocated to development activities following a standard platform process. mAb1 was excluded because of issues identified with respect to aggregation, viscosity, and integrity. mAb3 was dropped because of atypical findings in studies addressing *in vivo* fitness.

7.3.2.6 High-Throughput Early Selection Phase in Comparison to Detailed Characterization during the Final Profiling Phase

Approximately 90% of the potential candidates identified based on biological characteristics are eliminated in the early selection process. High-throughput assays are used to generate the data that result in elimination of this large number of candidates. It is therefore important to make sure that the assays used in the early selection process are predictive for their behavior during later-stage development.

Since the large number of candidates makes the use of high-throughput assays a must, it cannot be expected that these assays will provide a perfect prediction of late-stage behavior. The primary aim of the selection phase assays is to de-select the bad candidates while avoiding de-selection of developable candidates. Therefore, cutoff criteria for selection phase assays are set to rather underpredict badly behaving candidates in order to avoid overpredicting, which would cause the de-selection of acceptable candidates. The later profiling phase will provide in-depth characterization with a larger coverage, which will provide the second filter to remove any remaining problematic candidates.

The knowledge space covered during the selection and profiling space is shown in Figure 7.5. Fast aggregation (aggregation during expression and purification, in contrast to slow aggregation, which occurs upon storage and stress conditions), conformational stability, isoelectric point, hydrophobicity, and Fab interface stability are assessment criteria in the selection phase. These characteristics are confirmed and additional criteria covered during the profiling phase.

The correlation of candidate risk category assignment in the selection and profiling phases is shown in Figure 7.6. In the selection phase, 63% of the candidates which were later also assessed in the profiling phase, were assigned to the low-risk, 30% to the medium-risk, and 7% to the high-risk category, respectively. Detailed characterization of candidates during the profiling phase resulted in the assignment of 54% to the low-risk, 23% to the medium-risk, and 23% to the high-risk category. Thus, 74% of the risk category assignments during the selection phase were confirmed in the profiling phase. Twenty-six percent of the developability risk category assignments determined during the selection phase were adjusted based on data generated during the profiling phase. Seven percent of candidates assigned as high

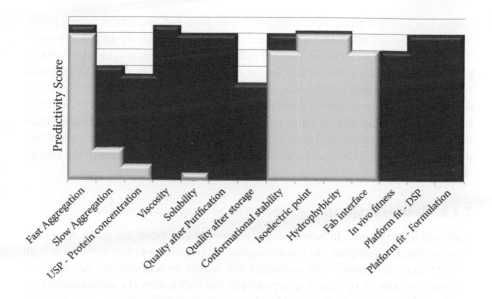

FIGURE 7.5 Knowledge space covered during the high-throughput selection phase (yellow bars) and the more comprehensive and in-depth profiling phase (brown bars). (Image reproduced with permission from *American Pharmaceutical Review.* Copyright 2014 CompareNetworks.)

		Low	Medium	High
	High	9%	7%	7%
Profiling Phase	Medium	5%	18%	0%
	Low	49%	5%	0%
		Low	Medium	High
Risk category			Selection Phase	

FIGURE 7.6 Correlation of developability risk ranking during the selection and profiling phases. Seventy-four percent of candidate risk assignments during the selection phase are confirmed in profiling (sum of gray squares). Additional information collected during the profiling phase leads to changes of the risk category in 26% of the cases. (Image reproduced with permission from *American Pharmaceutical Review.* Copyright 2014 CompareNetworks.)

risk in the selection phase were confirmed to be high-risk candidates during profiling, indicating that selection does, in general, not overestimate developability risks. The principal reasons for developability risk category adjustments after the profiling phase are the consequences of expression of candidates in the CHO cell line used for manufacturing and purification, which follows a process designed to evaluate suitability for platform downstream processing in the profiling phase. In addition, in profiling, the candidates are also characterized at high protein concentrations and after being subjected to stress conditions. *In vivo* fitness liabilities are only identified during the profiling phase, and therefore also contributed to changes in the overall risk assignment between the selection and profiling phases.

7.3.3 IMPROVING QUALITY OF CANDIDATES BY GENETIC ENGINEERING

Antibody concentrations in human serum are mostly well below the range of concentrations used for treatment. As a consequence, aggregation and low stability at high concentration may appear with antibodies that are not problematic in a natural environment, for example. It should be appreciated that only a subset of antibodies has the biophysical properties ideally suited for therapeutic applications.[18]

It may therefore be necessary to genetically engineer candidates in order to achieve satisfactory developability.[19] The thermodynamic stability of human antibodies can be increased through well-established grafting and consensus strategies.[20] Protein aggregation (colloidal stability) remains a more persistent problem. However, strategies for increasing the colloidal stability of human antibody V_H and V_L domains have recently become available.[21] Candidates that have been subjected to genetic engineering to decrease risk for high viscosity, low solubility, or aggregation propensity, for example, need to be reanalyzed for both their biological properties and developability risks.

7.4 DEVELOPABILITY ASSESSMENT OF NON-ANTIBODY PROTEIN FORMATS

The knowledge space for antibody developability has steadily been growing over the past years, with large numbers of candidates being assessed. Inclusion/exclusion criteria have been defined and are constantly refined. The availability of protein A chromatography for fast- and high-throughput quantification and purification has considerably helped to assess developability of antibody candidates. Commercially available anti-C_H1, anti-lambda, or anti-kappa affinity products help to quantify and purify other formats, like Fab fragments, single-chain antibodies, and nanobodies. Tailor-made solutions for quantification and purification need to be developed in other cases. Inclusion/exclusion criteria are also much less advanced for nonantibody formats because the number of candidates are small and comprehensive data sets are missing. Nevertheless, assessment of expression titer, feasibility of downstream processing, aggregation propensity during processing and upon storage, stability, heterogeneity, solubility, viscosity of a given therapeutic drug candidate, and possibly variants thereof provide valuable insights into developability risks. Knowledge of developabilty risks early in a project will protect from costly surprises.

7.5 ORGANIZATIONAL SETTING

A research organization is successful if a project shows positive data in proof-of-concept (POC) studies in humans. To achieve this, the organization covers discovery and pre-clinical research efforts where the project flow is very dynamic. New experimental findings can change the fate of a project literally overnight. The support of development functions is needed to prepare a drug under good manufacturing practice (GMP) conditions for toxicological studies and eventually clinical trials. Development functions operate in a regulated environment where typically the project flow is a lot more predictable. Prerequisites, expectations, and performance indicators of research and development organizations therefore naturally differ from each other.

Developability assessment takes place at the interface of the two organizations. It is essential that scientists responsible for developability assessment understand the different requirements of both organizations well, and are accepted by both research and development colleagues. This makes it less important where the multidisciplinary teams in charge of developability assessment are organizationally located.

The clear advantage of an organizational unit dedicated to developability assessment in contrast to functions across the organization lies in the ability to collect learnings across all projects and aspects. It ensures that candidates are assessed in a systematic way across projects and not just within a project. It provides a truly integrated approach in bringing together all aspects—expression, purification, physicochemical characterization, formulation, *in vivo* compatibility, immunogenicity risk—to enable an optimal selection of the lead molecule.

7.6 OUTLOOK

The concept of carrying out developability assessments to identify or optimize lead candidates is becoming increasingly accepted in the industry.[22] Performing developability assessment of biologic candidates can tell whether a candidate can be produced and formulated as a drug, is compatible with *in vivo* conditions, or might have safety liabilities. The cost for developing a drug is a billion dollar effort, with more than half of the cost arising after the lead molecule is selected.[23, 24] It therefore makes sense to select candidates with as little liabilities as possible in the early phase of the drug development process, before costly development efforts start. Moreover, developability assessment also allows us to identify areas of increased resource needs and better predict overall development efforts and product costs. Dropping or timely reengineering of candidates with developability liabilities lessens later development efforts. As depicted in Figure 7.7, relatively small profiling efforts by the Integrated Biologics Profiling Unit before initiation of technical development activities de-risk the development process by rejecting candidates whose development is expected to require excessive efforts, thus improving productivity, gaining speed in getting the drug to patients, and providing best-in-class drugs.

The more traditional approach to engineer later stages of manufacturing and pre-clinical development to fit the drug product can result in immense costs, prolong the time to reach the clinic, and increase the cost of goods for the whole lifetime of the drug.

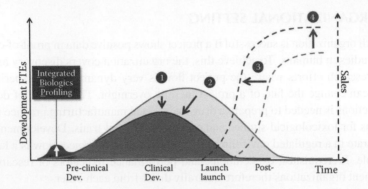

FIGURE 7.7 Effect of small profiling efforts by the Integrated Biologics Profiling Unit before initiation of resource-intensive technical development activities. (1) De-risked development process, (2) improved productivity, (3) accelerated development process, and (4) best-in-class drugs.

Of course, performing developability assessments in *in vitro* assays has its costs as well. Using *in silico* methodologies for developability assessments further lowers costs, increases throughput, and shortens timelines in comparison to the *in vitro* approach. It is for this reason that computational developability methodologies are built in-house, and commercially available propositions are evaluated and validated on a regular basis. As described, potential sites for post-translational modifications are first identified *in silico* and subsequently tested analytically *in vitro*. Determination of the isoelectric point, an important parameter for downstream processing of platform fit, is an example of a value that today is only determined *in silico*. With increasing predictability and reliability of *in silico* methodologies, the importance of these methodologies for developability assessment will continue to grow.

ACKNOWLEDGMENTS

The authors are very grateful to the current and former members of the Integrated Biologics Profiling Unit for their efforts during the setup and operation of integrated biologics profiling at Novartis. We also thank them for their scientific input, valuable discussions, and suggestions during the creation of this chapter. Moreover, we would like to thank our colleagues in partner organizations in research and development at Novartis for their continuous support.

REFERENCES

1. Jostock, T., and Knopf, H. P. 2012. Mammalian stable expression of biotherapeutics. *Meth. Mol. Biol.* 899: 227–238.
2. Meyer, H., and Schmidhalter, D. 2012. Microbial expression systems and manufacturing from a market and economic perspective. In *Innovation in Biotechnology*, ed. E. C. Agbo, 211–250. http://www.intechopen.com/books/innovations-in-biotechnology/microbial-expression-systems-and-manufacturing-from-a-market-and-economic-perspective

3. Zurdo, J. 2013. Developability assessment as an early de-risking tool for biopharmaceutical development. *Pharm. Bioprocess.* 1: 29–50.
4. Yang, X., Xu, W., Dukleska, S., Benchaar, S., Mengisen, S., Antochshuk, V., Cheung, J., Mann, L., Babadjanova, Z., Rowand, J., Gunawan, R., McCampbell, A., Beaumont, M., Meininger, D., Richardson, D., and Ambrogelly, A. 2013. Developability studies before initiation of process development. Improving manufacturability of monoclonal antibodies. *mAbs* 5: 787–793.
5. Saxena, V., Panicucci, R., Joshi, Y., and Garad, S. 2009. Developability assessment in pharmaceutical industry: An integrated group approach for selecting developable candidates. *J. Pharm. Sci.* 98: 1962–1979.
6. Jenkins, N., Murphy, L., and Tyther, R. 2008. Post-translational modifications of recombinant proteins: Significance for biopharmaceutical. *Mol. Biotechnol.* 39: 113–118.
7. Geisse, S., and Fux, C. 2009. Recombinant protein production by transient gene transfer into mammalian cells. *Meth. Enzymol.* 463: 223–238.
8. Rombach-Riegraf, V., Karle, A. C., Wolf, B., Sorde, L., Koepke, S., Gottlieb, S., Krieg, J., Djidja M.-C., Baban, A., Spindeldreher, S., Koulov, A., and Kiessling A. 2014. Aggregation of human recombinant monoclonal antibodies influences the capacity of dendritic cells to stimulate adaptive T-cell responses *in vitro*. *PLoS ONE* 9: e86322.
9. Sahin, E., and Roberts, C. J. 2012. Size-exclusion chromatography with multi-angle light scattering for elucidating protein aggregation mechanisms. *Methods Mol. Biol.* 899: 403–423.
10. Phillips, K., and Hernandez de la Pena, A. 2011. The combined use of the Thermofluor assay and ThermoQ analytical software for the determination of protein stability and buffer optimization as an aid in protein crystallization. *Curr. Protoc. Mol. Biol.* 94: 10.28.1–10.28.15.
11. Hanke, A. T., and Ottens, M. 2014. Purifying biopharmaceuticals: Knowledge-based chromatographic process development. *Trends Biotechnol.* 32: 210–220.
12. Lorenz, T., Fiaux, J., Heitmann, D., Gupta, K., Kocher, H., Knopf, H.-P., and Hartmann, S. 2014. Developability assessment of biologics by integrated biologics profiling. *Am. Pharm. Rev.* http://www.americanpharmaceuticalreview.com/Featured-Articles/167439-Developability-Assessment-of-Biologics-by-Integrated-Biologics-Profiling/.
13. Li, L., Kumar, S., Buck, P. M., Burns, C., Lavoie, J., Singh, S. K., Warne, N. W., Nichols, P., Luksha, N., and Boardman, D. 2014. Concentration dependent viscosity of monoclonal antibody solutions: Explaining experimental behavior in terms of molecular properties. *Pharm. Res.* DOI: 10.1007/s 11095-014-1409-0.
14. Ghetie, V., and Ward, E. S. 2000. Multiple roles for the major histocompatibility complex class I related receptor FcRn. *Ann. Rev. Immunol.* 18: 739–766.
15. Roopenian, D. C., and Akilesh, S. 2007. FcRn: The neonatal Fc receptor comes of age. *Nat. Rev. Immunol.* 7: 715–725.
16. Strohl, W. R. 2009. Optimization of Fc-mediated effector functions of monoclonal antibodies. *Curr. Opin. Biotechnol.* 20: 685–691.
17. Bertone, P., and Snyder, M. 2005. Advances in functional protein microarray technology. *FEBS J.* 272: 5400–5411.
18. Ewert, S., Huber, T., Honegger, A., and Plückthun, A. 2003. Biophysical properties of human antibody variable domains. *J. Mol. Biol.* 325: 531–553.
19. Strohl, W. R., and Strohl, L. M. 2012. Development issues: Antibody stability, developability, immunogenicity, and comparability. In *Therapeutic Antibody Engineering: Current and Future Advances Driving the Strongest Growth Area in the Pharmaceutical Industry*, 377–403. Oxford: Woodhead Publishing.
20. Rouet, R., Lowe, D., and Christ, D. 2014. Stability engineering of the human antibody repertoire. *FEBS Lett.* 588: 269–277.

21. Dudgeon, K., Rouet, R., Kokmeijer, I., Schofield, P., Stolp, J., Langley, D., Stock, D., and Christ, D. 2012. General strategy for the generation of human antibody variable domains with increased aggregation resistance. *Proc. Natl. Acad. Sci. USA* 109: 10879–10884.

22. Lowe, D., Dudgeon, K., Rouet, R., Schofield, P., Jermutus, L., and Christ, D. 2011. Aggregation, stability, and formulation of human antibody therapeutics. *Adv. Protein Chem. Struct. Biol.* 84: 41–61.

23. Paul, S. M., Mytelka, D. S., Dunwiddie, C. T., Persinger, C. C., Munos, B. H., Lindborg, S. R., and Schacht, A. L. 2010. How to improve R&D productivity: The pharmaceutical industry's grand challenge. *Nat. Rev. Drug Discov.* 9: 203–214.

24. DiMasi, J. A., and Grabowski, H. G. 2007. The cost of biopharmaceutical R&D: Is biotech different? *Manage. Decis. Econ.* 28: 469–479.

8 Best Practices in Developability Assessments of Therapeutic Protein Candidates in the Biopharmaceutical Industry

Nicolas Angell,* Randal R. Ketchem,†
Kristine Daris,‡ Jason W. O'Neill,†
and Kannan Gunasekaran†

CONTENTS

8.1 INTRODUCTION

Various stages involved in discovery and development of a typical drug are shown in Figure 8.1. Developability assessment (DA) processes are used during the candidate screening, optimization, and lead molecule selection. Developability assessments typically involve examining several candidate molecules for attributes that can impact

* Product Attribute Sciences, Amgen, Inc., Thousand Oaks, California
† Therapeutic Discovery, Amgen, Inc., Thousand Oaks, California
‡ Drug Substance Development, Amgen, Inc., Thousand Oaks, California

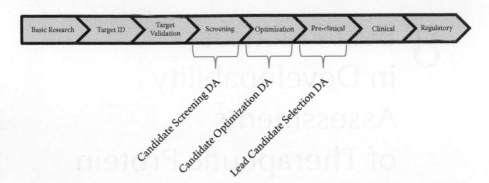

FIGURE 8.1 Stages involved in a typical therapeutic candidate development. A developability assessment (DA) process is utilized during the candidate screening, optimization, and lead molecule selection.

production process, formulation, safety, and efficacy, and then selecting a single lead molecule that delivers to the target expectations.[1-3] The primary focus of the developability assessment is to instill those product quality attributes that lead to favorable safety and efficacy. This means the selected candidate should augment the biological activity assessments in addition to meeting all of the desired developability attributes. While biological activity always trumps the developability assessments, there is also the need to ensure adequate productivity from expression and purification to meet the future commercial demand and provide product stability for processing and storage. The developability attributes typically examined include structural stability, chemical stability, transport stability, photo-stability, viscosity, *in vivo* stability, and ability to deliver to process expectations and meet the target product profile. Often the developability assessments begin with multiple candidates and end with identification of a lead candidate(s) that either has all the desired attributes or has unfavorable aspects that can be remediated through further formulation or process development or optimization. Frequently, due to the pressure to reach the market quickly, the goal becomes achieving the right balance between the bioactivity and developability attributes.

Throughout the developability assessment process, various questions are addressed. For example, questions relating to productivity or yield, purity, cost, chemical and physical stability, fit to established standard process, suitability to route of administration, serum half-life, specificity, immunogenicity, safety, and efficacy are often addressed or considered. The assessments begin with examination of the primary sequence using computational tools that facilitate the rapid analysis and triaging of a large number of sequences. However, while some of the attributes, such as potential chemical modification sites, can be predicted through analysis of the primary sequences, we still lack the ability to predict a wide range of issues that occur during process development and formulation, such as solubility, viscosity, or cell productivity. Similarly, there is not a single experiment or study that could rapidly address all of the above aspects. For these reasons, it becomes essential to begin the developability assessments with multiple candidates. The ability to quickly eliminate or identify problematic candidates depends on having a well-established target product profile (TPP) in hand. For example, the route of administration or dosing can determine if a viscosity assessment is necessary. If a formulation is required, then those molecules that have abnormally high viscosities can be eliminated or targeted for further mitigation strategies if they are lead candidates from a biological perspective. While developability assessments do not always result in final candidates that have all of the desired attributes, the vast amount of data collected during the developability assessment can potentially help guide process development or formulation optimization and allow for accurate resource and timeline planning, and in the worst-case scenario, trigger reengineering of the candidates to deliver to the TPP.

In this chapter, we attempt to describe the best practices in developability assessments of therapeutic protein candidates, primarily antibodies. The recent advances in *in silico* sequence analysis and high-throughput engineering provide us with an opportunity to evaluate large numbers of candidates and also to implement knowledge gained from previous assessments. The sequence analysis aspects range from assigning a residue numbering scheme to the primary sequence to predicting specific chemical modification sites. The engineering is primarily driven by past empirically derived knowledge around a protein sequence and its relation to structure attributes. It can be as simple as mutating an Asn residue to Gln in order to prevent glycosylation, to grafting complementarity determining regions (CDRs) onto a different variable domain framework in order to avoid use of a rare subtype. The next section addresses the important question of productivity and cost. Having sequence-diverse candidates in the therapeutic protein panel brings us the luxury to evaluate multiple candidates in a given established consistent expression and purification process. Aside from productivity, a range of other attributes, such as cell density and viability, must be considered when screening candidate molecules. The primary goal of the purification process is to efficiently remove contaminants, such as host cell proteins and aggregates, with minimal impact to the final yield of the product. However, a detailed analysis of the purification procedures may uncover issues related to certain candidate molecules that are prone to in-process aggregation or undergo amino acid modification, such as Trp oxidation or Asp deamidation. The biophysical analysis is the key part of the developability assessments. The initial product quality profile, established by the biophysical analysis, together with the target product profile (TPP), determines subsequent stability and formulation studies. For this reason, having the TPP at the early stage, as early as the engineering phase, is critical.

8.2 SEQUENCE ANALYSIS AND ENGINEERING: DEVELOPABILITY STRATEGIES

Protein engineering was first coined as a phrase in 1983 by Kevin Ulmer.[4] Simply stated, protein engineering is the design of *de novo* proteins with desired structure and function by the substitution, addition, or deletion of amino acids. The direction behind protein engineering can arise from a need to modulate a particular function, either positively or negatively; to shift its function; to modify the behavior of the protein within its function, such as by altering its half-life or tissue distribution; or to modify the behavior of the protein in its environment, such as its thermal stability or propensity to chemical modification. Implementation of this level of engineering can be accomplished through multiple means, but these means are most broadly

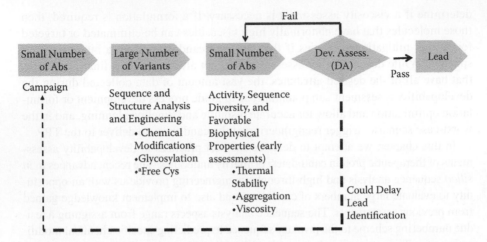

FIGURE 8.2 Typical engineering and developability assessment process. Having engineered variants that have desirable biological activity and biophysical properties and pass minimal stability requirements helps to quickly replace the lead molecule in case the lead fails in the extensive developability assessment process. The other option of having and progressing with only the lead molecule in the entire process is risky since the failure could lead to a new campaign and hence significant delay in the candidate development.

described as (1) rational design and (2) library approaches. While these two methods can, and are, certainly used together, here we focus on the area of protein engineering through rational design. More specifically, we utilize the analysis of antibody sequences and structures to discover particular residues and regions such that their modification leads to manufacturable, therapeutically active molecules.

Parental antibody sequences are evaluated utilizing computational analysis methods coordinating both sequence and structure (Figure 8.2). With a panel of antibodies exhibiting desirable binding or activity attributes, the sequences are first aligned structurally using the AHo numbering system,[5] ensuring that individual residues are in structurally aligned, externally consistent positions. Not only does this have the benefit of imparting structural meaning to individual residues, but also it allows for a uniform numbering and alignment across multiple Fv sequences without having to align the sequences to each other. The AHo system also provides structure-based gapping for the CDR residues. From this alignment, a sequence-based clading is performed that illustrates diversity across the panel and also allows for the evaluation of closely related antibodies, termed siblings, on a sequence level to aid analysis and potential engineering. Since in the panel all bind the target of interest and have been shown to be active in the screening assay, both sequence similarities and differences can be utilized in directing the optimizations to obtain multiple active therapeutics that will pass our more stringent, broader, later-stage developability assessments (Figure 8.3).

The clading is then used to select individual antibodies to represent the clades, and a structure model of the Fv or Fab is built for each using the Antibody Modeler method within the Molecular Operating Environment (MOE) from the Chemical Computing Group (Montreal, Quebec). Structure templates for the frameworks are selected from a single experimental structure in order to best maintain a proper Fv

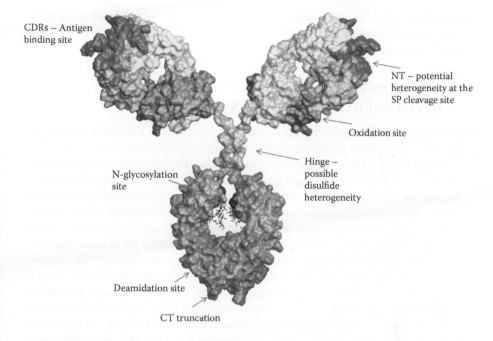

CDRs – Antigen binding site

NT – potential heterogeneity at the SP cleavage site

Oxidation site

Hinge – possible disulfide heterogeneity

N-glycosylation site

Deamidation site

CT truncation

FIGURE 8.3 Modeled structure of an antibody. Fab and Fc crystal structures were utilized to model the full antibody structure. The N-terminus is located near the CDRs, the antigen binding site. Some of the typical hot spot regions and potential heterogeneities are highlighted.

interface orientation. Individual CDR templates are chosen based on CDR length, sequence identity, and structural characteristics such as resolution. Care is also taken in the selection of CDR templates in which residues differ between the query sequence and the template in an effort to make reasonable, structure-based template decisions. Multiple initial structures are built, each with multiple side chain models.

Computational analysis is first performed on the individual antibody sequences to guide engineering of non-standard cysteines and N-linked glycosylation sites. If either of these hot spots exists in an antibody, it is modified into multiple variants to attempt to remove the site without impacting the binding or activity of the antibody. If function is lost in all of the variants for a particular site, the parental antibody should not be taken forward. One exception to this is if non-standard cysteines are shown to form a stable disulfide bond, and even then, a variant without the cysteines is attempted and preferred. Both sequence and structure play a role in the engineering of these hot spots, leading to a consideration of solvent exposure, atomic interactions, and sibling residue information in variant design.

Further computational analysis is then performed to identify deamidation sites, isomerization sites, and covariance violations.[6] While these hot spots are used to focus efforts in variant creation, if a particular hot spot cannot be removed by variant engineering, resulting in a maintained activity, the individual hot spot could remain in the final antibody. Deamidation and isomerization events in an antibody can lead to a loss of function in the antibody in which they occur. While it is not known *a priori* if the site will modify and if the modification will impact function, a functional variant without the hot spot avoids the need for a specific stress test for that site in later developability assessments.

8.3 MOLECULE PRODUCTION PROCESS: UPSTREAM DEVELOPABILITY ASSESSMENTS

The developability assessment process aims to identify from the outset a commercially viable therapeutic protein candidate that displays appropriate bioactivity. The focus of this section is how developability assessments during upstream development can ensure suitable productivity and attributes to meet future commercial demand.

A good developability strategy increases the potential to select candidate molecules that have overall greater productivity, in contrast to identification based on biological activity alone. There are many strategies to improve productivity, such as media optimization during process development. However, screening a set of candidate molecules with sequence diversity for expression early during the developability assessments can guide selection of higher-yielding candidate molecules (Figure 8.4). During this screen, molecules are compared using the same production process to directly visualize the impact of the molecule on expression. Appropriate expression levels must be balanced against a wide range of other factors, which can mean a desired upstream titer cannot always be reached. Finally, by capturing all of the expression data associated with each candidate, an enhanced understanding of impact of the candidate sequence on expression can guide future programs and success rates.

Aside from productivity, a range of other attributes must be considered when screening for a potential lead candidate molecule. These include doubling time of the candidate cell lines, peak cell densities, and ending culture viability during a fed batch production process. Candidates that have longer doubling times, low productivities, or poor growth using a representative process should be flagged as potential problems. Low titers are particularly a concern if the indication requires a high therapeutic dose, as they can have an impact on how the cost of goods and manufacturing (COGM) (Figure 8.5) factors into the overall lead candidate selection. It is also critical to understand key product quality attributes, such as levels of

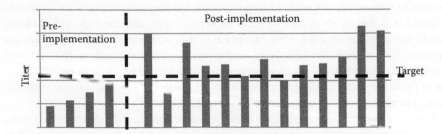

FIGURE 8.4 Titers were assessed using the same fed batch production process for candidate molecules pre-implementation vs. post-implementation of an upstream developability assessment. Post-implementation, the candidate molecules have greater potential to meet the target requirements for titer.

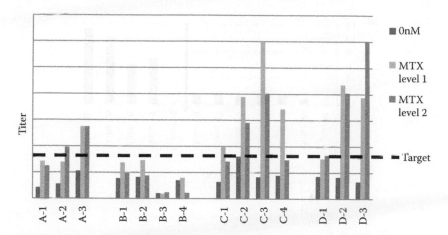

FIGURE 8.5 Titers from multiple transfections unamplified and amplified for four different candidate molecules (A–D) were assessed using a fed batch production assay. Screening candidates for expression during upstream development can help identify those that have low expression. Based on this information and other data generated for this project, candidate C was chosen as the lead molecule.

aggregates and other high-molecular-weight (HMW) species produced during cell culture. Certain candidate molecules may have a propensity for HMW species and can be flagged or eliminated during screening (Figure 8.6). HMW trends correlate between parental cell lines and clones, as well as scales, so they are a good predictor in early screening for flagging potential issues (Figure 8.7). High levels of HMW species can impact yield and be a potential immunogenicity concern. In addition,

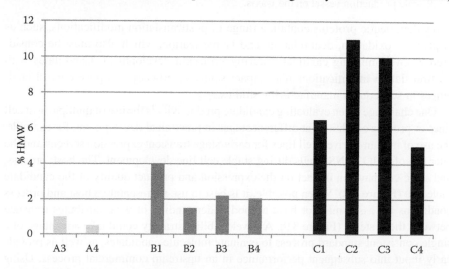

FIGURE 8.6 The percent high molecular weight for multiple amplified parental cell lines from three different candidate molecules (A–C) was assessed from protein A–purified conditioned media generated during a fed batch production assay. Screening for percent HMW during upstream development can identify candidates with high levels of aggregates.

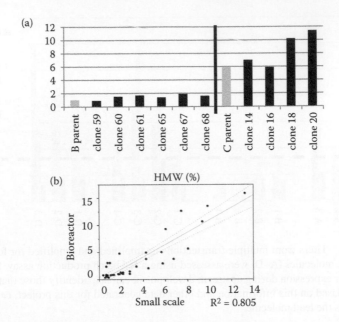

FIGURE 8.7 (a) The % high molecular weight for the parental cell line (B and C) and the clones resulting from the parental cell lines were assessed from Protein A purified conditioned media generated during a fed batch production assay. The HMW trend is the same for the parental cell lines and the clones derived from them later in development. This trend also correlates across scales, as shown by the correlation plot in Figure 8.7(b). The % HMW correlation has a high R^2 value of 0.805 for the bioreactor and is shown on the y-axis vs. the small-scale production vessel on the x-axis.

many therapeutic proteins contain a range of posttranslation modifications, such as N-glycans, oxidation, deamidation, and isomerization, which also must be considered when comparing candidates during upstream development. Candidates with posttranslation modifications that impact safety or efficacy can pose control challenges in subsequent manufacturing studies.

One challenge when evaluating candidate productivity is the use of multiple host cell lines or process conditions at different stages in preclinical development, for example, the use of human-derived cell lines for early-stage transient expression screens and the later use of CHO or NS0 cells during stable cell line development. The host, process, and scale can have an impact on the expression and product quality of the candidate molecule (Figure 8.8). When possible, it is best to use representative host and process conditions for assessment or have a good understanding of what attributes translate between the systems (Figure 8.9). A developability strategy could take advantage of a single consistent standard process to evaluate molecule candidates, as well as provide early input into subsequent performance in an upstream commercial process. Using these screens, one could observe expression and quality attribute differences between different candidates and determine if they meet the desired target values.

Lastly, another important consideration during upstream development when comparing a panel of candidate molecules is the impact of signal peptides. It has been

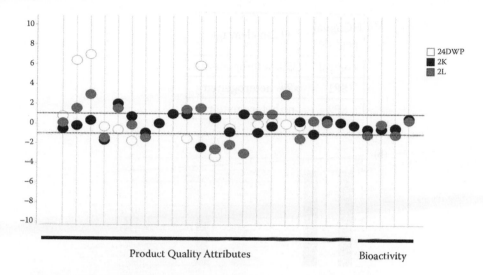

FIGURE 8.8 Process scale can have an impact on PQ as shown in the graph. Small-scale 24 deep well plates (open circles), 2L bioreactors (gray circles), and 2000L commercial-scale reactors (black circles) have differences outside of assay variance for some PQ parameters such as glycosylation, which can be affected by pH as an example. Other PQ parameters such as bioactivity or %HMW are less impacted by process scale.

host	molecule	% glycosylation
transient	A	75-80%
transient	B	85%
stable 1	A	90%
stable 1	B	90-95%
stable 2	A	85-90%
stable 2	B	95%

FIGURE 8.9 Host cell lines can have an impact on product quality. In this project example one transient and two stable host cell systems were used to generate cell lines for two different molecules. The resulting amount of glycosylation was different depending on the host. This highlights the importance of using the most representative system for making development decisions.

shown that different signal peptides, even on the same molecule, can have a signifi-cant effect on both the heterogeneity of the product and the expression of the molecule (Figure 8.10c and d). One way to minimize this issue is to implement preferred signal peptides that cleave cleanly and have no or minimal impact on expression for differ-ent molecules. Tools such as Signal 3P software can be used to predict signal peptide

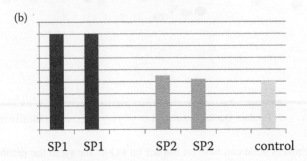

FIGURE 8.10 (a) and (b). Signal 3P software was used to predict the cleavage of control and SP1 signal peptides. The SP1 signal peptide is predicted to cleave better than the control signal peptide. In Figure 8.10(a) the table shows the actual heterogeneity observed using peptide mapping for the heavy chain signal peptide cleavage. The two alternate signal peptides, SP1 and 2, used cleaved more cleanly than the control signal peptide. In Figure 8.10(b) upon expression, cell lines using signal peptide SP1 also showed an increase in titer compared to signal peptide SP1 and the control.

cleavage (Figure 8.5a and b); however, this should be confirmed by other methods, such as mass spectrometry on material produced from a representative system and process.

8.4 MOLECULE PRODUCTION PROCESS: DOWNSTREAM DEVELOPABILITY ASSESSMENTS

Developability assessments of a candidate therapeutic protein panel require a ranked comparison of product attributes such as biophysical stability and functional readouts, which rely on a robust and self-consistent process to evaluate both purification and product quality. Developability determination on a panel of candidates also relies on a purification procedure that produces a high-quality product that isolates a homogenous, monodisperse, nonaggregated protein at high yields. Comparison and ranking of candidates' developability is highly dependent not only on a final purified behavior, but also on a robust analytical assessment throughout a purification process.

Initial candidate selection strategies could include core sequence-based parameters of the protein, such as class (e.g., Fc containing), MW, pI, and glycosylation, and additionally, the chemical hot spots may influence the purification and selection process (Table 8.1). As an example, knowledge of the isoelectric point (pI) of the protein can directly influence modeling of purification processes.[7] Selection of candidates having gone through hot spot repair may also be desirable.[1] However, purification

TABLE 8.1
Core Protein Parameters Identified Ahead of Candidate Development

Initiating Protein Characteristics	Tools and Examples
Protein class	Ab, non-Ab candidate, affinity tag
MW, extinction coefficients	Calculated, MS
Glycosylation (N-, O-) calculated	Calculated, MS, RP-HPLC
pI	Calculated, ceIEF
Titer	BiaCore, ELISA, SDS-PAGE, CE-SDS
Hot spots in AA sequence	*In silico* examples: non-standard Cys, CDR N-Glycosylation

Note: RP-HPLC, reversed-phase high-performance liquid chromatography; Capillary Electrophoresis Isoelectric Focusing (ceIEF); ELISA, enzyme-linked immunosorbent assay.

processes between modified candidate siblings should be closely compared to determine if the problematic behavior was resolved.

Product quality attributes can be impacted by a variety of influences, such as expression titer, and through purification procedures such as pH ranges, resin loading, resin type, and buffer conditions (pH, salts). While expression titer is important as an initiating criterion for candidate selection, those that exhibit the highest expression are also required to exhibit good purification efficiency, low aggregation levels, and high stability and demonstrate function.

At each step of the purification, suites of analytical processes are paired to sort problem candidates from promising candidates that can survive a robust purification process. These in-process analytical tools are used to follow the behavior of candidates, guide future purification procedure development, and rank candidates (Table 8.2). Core standard in-process tools, such a reduced and nonreduced capillary electrophoresis with sodium dodecyl sulfate (CE-SDS), as well as sodium dodecyl sulfate polyacrylamide gel electrophoresis (SDS-PAGE), and analytical size exclusion high-performance liquid chromatography (SE-HPLC),[8] are often paired to monitor gross levels of contaminant levels and to determine those chromatography steps or fractions yielding the highest purity and monodispersity. During chromatography SE-HPLC can also be paired with in-line multiangle light scattering to further refine the boundaries of aggregate and monodispersed species, and thus aid protein pooling analysis.[8] Utilization of CE-SDS, imaged Isoelectric Focusing (CE-IEF), and reversed-phase (RP) methods early on in the purification process can also highlight, for instance, if pairings of the heavy chain and light chain, clips, and amino acid modifications are within acceptable parameters.[9,32] Mass spectrophotometry techniques can be used in-process to verify the mass of target therapeutics (purity, clips, amino acid modifications) or help quantitate the burden of host cell proteins.[10]

The goal of purification is to efficiently remove from the product contaminants, host cell proteins (HCPs), aggregates (HMW), and clipped forms (LMW) with minimal impact to final yields. Determining developability standardizing purification procedures and analytics across a grouping of similar candidate classes provides the opportunity for head-to-head comparisons. As an example, protein A will be

TABLE 8.2

Key Analytical Tools and Criteria Utilized to Determine Product Quality through Manufacturing Assessment Purification, and to Rank Order a Candidate Developability Profile

In-Process Analytics and Readouts	In-Process Tool	Target
Elution condition and concentration	Resin load capacity	Capacity before breakthrough
In-process chromatography impact on aggregate levels	SE-HPLC (main peak, HWM, LMW), multiangle light scatter (MALS)	High purity[a]
Elution volume (column volumes)	Chromatography profile	Broad elution profiles may indicate aggregation or multiple isoforms
pH 3.5 viral inactivation 1 h hold	HMW% (protein A elution vs. low-pH viral inactivation)	HMW% unchanged
In-process formulation impacts to protein quality	Intact or peptide mass evaluation of degradation, oxidation, isomerization, deamidation, Trp oxidation	No alteration, or modification control
Determination of covalent LC/HC ratio, clips	nrCE-SDS, reversed phase	Very low unpaired %[a]
Host cell protein	ELISA, MS	Very low ppm[a]
Endotoxin	Kit	Very low EU/mg[a]

[a] Exact criterion or specification depends on the target product profile.

Note: LC/HC, liquid chromatography–mass spectrometry; nrCE-SDS, non-reduced Capillary Electrophorsis-SDS.

standardly utilized for antibody capture,[11] and then paired with an ion exchange or hydrophobic interaction chromatography (HIC) steps for the host cell protein and aggregate removal[12] (Figure 8.11). A detailed analysis of the purification and processing step outcomes may uncover issues related to certain candidate molecules, such as those that are prone to in-process aggregation during a low-pH viral inactivation hold or have already undergone amino acid modification, such as Trp oxidation or Asp deamidation, during production. Some purification process steps have been observed to impact aggregation levels, including a low-pH hold post–protein A elution, which, though a convenient step to reduce viral burden, can carry a risk of aggregation.[13,14] There have also been reports that utilizing certain strong cation exchange resins (CEX) may lead to additional aggregation for some antibody classes.[8] Tracking these in-process analytics data can also aid in altering purification processes to ameliorate these types of issues.

The developability of a molecule is determined by combining knowledge of sequence-based parameter and purification outcomes of yields and analytical assessments. Table 8.3 demonstrates an example manufacturing assessment readout of a candidate set. Here candidates are rank-ordered on purification quality determination (yields vs. aggregation): A > B = C > D > E = F. Yield assessments post–protein A

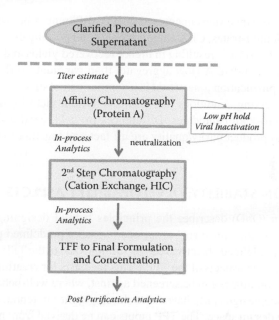

FIGURE 8.11 Standard antibody purification path. Here, in-process purification analysis is placed to stepwise query the titer, purification pooling, process step yields, aggregate, and cell host protein (CHOP). Generally, clarified supernatants (e.g., mammalian cell expression) will be loaded to an affinity step that has been determined to enrich the target protein and remove the bulk of host proteins. The example here is utilization of a high-capacity protein A to isolate an antibody. A low-pH hold is a typical in-process step to achieve viral burden reduction, and this step is monitored for impact product on aggregate levels. A second step of chromatography, for example, cation exchange, attempts to reduce further aggregate and cell host protein levels. A final buffer exchange and concentration using tangential flow filtration (TFF) strategies is another point where analytical strategies are implemented to, for instance, monitor aggregate levels.

TABLE 8.3

Downstream Purification Process MA Determination of Candidate Developability

Parameter	A	B	C	D	E	F
ProA elution pool CV (target < 3)	1.6	2.3	1.6	2.1	1.8	2
ProA elution pool HMW% (target < 4%)	2.9	2.6	5.4	2.2	3.5	4.2
ProA step recovery % (target > 90%)	99	100	100	94	100	92
Neutralized elution pool HMW% increase (target < 0.2%)	0.1	−1.1	−0.1	0.2	NS	4.9
CEX step recovery % (target > 75%)	77.4	71.3	71.3	62.2	46.8	49.4
CEX elution pool HMW% (target ≤ 2%)	0.2	2	0.8	0.2	0.9	1.1

Overall purification outcome ranking example: A > B = C > D > E = F. NS, not significant.

and CEX Cation Exchange steps uncovered quality differences between candidates. As this example demonstrates, CEX elution profiles and pooling criteria require low aggregate levels (SE-HPLC), resulting in higher recovered yields and a higher developability level for candidate A (low aggregate) than candidate F (high aggregate). The output of the purification assessment is a comparison of candidate performance under similar processing conditions that results in a final yield for the given target product quality across all candidates. Differences in yield, candidate performance in the process, and product purity/quality are all factored together to rank candidate molecules.

8.5   SOLUTION STABILITY: DEVELOPABILITY ASPECTS

Quality by design (QbD) describes the principles behind designing and developing formulations and manufacturing processes to ensure predefined product quality objectives. These predefined objectives can be summarized in the TPP. Understanding the formulation-related aspects of the TPP forms the basis for creating a formulation design space that candidates can be screened against, where well-behaved molecules that fit well into the design space have more developability potential than those that fall outside of the design space. The TPP inputs can be derived from many aspects of therapeutic development, where most are related to safety and efficacy (such as product quality and therapeutic dose), but many may address the commercial aspects, such as patient use, storage, and desired shelf life. When all of the aspects of the TPP are defined, a desired formulation design space can be created.

In practice, it is best to avoid those candidates that may present significant challenges during commercialization. In reality, the biology may dictate the need for nonideal or poorly behaved molecules, with respect to the formulation design space. Early identification of the potential liabilities can often lead to further engineering that may mitigate or limit these liabilities or, at a minimum, provide a risk-balanced perspective for moving a nonideal candidate forward into clinical development and allow for adjustment of timelines or resources to facilitate clinical development.

A developability assessment focuses on how well the candidate(s) fitting a formulation design space can be divided into four aspects: (1) product quality, (2) solubility, (3) stability, and (4) deliverability. The flow of the section mimics the workflow of the assessment, where failure to achieve the target product profile for each aspect could result in elimination of the candidate from subsequent testing.

8.6   MOLECULE SOLUTION PROPERTIES: FORMULATION DEVELOPABILITY ASSESSMENTS

Establishing an initial product quality profile for each of the candidates is of first importance, as it is used to establish the identity of the candidate, the level and type of post-translational modifications, and the purity of the material with regard to size variants (Table 8.4). Each of these aspects can have an impact on subsequent studies and may, in itself, be a quick determination of a candidate's ability to move forward in the assessment process. Incorrect candidates, candidates with heavily modified

TABLE 8.4

Product Quality Assays, Their Purpose, and the Potential Risk or Liability That the Assays Can Help to Identify

Assay Overview	Purpose	Potential Risk or Liability
LC/MS of intact molecule	Confirm the molecular weight of candidates. Assess truncations, clips, covalent dimers, half molecules, unusual glycosylation.	Product heterogeneity. Additional time/effort to characterize and develop quality assays.
LC/MS of reduced and intact deglycosylated molecules	Confirm the molecular weights of HC and LC. Obtain information on glycosylation and glycation.	Product heterogeneity. Additional time/effort to characterize and develop quality assays.
SE-HPLC	Oligomers and pre- and post-peak shoulders.	Higher than expected amount of such species may require additional process development time and effort.
SDS-PAGE, rCE-SDS, nrCE-SDS	Unexpected peaks, which could suggest truncations, dimers, oligomers, partial molecules, etc.	Product heterogeneity that may require additional process development time and effort to resolve.
Peptide mapping	Isomerization, oxidation, or deamidation in CDRs. N- and C-terminal modifications, signal peptide, unusual glycosylation, glycation.	Chemical stability, heterogeneity.

amino acid residues, and heavily aggregated candidates should be flagged, as each could have a detrimental impact on the validity of further studies. At a minimum, a candidate product quality profile should align with the expectations set forth in the TPP. If a candidate does not meet the established minimum requirements outlined in the TPP, the candidate may be considered for elimination from the assessment. Lastly, the product quality data provide a baseline (time = 0) for subsequent studies.

The identity of a candidate can typically be established by performing mass spectrometry (MS) analysis of the intact molecule and comparison of the predicated molecular weight to that of the observed. Additional analysis can include intact MS analysis of the deglycosylated or reduced and alkylated molecule to assess the candidate(s) without N-linked glycans (mAbs) or analysis of individual peptide chains that are covalently linked through disulfides, as in the case of IgGs. This can aid in further confirmation of the candidate identity by minimizing the glycan heterogeneity or by identification of the individual heavy and light chains.[15] An inability to establish the specific identity (established from the predicted amino acid sequence) of a candidate should result in elimination of the candidate from the assessment. A peptide map can be used to further confirm identity by using one or more enzymes to provide 100% sequence coverage. As indicated above, the initial ($t = 0$) peptide map is used to interrogate the purity of the candidate with regard to amino acid modifications and establish any sequence heterogeneity, such as N-terminal extensions from signal peptide or C-terminal lysine variants that may not be directly predicted from the sequence alone.[16] Finally, the biophysical state of the protein is assessed with regard to size variants in order to establish the purity of each of the candidates. Assays typically include size exclusion chromatography (SEC) to assess high and low molecular weight species (HMWS and LMWS),[17] as well as reduced capillary electrophoresis with SDS (rCE-SDS)[18] to evaluate covalent aggregates and clipped species. The results of these assays will be used during later studies as a $t = 0$ data set. Heavily aggregated or clipped candidates may be flagged for further analysis, additional purification, or elimination based on the nature and extent of the aggregation and the desired TPP.

Of primary importance, after establishing the correct identity of a candidate, is the solubility of the candidate.[19,20] Failure to achieve a target product profile concentration may limit a candidate's ability to become a viable therapeutic candidate.[20,21] Solubility, for the purposes of ranking and identifying problematic candidates, can easily be determined by visual observations (haziness, precipitation, gelling, etc.), as well as UV absorbance spectroscopy (A280) after filtration, where comparison to a nonfiltered sample may indicate a loss of material due to a lack of solubility. In addition to the solubility assessment, the stability at the given concentration(s) can be assessed under varying stresses and then compared to the appropriate controls (Table 8.5).

The stability of a candidate at the target product profile concentration provides a perspective of how it may behave during long-term storage and transportation stresses. Additionally, if a candidate progresses into clinical development, the assessment of a candidate's physical and chemical stability under a given formulation condition or under a given stress can also be used to identify program risks that may necessitate a longer development time or resource burden during formulation development. Stability under elevated temperatures (e.g., 37°C or 25°C), as well as at a target clinical storage temperature (typically 4°C for liquid therapeutic proteins) can be used to ascertain biochemical and biophysical liabilities, such as aggregation, particulation, clipping, changes in charge profile, and oxidations, under the target product profile formulation conditions (Table 8.6). The techniques employed typically include SEC, ion exchange chromatography (IEX), dynamic light scattering (DLS), reduced and

---

TABLE 8.5

Concentration Study, Its Purpose, and the Potential Risk or Liability That the Study Can Help to Identify

Study/Assay Overview	Purpose	Potential Risk or Liability
The mAb candidates are concentrated in u standard buffer. Solubility is determined by visual observation. The stability and solubility of the candidates at high concentrations are compared after 1 week at 4°C using SE-HPLC and visual observation.	Determine solubility in a standard formulation. Assessment of stability at high concentrations.	Insolubility or instability or inability to reach a target concentration of candidates, especially antagonist candidates, in the standard formulation would require that additional time and resources are available to develop a non-standard formulation.

TABLE 8.6
Stability Studies, Their Purpose, and the Potential Risk or Liability That the Studies Help to Identify

Study/Assay Overview	Purpose	Potential Risk or Liability
Stability in standard formulation is compared after incubation at 4°C, 40°C, and 25°C using pH, visual, SE-HPLC, DLS, rCE-SDS, nrCE-SDS, and LC/MS analyses, including intact, intact deglyco, reduced LC/MS, and peptide mapping.	Ensure stability (physical and chemical) in the standard formulation. Monitor the change between control and stress samples for truncations, clips, covalent dimers, half molecules, glycation, isomerization, oxidation, and deamidation.	Unstable candidates incur a greater resource burden during formulation development.
Stability after light exposure is assessed by visual inspection, SEC, pH, and LC/MS analyses, including intact, intact deglyco, reduced LC/MS, and peptide mapping.	Identify chemical modification sites in CDRs that are sensitive to light.	APX formation and modifications in CDRs. May require additional bioactivity assessment if degradation is observed.
The solubility and stability of the candidates are monitored as the candidate transitions from the formulation pH to a neutral pH using pH, OD, visual observations, DLS, concentration by UV, SE-HPLC, and LC/MS analyses, including intact, intact deglyco, reduced LC/MS, and peptide mapping.	Solubility and stability after transition to neutral pH. Identify chemical modifications in regions that are sensitive to pH neutral conditions.	Identify candidates that may have challenges during subcutaneous delivery.
The propensity for agitation-induced particulation of the candidates is assessed in the standard formulation with and without surfactant. Samples are analyzed after agitation by visual inspection, OD, DLS, and SE-HPLC.	Assessment of stability after agitation-induced particulation.	Candidates that are susceptible to agitation-induced particulation may be prone to particle formation during handling, transport, or prolonged exposure to a hydrophobic interface (e.g., air).

Note: APX 2-Amino-3 H-phenoxazine-3-one; OD, Optical Density.

nonreduced CE-SDS, and peptide mapping.[22] Additional stress studies may be used to target specific biochemical liabilities, if present, such as susceptibility of amino acids to oxidation upon light exposure that may be encountered during production or fill-finish operations.[23,24] Candidates can also be screened to determine their ability to resist typical processing, delivery, or transportation stresses, such as freeze and thaw,[24] sensitivity toward changes in pH associated with subcutaneous injection,[25] and resistance toward agitation.[26,27]

TABLE 8.7

Delivery Study, Its Purpose, and the Potential Risk or Liability That the Study Helps to Identify

Study/Assay Overview	Purpose	Potential Risk or Liability
Viscosity of the candidates is measured in the standard formulation.	Ensure tolerable viscosity in standard formulation, assessment of stability at high concentrations. Predict ease of manufacturing unit operations, UF/DF.	Insolubility or instability or inability to reach a target concentration of candidates, especially antagonist candidates, in the standard formulation would require that additional time and resources are available to develop a non-standard formulation. High viscosity can cause difficulty in the UF/DF process and may hinder the use of prefilled syringe (PFS) and commercial automated delivery devices.

The viscosity of a candidate at a target product profile concentration can be used to identify molecules that cannot easily meet process or delivery device limitations without incurring additional time or resource expenditures during development[28,29] (Table 8.7). Viscosity determination can be done using cone and plate rheometers,[30] as well as DLS.[31]

8.7   CONCLUSIONS

The primary focus of the developability assessments is to ensure manufacturability, safety and efficacy. Various attributes are examined during the developability assessments. Elimination of candidates based on the attributes examined during the process depends on having the TPP in hand. Sequence analysis can be utilized to screen a large number of candidates and identify candidates with sequence diversity and fewer chemical hot spots for further engineering and developability assessments. While we currently lack the ability to predict most issues using computational tools, the future is likely to see significant advancements in the development of tools. The vast experimental data collected during the developability assessments will provide learnings and understandings that will help engineer candidates up front with better quality and safety.

REFERENCES

1. Yang, X., Xu, W., Dukleska, S., Benchaar, S., Mengisen, S., Antochshuk, V., Cheung, J., Mann, L., Babadjanova, Z., Rowand, J., Gunawan, R., McCampbell, A., Beaumont, M., Meininger, D., Richardson, D., and Ambrogelly, A. (2013). Developability studies before initiation of process development: Improving manufacturability of monoclonal antibodies. *MAbs* 5, 787–94.
2. Zurdo, J. (2013). Developability assessment as an early de-risking tool for biopharmaceutical development. *Pharm Bioprocess* 1, 29–50.

3. Saxena, V., Panicucci, R., Joshi, Y., and Garad, S. (2009). Developability assessment in pharmaceutical industry: An integrated group approach for selecting developable candidates. *J Pharm Sci* 98, 1962–79.

4. Ulmer, K. M. (1983). Protein engineering. *Science* 219, 666–71.

5. Honegger, A., and Pluckthun, A. (2001). Yet another numbering scheme for immunoglobulin variable domains: An automatic modeling and analysis tool. *J Mol Biol* 309, 657–70.

6. Kannan, G. (2012). Method of correlated mutational analysis to improve therapeutic antibodies. Google Patents.

7. Adhikari, S., Manthena, P. V., Sajwan, K., Kota, K. K., and Roy, R. (2010). A unified method for purification of basic proteins. *Anal Biochem* 400, 203–6.

8. Hong, P., Koza, S., and Bouvier, E. S. (2012). Size-exclusion chromatography for the analysis of protein biotherapeutics and their aggregates. *J Liq Chromatogr Relat Technol* 35, 2923–50.

9. Stackhouse, N., Miller, A. K., and Gadgil, H. S. (2011). A high-throughput UPLC method for the characterization of chemical modifications in monoclonal antibody molecules. *J Pharm Sci* 100, 5115–25.

10. Zhang, Q., Goetze, A. M., Cui, H., Wylie, J., Trimble, S., Hewig, A., and Flynn, G. C. (2014). Comprehensive tracking of host cell proteins during monoclonal antibody purifications using mass spectrometry. *MAbs* 6, 659–70.

11. Shukla, A. A., Hubbard, B., Tressel, T., Guhan, S., and Low, D. (2007). Downstream processing of monoclonal antibodies: Application of platform approaches. *J Chromatogr B Analyt Technol Biomed Life Sci* 848, 28–39.

12. Shukla, A. A., and Thommes, J. (2010). Recent advances in large-scale production of monoclonal antibodies and related proteins. *Trends Biotechnol* 28, 253–61.

13. Miesegaes, G., Lute, S., and Brorson, K. (2010). Analysis of viral clearance unit operations for monoclonal antibodies. *Biotechnol Bioeng* 106, 238–46.

14. Arosio, P., Rima, S., and Morbidelli, M. (2013). Aggregation mechanism of an IgG2 and two IgG1 monoclonal antibodies at low pH: From oligomers to larger aggregates. *Pharm Res* 30, 641–54.

15. Bondarenko, P. V., Second, T. P., Zabrouskov, V., Makarov, A. A., and Zhang, Z. (2009). Mass measurement and top-down HPLC/MS analysis of intact monoclonal antibodies on a hybrid linear quadrupole ion trap-orbitrap mass spectrometer. *J Am Soc Mass Spectrom* 20, 1415–24.

16. Kotia, R. B., and Raghani, A. R. (2010). Analysis of monoclonal antibody product heterogeneity resulting from alternate cleavage sites of signal peptide. *Anal Biochem* 399, 190–95.

17. Diederich, P., Hansen, S. K., Oelmeier, S. A., Stolzenberger, B., and Hubbuch, J. (2011). A sub-two minutes method for monoclonal antibody-aggregate quantification using parallel interlaced size exclusion high performance liquid chromatography. *J Chromatogr A* 1218, 9010–18.

18. Rustandi, R. R., Washabaugh, M. W., and Wang, Y. (2008). Applications of CE SDS gel in development of biopharmaceutical antibody-based products. *Electrophoresis* 29, 3612–20.

19. Arakawa, T., and Timasheff, S. N. (1985). Theory of protein solubility. *Methods Enzymol* 114, 49–77.

20. Shire, S. J., Shahrokh, Z., and Liu, J. (2004). Challenges in the development of high protein concentration formulations. *J Pharm Sci* 93, 1390–402.

21. Goswami, S., Wang, W., Arakawa, T., and Ohtake, S. (2013). Developments and challenges for mAb-Based therapeutics. *Antibodies* 2, 452–500.

22. Ha, S., Wang, Y., and Rustandi, R. R. (2011). Biochemical and biophysical characterization of humanized IgG1 produced in *Pichia pastoris*. *MAbs* 3, 453–60.

23. Qi, P., Volkin, D. B., Zhao, H., Nedved, M. L., Hughes, R., Bass, R., Yi, S. C., Panek, M. E., Wang, D., Dalmonte, P., and Bond, M. D. (2009). Characterization of the photo-degradation of a human IgG1 monoclonal antibody formulated as a high-concentration liquid dosage form. *J Pharm Sci* 98, 3117–30.

24. Radmanovic, N., Serno, T., Joerg, S., and Germershaus, O. (2013). Understanding the freezing of biopharmaceuticals: First-principle modeling of the process and evaluation of its effect on product quality. *J Pharm Sci* 102, 2495–507.

25. Laursen, T., Hansen, B., and Fisker, S. (2006). Pain perception after subcutaneous injections of media containing different buffers. *Basic Clin Pharmacol Toxicol* 98, 218–21.

26. Thirumangalathu, R., Krishnan, S., Ricci, M. S., Brems, D. N., Randolph, T. W., and Carpenter, J. F. (2009). Silicone oil- and agitation-induced aggregation of a monoclonal antibody in aqueous solution. *J Pharm Sci* 98, 3167–81.

27. Fesinmeyer, R. M., Hogan, S., Saluja, A., Brych, S. R., Kras, E., Narhi, L. O., Brems, D. N., and Gokarn, Y. R. (2009). Effect of ions on agitation- and temperature-induced aggregation reactions of antibodies. *Pharm Res* 26, 903–13.

28. Shieu, W., Torhan, S. A., Chan, E., Hubbard, A., Gikanga, B., Stauch, O. B., and Maa, Y. F. (2014). Filling of high-concentration monoclonal antibody formulations into pre-filled syringes: Filling parameter investigation and optimization. *PDA J Pharm Sci Technol* 68, 153–63.

29. Cheng, W., Joshi, S. B., Jain, N. K., He, F., Kerwin, B. A., Volkin, D. B., and Middaugh, C. R. (2013). Linking the solution viscosity of an IgG2 monoclonal antibody to its structure as a function of pH and temperature. *J Pharm Sci* 102, 4291–304.

30. Patapoff, T. W., and Esue, O. (2009). Polysorbate 20 prevents the precipitation of a monoclonal antibody during shear. *Pharm Dev Technol* 14, 659–64.

31. He, F., Becker, G. W., Litowski, J. R., Narhi, L. O., Brems, D. N., and Razinkov, V. I. (2010). High-throughput dynamic light scattering method for measuring viscosity of concentrated protein solutions. *Anal Biochem* 399, 141–43.

32. Sosic Z, et al. (2008). Application of imaging capillary IEF for characterization and quantitative analysis of recombinant protein charge heterogeneity. *Electrophoresis* 29(21), 4368–4376.:

9 Best Practices in Assessing the Developability of Biopharmaceutical Candidates

Juan C. Almagro* and Alessandro Mascioni*

CONTENTS

---

* Centers for Therapeutic Innovation, Pfizer Inc., Boston, MA

9.1   INTRODUCTION

Biologics are gaining momentum in all major pharmaceutical and biotechnology company portfolios, with a wide range of products being developed that include antibodies, growth factors, enzymes, peptides, and toxins.[1] The successful development of biologics depends not only on the ability of the drug to alter a biological pathway, and thus produce a therapeutic effect, but also on the competence of drug developers to formulate and manufacture the drug. Undesirable biophysical and biochemical properties, such as poor solubility and stability, propensity to aggregate, and chemical degradation, can impact efficacy[2] and generate unwanted toxic reactions, such as immunogenicity,[3] leading to failure of the drug during late-stage development and clinical trials. Therefore, transitioning molecules selected during early discovery with suboptimal biophysical and biochemical properties to late development increases attrition rate and imposes expensive and time-consuming formulation and development campaigns.

For these reasons, and as an increasing number of biologics have been reaching the market and many others have failed in development and clinical trials, biologic drug development has undergone a paradigm shift in recent years. In the new model, more knowledge-based design principles and experimental screening to select developable molecules are implemented as early as possible in the discovery process, as opposed to fixing "badly behaved" leads in the formulation and late-stage development. Obviously, design and early-on selection of drug candidates suitable for further development, formulation, and manufacturing de-risk late-stage failures, with the consequent reduction in attrition rates, lowering development costs and shortening the timeline to produce the drug.

The term *developability* has been coined to encompass the design principles and experimental assessment of the characteristics a molecule should meet to be further developed or manufactured, formulated, and stabilized in order to achieve the desired therapeutic effects. Developability is affected by a number of factors, including the intrinsic biophysical and biochemical properties of the molecule, as well as extrinsic parameters, such as ionic strength, pH, and formulation additives.

In this chapter, we describe typical developability parameters and tests performed during the biologic discovery and optimization campaigns. It should be noted that not all the analytical methods described here are routinely included in developability assessment. In fact, tests of increasing sophistication are performed as the candidate selection progresses, with late-stage candidates being subjected to more in-depth scrutiny. We also describe predictive methods that have been recently developed to identify developability liabilities. Such methods have great potential in assisting the design of more developable molecules and in complementing experimental developability methods to select molecules with a higher probability of success in late development.

9.2   DEFINITION OF DEVELOPABILITY ASSESSMENT
AND THE DRUG DISCOVERY PROCESS

The developability assessment is conducted at several stages of the discovery and optimization processes and consists of a set of experimental and predictive methods intended to assess both the intrinsic biophysical properties of a potential drug and its ability to fit in the production processes compatible with the available manufacturing platforms. Although the developability assessment for biologic development is not yet harmonized across the industry, there has been a convergence toward general best practices, in particular for therapeutic antibodies, which are a very significant proportion of the portfolio of biologics in current development.[4]

Figure 9.1 depicts a typical antibody drug development process.[5] It starts (top of the figure) with exploratory research, which is focused on target validation and assay development. This initial phase is followed by discovery and optimization of potential leads, which are developed into potential drug candidates during the development phase, before a drug candidate is submitted to the regulatory agencies for approval of investigational new drug (IND) status. Early discovery (middle of the figure), that is, where the future drug candidate is engineered, focuses on three main steps: design, screening, and selection. During these processes *in vitro* and *in vivo* assays are performed to assess binding to the target, cross-reactivity with orthologs, neutralization, and other parameters relevant to the biological activity. In addition, four developability parameters are assessed: expression yield, chemical stability, physical stability, and solubility.

These parameters are interrelated. For instance, chemical instabilities such as oxidation or clipping sites result in sample heterogeneity and eventually can impact the physical stability or lead to low solubility or aggregation. Poor physical stability can expose side chains prone to oxidation or degradation, eventually leading to aggregation when these residues are degraded. Nevertheless, each of these parameters can be measured independently (Table 9.1), and molecules not meeting the

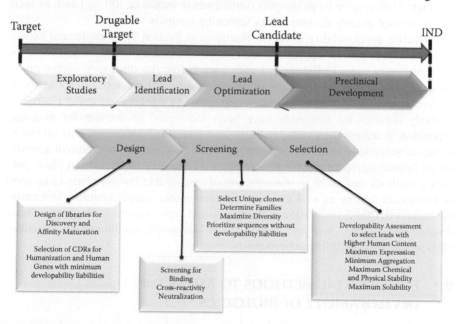

FIGURE 9.1 Drug discovery and developability assessment.

TABLE 9.1
Developability Parameters, Target Values, and Methods to Engineer Molecules with Developability Issues

Parameter	Target Value	Engineering Method
Physical stability (Tm)	60°C (C_H2) or higher	Engineer stabilizing mutations, framework replacement
Chemical stability	Minimize solvent exposure of residues prone to degradation	Remove unpaired Cys, N-glycosylation sites and exposed Met and Trp
Solubility	IV = 50 mg/ml; sub-Q > 100 mg/ml	Select well-behaved V regions, remove hydrophobic patches
Aggregation	<5% of High Molecular Mass Species (HMMS)	Select well-behaved V regions, remove hydrophobic amino acids

success criteria are deprioritized. Alternatively, if leads are identified with promising biological activity but do not perform well during the developability assessment, they are targeted for developability enhancement during the optimization phase by the methods described in the last column of the table.

To assess developability experimentally as early as possible in the discovery process, screening of biophysical and biochemical characteristics of the engineered molecules has to be redesigned to work with small amounts of protein and be able to handle as many variants as possible. Therefore, it is desirable that developability assessment methods be amenable to high-throughput screening. Unfortunately, some techniques still require large samples (sometimes in excess of 100 mg) and, as such, cannot be used in early developability screening methods.

Predictive developability methods, discussed in Section 9.4, complement experimental methods that are applied in each step of the process (boxes underneath the early discovery steps in Figure 9.1) and, even before the discovery campaign, in the design phase. For instance, the initial phage display libraries used for antibody discovery were designed to maximize expression in *Escherichia coli*.[6] Currently, antibody libraries for discovery have been redesigned to account for maximal expression in mammalian cells, stability, and solubility.[7] Moreover, humanization, as the cornerstone of antibody engineering methods to prepare non-human antibodies for human therapeutic settings, has evolved to include in the design phase predictive methods that assist in the selection of developable human genes to be used as framework donors as well as choosing developable complementarity determining regions (CDRs).[8] Integration of predictive and experimental methods provides a detailed assessment of well and badly behaved therapeutic candidates and expedites the discovery and optimization processes.

9.3   EXPERIMENTAL METHODS TO ASSESS THE DEVELOPABILITY OF BIOLOGICS

Table 9.2 summarizes experimental methods applied during discovery and optimization campaigns of biologics. As mentioned above, methods of increased complexity or

TABLE 9.2
Analytical Assays Typically Included in Developability Assessment Campaigns

Purpose	Test
Purity	PAGE, cGE, analytical SEC
Efficiency of purification scheme	
Biophysical properties: pI, extinction coefficient, charge distribution, titration curve, etc.	Sequence–structure analysis, IEF and cIEF
Aggregation level	Analytical SEC, DLS, analytical ultracentrifugation (AUC)
Expression level in the desired cell line	Protein A high-performance liquid chromatography (HPLC), biolayer interferometry
Thermal stability	DSC/DSF, Circular Dichroism (CD)
Unpaired cysteines	Free thiol assay
Solubility and viscosity	Concentration studies, rheological measurements, turbidity assessment
Charges heterogeneity	IEF, cIEF, capillary zone electrophoresis (CZE), analytical AEX, analytical CEX
Scouting for chromatographic conditions compatible with manufacturing platforms	AEX/CEX screening, resin selections
Protein heterogeneity: N- and C-terminal modifications, incomplete processing of signal peptide, glycosylation microheterogeneity, glycation, other proteolytic cleavages, and post-translational modifications	Intact MS, IEF, cIEF, cGE, Edman degradation
Stability studies: Sensitivity to temperature, pH, storage, mechanical stress, freezing–thawing, cryogenic lyophilization	Incubation at 2°C–8°C, 25°C, 37°C Incubation at low pH Freeze–thaw cycles Cryo-lyophilization cycles Agitation studies
Glycosylation analysis: Glycoform identification, glycosylation site mapping, glycan occupancy, glycation of primary amines	MS

Note: These tests are organized in approximate chronological order of use during different stages of candidate selection. The analyses at the bottom of the list are more demanding and tend to be performed during a later stage of development.

requiring large amounts of protein are performed during the later stage of candidate development, while simpler methods amenable to high-throughput applications are implemented during early candidate screening. In the following sections, we describe the mechanisms affecting developability and discuss experimental methods to assess them.

9.3.1 CHEMICAL STABILITY

Chemical stability of biologics has been the subject of extensive review.[9–11] Chemical degradation occurs through different mechanisms, such as side chain degradation or main chain clipping. Degradation of the side chains affects diverse amino acids.[12] For instance, oxidation can affect Met, Cys, His, Trp, and Tyr; hydrolysis affects Asp-Pro, Asp-Gly, and Asp-Ser; deamidation alters the chemical structure of Asn and Gln and isomerization of Asn and Asp; racemization impacts His, Asp, and Ser; and beta-elimination affects Cys, Ser, Thr, Phe, and Lys. Also, disulfide scrambling due to incorrect pairing of Cys residues and glycan desialylation impacts the stability of biologics.

Several of these modifications have been reported to cause partial or complete loss of activity, increase the tendency to aggregate, reduce the shelf life span of the product, and increase the sample heterogeneity. For antibodies, in addition to degradation of the side chain of residues involved in antigen binding, which can lead to a decrease or loss of efficacy, several residues in the constant regions have been documented to be potential degradation sites. These residues are summarized in Figure 9.2. As more therapeutic antibodies are designed and developed in the near future, it is expected that the list of observed modifications will grow.

9.3.1.1 Deamidation and Isomerization

Asn deamidation and Asp isomerization are common degradation processes that involve intramolecular reactions leading to aspartic and isoaspartic acids, usually in a 3:1 ratio. Although several factors affect these degradation pathways, high (>8) pH is recognized as a common cause of deamidation. Several studies on deamidation rates have been conducted on soluble peptides, although in full-length proteins

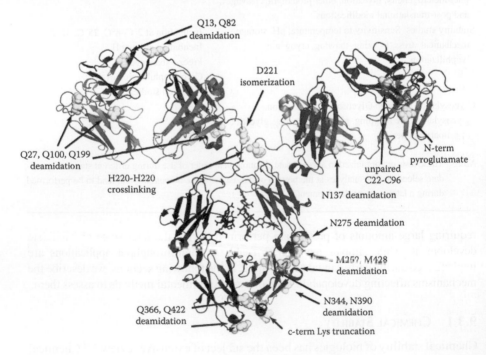

FIGURE 9.2 Documented sequence liabilities in the human IgG1.

the local tertiary structure has been shown to affect the deamidation rates considerably.[13] Gln deamidation follows a similar mechanism and has also been reported to strongly affect the tertiary structure and thus stability.[2]

Both Asn and Gln deamidation rates depend on the nature of the flanking amino acids.[14] For Asn, studies of soluble peptides indicate that the residues at the C-terminus of the Asn residue play a crucial role in the deamidation reaction. Asn and Asp, followed by Gly, His, Asp, Ser, or Ala, are considered potential deamidation sites, with NG being the most sensitive.[15] Predictive methods for the identification of deamidation have been devised, and reports suggest a prediction reliability of around 95%.[16] Gln deamidation is known to proceed at a slower rate with respect to Asn deamidation. Similarly to Asn, charged residues with low molecular hindrance on the N-terminus of Gln have been shown to accelerate the Gln degradation rate.[17]

For antibodies, deamidation occurring in CDRs has been shown to affect the potency of an anti-HER2 antibody.[18] Asn deamidation in the constant regions of heavy and light chains has also been reported for several antibody isotypes, with the most reactive Asn located in the sequence GFYPSDIAVEWESNGQPENNYK. Although human IgG1 contains 15 Asn in the constant region, only four of these residues have been shown to be sensitive to deamidation[19]—N275, N344 (QDWLN, WESN), and N390 (QPEN) in the heavy chain, and N137 (VCLLN) in the light chain—demonstrating the high sensitivity of the NG and NN sequences to deamidation.[19-21] The same Asn residues were also reported to undergo deamidation in IgG2 N386 (WTNN) and N423 (NWERN) in the C_H3, and N156 (ERQNGVLN) in the constant light chain.[22] In mouse IgG1, deamidation of N141 (SAAQTN) in the heavy chain and N161 (ERQN) in the light chain has been also reported.[23] Interestingly, *in vivo* deamidation has also been observed in animal studies following injection of monoclonal antibodies containing Asn in the CDRs.[24] Some degree of Gln deamidation has been reported for Q13 (EVQLVESGGGLVQPGR) and Q82 (NSLYLQMNSLR) on the variable heavy chain, Q27 (ASQGIR) and Q100 (APYTFGQGTK) on the variable light chain, Q199 on the light chain C_L, and Q366 (NQVSLTCLVK) and Q422 (SRWQQGN) in the C_H3.[17]

Similar to Asn deamidation, Asp residues in DG sequences can undergo isomerization to isoaspartyl residues.[25] Isomerization of Asp residues has been reported for DG sequences occurring in the CDRs,[26] as well as D221 in the hinge of IgG1.[27] Since deamidation results in a gain of negative charge on the side chain of Asp and Glu, the first consequence of deamidation is an increase in charge heterogeneity, which can be detected by analytical cation exchange chromatography (CEX) chromatography. However, the precise identification of the deamidation site can only be determined by mass spectrometry (MS) on tryptic peptides. For practical reasons, developability assessment packages usually include analytical CEX or capillary gel focusing (cGE) at an early stage to assess the overall charge heterogeneity, whereas accurate identification of each degradation species is left to a later stage of development.

9.3.1.2 Oxidation

The impact of Met oxidation on the development of biotherapeutics has been extensively documented. A representative example is alpha-1-antitrypsin (a1AT),[28] a member of the serpin family and inhibitor of serine proteases used in the treatment of hereditary emphysema. a1AT is known to undergo a complete inactivation at low pH due to oxidation of a crucial Met in the active site. For antibodies, Met oxidation was shown to lead to structural changes in the Fc, which affects the stability of the antibody as indicated by the reduced Tm of the C_H2.[10] As expected, oxidized samples displayed increased aggregation rates.

Human IgG1 contains two conserved Met residues in the Fc, M252 in the C_H2 and M428 in the C_H3, with an additional M358 present in some allotypes. Since these residues are located at the protein A and protein G binding sites (around the C_H2-C_H3 interface), their oxidations have been shown to alter the affinity for protein A and protein G.[29] Also, Met oxidation has been shown to affect the interaction with the neonatal Fc receptor and the Fc-gamma receptor.[30] In IgG2 and IgG3, in addition to M252 and M428, two additional Met residues, M358 and M397, with a lower degree of surface exposure, oxidize at a lower rate. In both IgG1 and IgG2, Met oxidation in the Fc was demonstrated to alter the binding to the Fc-gamma receptor,[30, 31] with a strong reduction in the antibody serum half-life.[32]

Met oxidation proceeds through the formation of a Met sulfoxide, which can be further oxidized irreversibly to a sulfone. Several factors can induce Met oxidation, including exposure to metal ions (like iron licking from the stainless steel container of bioreactors), light, peroxide impurities, or other sources of radicals. Although this may occur during storage and handling, a high risk of Met oxidation appears to happen during fermentation, where the iron concentration in the condition media can be relatively high.

Met residues in the binding site can be identified very early by inspection of the amino acid sequence and eliminated through sequence engineering. If the Met residue is critical for binding and thus cannot be mutated to another amino acid, oxidative degradation can be prevented through the use of antioxidants or chelating agents. For example, radical scavengers such as free Met, sodium thiosulfate, catalase, and platinum have been investigated in stabilizing the HER2 antibody.[33] Other chelating agents, like EDTA, diethylenetriaminepentaacetic acid (DTPA), and deferoxamine methylate, have also been used.

Trp represents another amino acid prone to oxidation. The indole ring of the Trp may oxidize when exposed to UV light, increasing the sample heterogeneity.[34] This chemical modification has been associated with a yellowing in commercial samples of therapeutic antibodies.[34] Since degradation of Trp residues does not affect the charge of the protein, detection of Trp degradation by-products is usually performed with hydrophobic interaction chromatography (HIC) or MS.[35] Most of the reports of Trp oxidation in recombinant antibodies involve the Trp in the CDRs.[36]

Trp residues in the CDRs, if exposed, can also affect the solubility of the molecule, leading to non-specific interactions.[37] In fact, the presence of Trp residues in the CDRs has been shown to generate polyreactive antibodies.[38] This is especially true when Trp and Tyr residues are combined in long CDRs.[39–42] Human IgG1 contains several conserved Trp residues in the framework regions: eight in the Fc, two in the C_H1, and two in the Fv, which form the core of the constant and variable domains. These residues are not solvent exposed and usually do not affect the solubility.

His residues can also undergo metal-catalyzed oxidation to 2-oxo-histidine,[43] although the degradation of His has received less attention in the literature.[44] Recently, oxidation and His-His cross-linking of His220 in the hinge of IgG1 has been reported.[45]

9.3.1.3 N- and C-Terminal Modifications

A number of N- and C-terminal modifications are known to occur in proteins. In antibodies, the N-terminal residues are frequently a Gln or Glu, which are often cyclized to pyroglutamate.[46] One of the most common sequence modifications in antibodies is the processing of the C-terminal Lys.[47,48] This modification is observed so frequently that it is a common practice in antibody engineering groups to remove the C-terminal Lys.

Other modifications of the N-terminal of antibodies can occur due to the incorrect processing of the signal peptide or other proteolytic cleavages occurring during cell culture and sample handling. Modifications of the N-terminal can be easily detected using automated Edman N-terminal sequencing, whereas C-terminal modifications are more difficult to characterize and require MS-based peptide mapping.[47] Alternatively, since most of these modifications alter the overall charge and size of the protein, they can be detected by analytical CEX, isoelectric focusing (IEF), or capillary isoelectric focusing (cIEF).

9.3.2 PHYSICAL STABILITY

Irreversible unfolding can alter the functional activity of proteins and lead to aggregation. During late-stage candidate selection, the stability of biotherapeutic candidates is often tested in different conditions of temperature, pH, mechanical stress, and diverse formulation buffers. These studies can be informative for downstream processing and help formulation groups to select drug candidates that perform well in the preferred formulation platforms.

The stability of biotherapeutics at low pH is particularly important for antibodies, since they are exposed to low pH during the purification with protein A affinity chromatography, and during the viral inactivation step.[49] The effect of low pH on antibody stability has been extensively investigated.[50–52] Stability studies at low pH are performed by incubating the samples at the desired pH for a few hours, followed by a series of analyses that include analytical size exclusion chromatography (SEC), polyacrylamide gel electrophoresis (PAGE), and activity assays to detect possible increases in aggregation and retention of protein activity.[53] Exposure to low pH is known to destabilize the antibody, causing aggregation and misfolding.[54]

Additional stability studies are performed to assess the resistance of drug candidates to storing and processing such freezing and thawing or cryogenic lyophilization. These experiments typically consist of subjecting the samples to several cycles of freezing and thawing or lyophilization and resolubilization, and analyzing the samples for aggregate content and biological activity.[55]

Finally, some developability assessment campaigns may involve testing the drug candidates for their sensitivity to mechanical stress. Proteins tend to aggregate at the air–water interface where foaming may occur under agitation. These studies are conducted by mechanically rotating or shaking the protein samples and measuring the formation of aggregates at different time intervals.[56]

9.3.2.1   Thermal Stability

The structural stability of biotherapeutics is typically estimated measuring the temperature-dependent transition points of the protein (Tm) in different formulation conditions. These experiments are usually performed with differential scanning calorimetry (DSC)[57] or differential scanning fluorimetry (DSF), which is more amenable to high-throughput settings.[58, 59] The thermal stability of the protein is considered a predictor of structural stability under the basic assumption that proteins exhibiting transitions at higher temperatures will also be more stable at physiological temperature and storage conditions.

Antibodies display two characteristic thermal transitions around ~70°C and ~80°C corresponding to the C_H2 and C_H3, respectively.[60] Although the Fab domain melting band may occasionally overlap with the bands of the C_H2 and C_H3, it often appears as a third band with a Tm that varies for each antibody. In developability campaigns, a minimum Tm threshold at 60°C is typically imposed as the pass criterion for the lowest thermal transition. Instead of using the peak maximum temperature (T_m), the temperature at which 1% of the protein has unfolded (T1%) can provide a more accurate estimate of stability.[61]

9.3.2.2   Long-Term Stability

Long-term stability is the ability of the drug to remain in solution under different storage conditions for a relatively long time (months). Therefore, variants of the drug are tested at different temperatures, typically 2°C–8°C, 25°C, and 37°C for several weeks. Samples are collected at different time points and analyzed for the formation of aggregates using analytical SEC and sodium dodecyl sulfate polyacrylamide gel electrophoresis (SDS-PAGE).[62] Functional assays can also be performed to ensure retention of the protein activity. These studies are time and sample consuming, and therefore are typically conducted during the late stage of candidate selection.

9.3.3   Solubility

Due to the sensitivity to enzymatic degradation of biological drugs, making oral delivery problematic, the vast majority of biologics are currently administered through subcutaneous/intramuscular injection or via intravenous infusion. Limitations in the volume of injection thus require a high concentration of protein. For example, while intramuscular injections have typical volumes around 5 ml, subcutaneous injections can only be performed with injection volumes below 1.5 ml.[63] Consequently, the need to formulate antibodies at concentrations above 100 mg/ml is not uncommon, although only a few commercial therapeutic antibodies have been successfully formulated at such high concentrations. While in rare cases multiple injections can be performed, this is not desirable. Intravenous infusion, on the other hand, can accommodate larger volumes at lower concentrations. However, this delivery route is not preferable for frequent dosing since it requires hospital visits.

From an experimental perspective, concentration experiments are performed by formulating the protein in the desired buffer and concentrating the samples stepwise using ultrafiltration devices. During the process, the aggregation level and the appearance of precipitate or phase separation is monitored by visual inspection and by assessing the sample turbidity with visible spectroscopy or dedicated devices.[64] Ideally, the formation of soluble particles would also be measured by light obscuration (LO) for micron-sized particles and nanoparticle tracking analysis (NTA) for submicron particles ranging from 40 to 1000 nm.[65]

9.3.3.1 Viscosity

Concentration studies are performed to test the feasibility to concentrate the protein drug while maintaining the sample viscosity below certain thresholds. Samples with high viscosity are undesirable since they may be difficult to operate with a syringe, requiring a larger needle gauge, ultimately increasing the discomfort of the patient. Although an empirical viscosity of 50 cP is commonly used as an upper limit for subcutaneous injections, a lower viscosity threshold of 15 cP has been recommended as the optimal limit of injectability for biologics.[66, 67]

9.3.3.2 Aggregation

Aggregation is one of the major concerns in the development of biotherapeutics, as high molecular mass species (HMMS) can be immunogenic.[68, 69] Aggregation may occur at different stages of the production process, for example, expression, purification, formulation, storage, and administration.[3] In the past decade, a large body of work has been dedicated to elucidating the mechanism of protein aggregation. However, protein aggregation is a complex process, proceeding through several possible mechanisms. It is affected by different factors, such as pH, temperature, and mechanical stresses, as well as the intrinsic instability in the protein fold.

Aggregates can be of a diverse nature.[70] In the simplest form, HMA may originate from the presence of hydrophobic or charged patches on the protein surface, while more complex forms of aggregation may involve partial or complete unfolding of the protein.[71] In general, protein aggregates can be soluble or insoluble (by evolving to larger particles). They can have covalent disulfide linkage and therefore appear in PAGE gels, or have noncovalent linkage and therefore be undetectable in SDS-PAGE. Aggregates can retain activity if the biologically active domain of the protein is not completely denatured in the aggregate particle.

Assessing the aggregation level of biotherapeutics is a regulatory requirement,[72] and it is part of every developability assessment campaign. Regulatory thresholds are imposed for visible and subvisible particles in IV injections.[64, 73] Although defined cutoff limits on soluble aggregates have not been established, it is customary to maintain aggregates below 5%.

The standard method to assess the amount of aggregation in protein samples is analytical SEC.[74] This method, however, is known to be prone to misleading results.[75, 76] Some arise from the possibility that part of the aggregates is retained by the column or other components of the instrumentation. Other misleading results may be due to the reversible nature of some aggregates, which dissolve in the buffer flow, leading to underestimation of the real degree of aggregation in the sample. In fact, although the stationary phase in SEC columns is intended to be chemically neutral, a certain degree of interaction with proteins is commonly observed, and both the chemical nature of the resin and the mobile phase affect the quality of the results.[77]

For this reason, the use of orthogonal techniques to validate the SEC results has been strongly recommended.[76] Analytical ultracentrifugation is broadly recognized to produce a more reliable estimation of the aggregate content.[78] Nevertheless, it requires specialized personnel and dedicated equipment. Alternatively, dynamic light scattering (DLS) measurements in cuvettes are exquisitely sensitive to the presence of high-molecular-weight species. DLS avoids the risk of underestimating the aggregate content derived from the dissolution of reversible aggregates in the buffer flow of SEC, and it is amenable for high-throughput applications.[79] Also, soluble particles that are filtered out by the SEC column due to their large size can be detected by DLS.

9.3.3.3 Unpaired Cysteines

Some accounts of unpaired cysteines in biotherapeutics have been reported.[80, 81] Testing for unpaired cysteines in developability assessment campaigns is not a common practice. However, we include here a brief discussion on this subject since the presence of unpaired or mispaired Cys residues can dramatically impact the protein folding, function, and stability.

Disulfide bonds are crucial structural elements to maintain the correct native fold of proteins. Unpaired cysteines are reactive centers that can lead to the formation of covalent aggregates and reduce the protein stability.[80] In some cases, scrambling of disulfide bonds can occur and lead to protein misfolding, such as reports for interferons.[82] Some cytokines require cysteine knots to maintain a correct folding.[83]

Depending on the species and isotype, antibodies contain different numbers of intra- and interchain disulfide bridges. Human IgG1 has 12 intrachain disulfide bonds and 4 interchain disulfide bonds, 2 in the lower hinge and 2 connecting the HC to the LC, which stabilize the IgG structure. In species such as rabbit and chicken, conserved cys residues are also found. A pair of Cys residues in the CDRs of human and mouse antibodies have also been reported, which is important for binding the target.[81]

Although properly folded IgG1 should not contain unpaired Cys residues, the presence of free thiols was reported in therapeutic antibodies.[84] In particular, unpairing of the disulfide bond C22–C96 in the Fab, closely positioned to the antigen binding site, was reported to lead to a reduction in the antibody potency.[85] Thus, although therapeutic antibodies containing Cys residues in the CDRs have been reported,[86] nonconserved Cys, especially in the CDRs, are considered a liability due to their reactivity and tendency to form covalent oligomers.

Unpaired Cys residues forming aggregates in antibodies can often be detected in nonreducing acrylamide gels where antibody fragments appear in a typical ladder pattern. Traditionally, unpaired cysteines can be detected using colorimetric free thiol assays based on the Ellman reagent, sometimes also referred to as DTNB.[87] Recently, more sensitive fluorimetric methods have been marketed that offer increase sensitivity.

Interestingly, the fact that the free thiol could be detected only after denaturation of the protein fold indicates that (1) the free cysteine is buried in the protein and (2) the antibody remains assembled noncovalently in spite of a missing disulfide bridge. This in turn suggests that the free thiol assay in native conditions may not be able to detect the presence of buried unpaired cysteines, and sample unfolding may be required in these assays. Although it is possible that traces of reducing impurities in the condition media or the formulation buffers may reduce these disulfide bonds, it would primarily affect the more sensitive and solvent exposed interchain bridges in the hinge region.[88] Therefore, the presence of buried unpaired cysteines could be attributed to disulfide bonds mispairing during the protein synthesis in the Endoplasmic Reticulum (ER).

Non-canonical disulfide bond structures have also been identified in other IgG Isotypes.[89] IgG2 antibodies contain two additional disulfide bonds in the hinge region and are known to undergo scrambling.[90] An interesting case is also represented by IgG4 antibodies, which display exceptionally weak disulfide bridges between the heavy chains and can undergo Fab–arm exchange *in vitro* and *in vivo*.[91]

9.3.4 GLYCOSYLATION

Glycosylation plays a crucial role in biopharmaceutical product development since it is generally regarded as a source of heterogeneity. Glycosylation can also affect the stability, solubility, half-life, and effector function of therapeutic antibodies. Importantly, the presence of a glycan in the CDRs can impair binding to the target with a consequent loss in biological activity. Therefore, elucidating the presence and nature of glycans in biopharmaceutical products is a regulatory requirement.

Antibodies have conserved N-glycosylation sites at N297 in the C_H2 domain, which is critical for the effector function of the antibody. Although the presence of N-glycans on the variable region is not uncommon, most of the human germ line sequences are not glycosylated on the variable region. These extra glycosylation sites are introduced by somatic mutations in about 20% of sequences.[92,93] Interestingly enough, an increased level of glycosylation in the variable region has been observed in some diseases.[94,95]

In principle, additional glycans at nonconserved sites may not degrade the biophysical properties of the protein or reduce its potency. In fact, glycans may improve the protein solubility since they consist of large hydrophilic moiety, and in nature they are known to mask hydrophobic surfaces.[96] In antibody engineering, successful attempts have been made to use glycans to mask hydrophobic patches and improve the protein solubility without affecting the affinity.[97] Glycosylation at nonconserved sites is also observed in commercial biotherapeutics like in the case of cetuximab, which carries an extra glycan on the variable region of the heavy chain.[98]

However, the presence of additional glycans in nonconserved spots is generally considered undesirable since it increases the analytical effort required for regulatory compliance. Antibody heterogeneity arises from the fact that glycans can have multiple chemical forms. The Fc glycans, for example, are predominantly biantennary structures bearing a core fucose on the first ring. These glycoforms may also have a bisecting N-acetylglucosamine, and a small fraction have a terminal sialic acid residue. Additional variability derives from truncations at the galactose rings leading to structures carrying two, one, or no galactose residue (referred to as G2, G1, and G0).

Although a detailed characterization of each antibody glycoform is not usually part of developability assessment campaigns, some degree of investigation may be required when the glycosylation is involved in the biological activity. For example, nonfucosylated antibodies are known to have antibody-dependent cellular cytotoxicity (ADCC) up to 100 times higher than their fucosylated counterparts.[99] Terminal sialylation was shown to negatively affect ADCC by decreasing the affinity for the Fc receptors,[100] but also positively increase anti-inflammatory activity of intravenous IgGs.[101] Also, the terminal sialylation affects the protein half-life and clearance from the body.[102] Finally, the lack of terminal galacatose, and high mannose have been respectively shown to reduce complement-dependent cytotoxicity (CDC)[103] and increase serum clearance.[104] Therefore, when certain biological activities such as ADCC and CDC are desired, a more thorough analysis of the glycan moiety is required, including an accurate estimation of the level of fucosylation and sialylation.

O-glycosylation at Ser and Thr is also often observed in biotherapeutics.[105] For example, the CTLA4-Ig Fc fusion protein is glycosylated at Ser129 and Ser139 in the hinge region.[106] Differently from N-glycosylation, which can be easily predicted from sequence analysis, consensus sequences for O-glycosylation have not been identified and predictive methods are limited to the identification of mucin-like domains.[107, 108]

A somewhat related glycan modification often observed in therapeutic proteins is the glycation of the ε-amino group in lysine and the N-terminal amine.[109] This modification is known to occur during the mammalian cell culture, where the concentration of glucose is particularly high. The addition of glucose to a primary amine leads to the loss of one positive charge, which can be detected by cIEF and IEC.[110]

In developability assessment, preliminary assessment of the degree of heterogeneity of the antibody glycoforms can be performed using cGE or HILIC, while a more detailed structural determination of the glycan antennary structure, the quantitative evaluation of the glycoform distribution, and the precise location of the glycosylation site can be assessed during the late stage of development using mass spectrometric methods.[111]

9.3.5 ELECTROSTATIC PROPERTIES AND SCREENING CHROMATOGRAPHIC CONDITIONS

A number of protocols are typically included in developability assessment to investigate the compatibility of the drug candidates with certain purification methods.

Different organizations adopt different manufacturing platforms in terms of purification and formulation methods, and these protocols are usually designed in collaboration between discovery and manufacturing groups.

These studies are performed by screening different chromatographic conditions of buffer, ionic strength, and pH, and eventually different chromatographic resins.[112] The purification efficiency is assessed with respect to the removal of undesired contaminants, with the most common impurities being high- and low-molecular-weight species, product-related impurities (aggregates and clipped products), host cell proteins, endotoxins, DNA, leached protein A, and media components. For antibodies, for example, protein A affinity chromatography is usually followed by polishing steps consisting of orthogonal methods such as anion exchange chromatography (AEX) or CEX in flow-through mode.[113,114] Therefore, a common step in developability assessment is determining if the antibody is amenable to purification in flow-through mode versus gradient mode, with the ion exchange chromatography (IEX) resins compatible with the organization's manufacturing platform.[115]

The use of some chromatographic methods imposes thresholds on the protein isoelectric point. This arises from the fact that the electrostatic properties of the protein (together with the pH and ionic strength of the sample) will determine the protein affinity for ion exchange resins. For example, a pI above 8.5 may be desirable to ensure solubility at neutral pH and allow the use of the AEX polishing step in flow-through mode. Developability assessment studies thus require characterizing the electrostatic properties of the protein in terms of isoelectric point, charge profile, and titration curves. This is usually done either theoretically from the protein sequence or, more accurately, experimentally by IEF or cIEF.

9.4   PREDICTIVE METHODS

Development of predictive methods can identify developability liabilities even before the synthesis of the molecule, and thus can save a significant amount of material and cut substantial costs in the drug development process. Some liabilities can be predicted by simple inspection of the amino acid sequence, but more often than not, reliable assessment of developability liabilities requires the three-dimensional (3D) structure.

A typical developability assessment campaign initiates with an analysis of the sequences with promising biochemical and biological profiles to identify potential developability liabilities (Figure 9.3). For antibodies, the Fv sequences derived from phage display campaigns or obtained by hybridoma technology are usually available before larger batches of antibodies are purified. Thus, sequence analysis can rapidly identify liabilities in the variable regions and the antigen binding site of the potential lead candidates.

Sequences with similar binding profiles and without apparent liabilities are prioritized over those with residues that can compromise further development. A more thorough analysis based on the 3D structure is reserved for molecules with promising binding profiles but carrying liable residues. The following sections describe the most common developability liabilities detected in the amino sequence, the state of

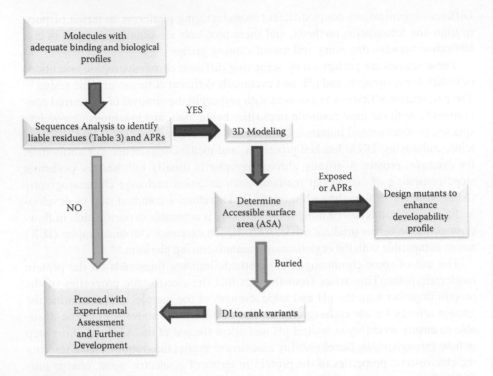

FIGURE 9.3 Workflow of the use of predictive methods during the discovery and optimization processes. Red arrows indicate a no-go. Green arrows indicate a go. Blue arrows indicate the direction of the process.

the art in antibody 3D modeling, and methods to predict aggregation, one of the key developability hurdles in biologic development, as it can lead to immunogenicity.

9.4.1 COMMON SEQUENCE LIABILITIES

Typical sequence liabilities have been discussed in previous sections and are summarized in Table 9.3. The amino acids listed in the table, if exposed to the solvent, can experience deamidation, isomerization, oxidation, and other chemical instabilities. Some residues, such as Trp, can lead to aggregation and should be flagged for further scrutiny. N-Glycosylation sites can easily be predicted from the sequence, in contrast to O-glycosylation sites that are difficult to predict. Also, we have listed under "other potential liabilities" a number of sequence motifs found in antibodies selected from phage display campaigns that have been found to cross-react with reagents frequently used in the selection process, such as biotin, protein A, or bovine serum albumin.

Analysis of the sequence alone can potentially overpredict chemical instabilities since liable residues can be buried in the protein and thus be shielded from degradation. In addition, hydrophobic patches leading to non-specific interactions and aggregation can be formed by residues distant in the sequence that are brought together by folding

TABLE 9.3

Sequence Liabilities Typically Scrutinized during the Sequence Analysis of Therapeutic Proteins

Sequence Liabilities	Liability	Reference
Chemical Stability		
NG, NS, NA, NH, ND	Deamidation	15, 17
DG, DP, DS	Isomerization	26
M, W	Oxidation and hydrophobicity (W)	17, 33
Glycosylation		
NXS/T; X = amino acid, not P	N-linked glycosylation	116
Aggregation		
C	Scrambled disulfide bonds, covalent oligomerization	81
APRs	Aggregation	117
Other Potential Liabilities		
Multiple W and Y in CDRs	Polyreactivity/non-specific binding	37–41
LLQG	Transglutaminase motif	
[RK]GD	Integrin binding motif	
HPQ, EPDQ[FY], DVEAW[LI], GD[FW] XF, PWXWL	Streptavidin	
WXPPF[KR]	Biotin	
WT[IL]XXHR	Protein A	
RPSP	Anti-IgE	
GLTFQ	Anti-IgM	
RT[IL][ST]KP	Anti-mIgG	
SS[IL]	Anti-IgG	
QSYP	Anti-cowIgG	
FHENSP	Plastic	
LPRWG	Albumin	
HHH	His tag	

(or unfolding). Therefore, the availability of the 3D structure is critical to weight up solvent exposure of degradable residues and the propensity of a protein to aggregate.

9.4.2 3D Structure and Modeling

Protein structure determination by x-ray crystallography is relatively robust, but it is resource-intensive, and in some cases the turnaround time is unpredictable, relative to computational modeling methods. In the absence of an experimental structure, the question is how a 3D model compares to an experimentally determined high-resolution structure, and therefore how reliable predictions and designs made based on models turn out to be.

To answer this question for antibody 3D modeling, one of the authors[118] recently compared seven modeling methodologies. The modeling methods including protocols developed by Accerlys, Inc., Chemical Computing Group (CCG), Schrödinger, Jeff Gray's lab at Johns Hopkins University, Macromoltek, Astellas Pharma/Osaka University, and Prediction of ImmunoGlobulin Structure (PIGS). These methodologies vary in many aspects, including degree of automation, criteria used for template selection, types of energy functions, and sampling algorithms. Nevertheless, all the assessed methods produced similar and reliable models of the Fv framework and, with few exceptions, for CDRs with canonical structures.

The comparison with a first modeling assessment[119] also indicated that progress has been made in the last 3 years at modeling the CDR-H3, although it still is an area of opportunity for improvement. The CDR-H3 is by far the most variable region of the Fv in terms of sequence, length, and conformation, and although some rules have been described to predict the base of the CDR-H3 loop,[120] no reliable rules to predict the conformation of its apical region strictly based on sequence patterns have been found (and probably do not exist due to the intrinsic dynamic properties of these long peptide chains). Therefore, 3D models of Fvs are reliable except at the CDR-H3.

9.4.3 AGGREGATION AND DEVELOPABILITY INDEX

Aggregation, as one of the main concerns for development of biologics, has been the subject of numerous studies, and in recent years, several methods have been developed to identify regions prone to aggregation (reviews in 117, 121). Some of the methods, such as TANGO,[122] Aggrescan,[123] Amylprep,[124] and Zyggregator,[125] have been developed based on the knowledge gained from amyloid formation of short peptides, called aggregation-prone regions (APRs). Other methods, such as Spatial Aggregation Propensity (SAP),[126, 127] predict aggregation-prone structural motifs via a combination of accessible surface area and hydrophobicity.

SAP has also been combined with estimation of the charge of the molecule to develop the so-called developability index (DI).[128] DI estimates the long-term physical stability, which is thought to be influenced mainly by the tendency of a molecule to aggregate based on solvent-exposed hydrophobic residues and the total net charge of the molecule. This latter parameter is calculated over the full sequence. The extent of hydrophobic patches in the surface of the molecule is estimated via SAP. Since DI provides a numerical output, this tool can be used to rank protein variants with diverse designs or Fvs after selection. As the charge of the molecules varies with the pH, DI can also be used to estimate aggregation in different pH conditions. It should be noted that DI does not consider other degradation pathways, such as oxidation, hydrolysis, and the presence of unpaired Cys residues and other liable residues. Thus, other liabilities need to be considered when designing or selecting a lead from a pool of drug candidates.

9.5 CONCLUSIONS AND FUTURE PERSPECTIVE

The developability assessment process is an evolving field where more experimental tests are being pushed upstream into early discovery. Ultimately, this will reduce the cost of development and avoid late-stage failures by transitioning to late-stage-only molecules that fit into certain standards of quality and are compatible with certain purification/formulation platforms. Currently, the inception of some developability tests in the early stages of development is hampered by the fact that certain techniques are not amenable to high-throughput multiplexing applications. In this context, the development of new automated instrumentation with high-throughput modalities can rapidly change the developability assessment practice.

Predictive methods, due to the speed and relatively low cost, have the potential to strongly improve the developability assessment process. Integration of predictive methods with high-throughput analytical tools will provide a detailed assessment of well- and badly behaved therapeutic candidates early on. Such information should expedite the discovery, formulation, and manufacturing processes. As the number of biotherapeutics in development grows, and novel scaffolds are introduced that differ from the traditional monoclonal antibody and Fc fusion modalities, the design of developability assessment protocols will also improve and lead to more robust methods, harmonized standards, and good practices across the industry. Currently, some accounts are available in the literature reporting developability strategies,[129–131] but most of the discussions on this subject are in rapid evolution and development.

REFERENCES

1. Sliwkowski MX, Mellman I. 2013. Antibody Therapeutics in Cancer. *Science* 341 (6151): 1192–98.
2. Liu H, Gaza-Bulseco G, Faldu D, Chumsae C, Sun J. 2008b. Heterogeneity of Monoclonal Antibodies. *Journal of Pharmaceutical Sciences* 97 (7): 2426–47.
3. Maas C, Hermeling S, Bouma B, Jiskoot W, Gebbink MF. 2007. A Role for Protein Misfolding in Immunogenicity of Biopharmaceuticals. *Journal of Biological Chemistry* 282 (4): 2229–36.
4. Zhong X, Neumann P, Corbo M, Loh E. 2011. Recent Advances in Biotherapeutics Drug Discovery and Development. In *Drug Discovery and Development: Present and Future*, 363–78. InTech.
5. Strohl WR, Strohl LM. 2012. *Therapeutic Antibody Engineering: Current and Future Advances Driving the Strongest Growth Area in the Pharmaceutical Industry.* Woodhead Publishing, Cambridge, UK.
6. Knappik A, Ge L, Honegger A, Pack P, Fischer M, Wellnhofer G, Hoess A, Wölle J, Plückthun A, Virnekäs B. 2000. Fully Synthetic Human Combinatorial Antibody Libraries (HuCAL) Based on Modular Consensus Frameworks and CDRs Randomized with Trinucleotides. *Journal of Molecular Biology* 296 (1): 57–86.
7. Ponsel D, Neugebauer J, Ladetzki-Baehs K, Tissot K. 2011. High Affinity, Developability and Functional Size: The Holy Grail of Combinatorial Antibody Library Generation. *Molecules* 6 (5): 3675–700.
8. Almagro JC, Kodangattil S, Li J. 2013. Humanized Antibodies. In *Making and Using Antibodies*, ed. GC Howard, R Matthew, MR Kaser, 395–421. Boca Raton, FL: CRC Press.
9. DeFelippis M, Harmon B, Huang L, Sukumar M. 2011. Stress Testing of Therapeutic Monoclonal Antibodies. In *Pharmaceutical Stress Testing: Predicting Drug Degradation*, ed. SW Baetsch, KM Alsante, 370–90. London: Taylor & Francis.

10. Liu D, Ren D, Huang H, Dankberg J, Rosenfeld R, Cocco MJ, Li L, Brems DN, Remmele RL Jr. 2008. Structure and Stability Changes of Human IgG1 Fc as a Consequence of Methionine Oxidation. *Biochemistry* 47 (18): 5088–5100.
11. Li S, Schöneich C, Borchardt RT. 1995. Chemical Instability of Protein Pharmaceuticals: Mechanisms of Oxidation and Strategies for Stabilization. *Biotechnology and Bioengineering* 48: 490–500.
12. Goolcharran C, Khossravi M, Borchardt RT. 2000. Chemical Pathways of Peptide and Protein Degradation. In *Pharmaceutical Formulation Development of Peptides and Proteins*, ed. S Frokjaer, L Hovgaard, 70–88. London: Taylor & Francis.
13. Robinson AB, McKerrow JH, Cary P. 1970. Controlled Deamidation of Peptides and Proteins: An Experimental Hazard and a Possible Biological Timer. *Proceedings of the National Academy of Sciences of the United States of America* 66 (3): 753–57.
14. Robinson NE, Robinson AB. 2001. Molecular Clocks. *Proceedings of the National Academy of Sciences of the United States of America* 98 (3): 944–49.
15. Wakankar AA, Borchardt RT. 2006. Formulation Considerations for Proteins Susceptible to Asparagine Deamidation and Aspartate Isomerization. *Journal of Pharmaceutical Sciences* 95 (11): 2321–36.
16. Robinson NE. 2002. Protein Deamidation. *Proceedings of the National Academy of Sciences of the United States of America* 99 (8): 5283–88.
17. Liu H, Gaza-Bulseco G, Chumsae C. 2008a. Glutamine Deamidation of a Recombinant Monoclonal Antibody. *Rapid Communications in Mass Spectrometry* 22 (24): 4081–88.
18. Harris RJ, Kabakoff B, Macchi FD, Shen FJ, Kwong M, Andya JD, Shire SJ, Bjork N, Totpal K, Chen AB. 2001. Identification of Multiple Sources of Charge Heterogeneity in a Recombinant Antibody. *Journal of Chromatography B* 752 (2): 233–45.
19. Chelius D, Rehder DS, Bondarenko PV. 2005. Identification and Characterization of Deamidation Sites in the Conserved Regions of Human Immunoglobulin Gamma Antibodies. *Analytical Chemistry* 77 (18): 6004–11.
20. Harris RJ, Wagner KL, Spellman MW. 1990. Structural Characterization of a Recombinant CD4-IgG Hybrid Molecule. *European Journal of Biochemistry/FEBS* 194 (2): 611–20.
21. Wang L, Amphlett G, Lambert JM, Blättler W, Zhang W. 2005. Structural Characterization of a Recombinant Monoclonal Antibody by Electrospray Time-of-Flight Mass Spectrometry. *Pharmaceutical Research* 22 (8): 1338–49.
22. Kroon DJ, Baldwin-Ferro A, Lalan P. 1992. Identification of Sites of Degradation in a Therapeutic Monoclonal Antibody by Peptide Mapping. *Pharmaceutical Research* 9 (11): 1386–93.
23. Perkins M, Theiler R, Lunte S, Jeschke M. 2000. Determination of the Origin of Charge Heterogeneity in a Murine Monoclonal Antibody. *Pharmaceutical Research* 17 (9): 1110–17.
24. Huang L, Lu J, Wroblewski VJ, Beals JM, Riggin RM. 2005. *In Vivo* Deamidation Characterization of Monoclonal Antibody by LC/MS/MS. *Analytical Chemistry* 77 (5): 1432–39.
25. Cacia J, Keck R, Presta LG, Frenz J. 1996. Isomerization of an Aspartic Acid Residue in the Complementarity-Determining Regions of a Recombinant Antibody to Human IgE: Identification and Effect on Binding Affinity. *Biochemistry* 35 (6): 1897–1903.
26. Sreedhara A, Cordoba A, Zhu Q, Kwong J, Liu J. 2012. Characterization of the Isomerization Products of Aspartate Residues at Two Different Sites in a Monoclonal Antibody. *Pharmaceutical Research* 29 (1): 187–97.
27. Hambly DM, Banks DD, Scavezze JL, Siska CC, Gadgil HS. 2009. Detection and Quantitation of IgG 1 Hinge Aspartate Isomerization: A Rapid Degradation in Stressed Stability Studies. *Analytical Chemistry* 81 (17): 7454–59.

28. Griffiths SW, Cooney CL. 2002. Relationship between Protein Structure and Methionine Oxidation in Recombinant Human R1-Antitrypsin. *Biochemistry* 41 (20): 6245–52.
29. Gaza-Bulseco G, Faldu S, Hurkmans K, Chumsae C, Liu H. 2008. Effect of Methionine Oxidation of a Recombinant Monoclonal Antibody on the Binding Affinity to Protein A and Protein G. *Journal of Chromatography B* 870 (1): 55–62.
30. Bertolotti-Ciarlet A, Wang W, Lownes R, Pristatsky P, Fang Y, McKelvey T, Li Y, Li Y, Drummond J, Prueksaritanont T, Vlasak J. 2009. Impact of Methionine Oxidation on the Binding of Human IgG1 to Fc Rn and Fc Gamma Receptors. *Molecular Immunology* 46 (8–9): 1878–82.
31. Pan H, Chen K, Chu L, Kinderman F, Apostol I, Huang G. 2009. Methionine Oxidation in Human IgG2 Fc Decreases Binding Affinities to Protein A and FcRn. *Protein Science* 18 (2): 424–33.
32. Wang W, Vlasak J, Li Y, Pristatsky P, Fang Y, Pittman T, Roman J, Wang Y, Prueksaritanont T, Ionescu R. 2011. Impact of Methionine Oxidation in Human IgG1 Fc on Serum Half-Life of Monoclonal Antibodies. *Molecular Immunology* 48 (6–7): 860–66.
33. Lam XM, Yang JY, Cleland JL. 1997. Antioxidants for Prevention of Methionine Oxidation in Recombinant Monoclonal Antibody HER2. *Journal of Pharmaceutical Sciences* 86 (11): 1250–55.
34. Boyd D, Kaschak T, Yan B. 2011. HIC Resolution of an IgG1 with an Oxidized Trp in a Complementarity Determining Region. *Journal of Chromatography B* 879 (13–14): 955–60.
35. Haverick M, Mengisen S, Shameem M, Ambrogelly A. 2014. Separation of mAbs Molecular Variants by Analytical Hydrophobic Interaction Chromatography HPLC: Overview and Applications. *mAbs* 6 (4): 852–58.
36. Hensel M, Steurer R, Fichtl J, Elger C, Wedekind F, Petzold A, Schlothauer T, Molhoj M, Reusch D, Bulau P. 2011. Identification of Potential Sites for Tryptophan Oxidation in Recombinant Antibodies Using Tert-Butylhydroperoxide and Quantitative LC-MS. *PLoS ONE* 6 (3): e17708.
37. Bethea D, Wu SJ, Luo J, Hyun L, Lacy ER, Teplyakov A, Jacobs SA, O'Neil KT, Gilliland GL, Feng Y. 2012. Mechanisms of Self-Association of a Human Monoclonal Antibody CNTO607. *Protein Engineering, Design and Selection* 25 (10): 531–37.
38. Mian IS, Bradwell AR, Olson AJ. 1991. Structure, Function and Properties of Antibody Binding Sites. *Journal of Molecular Biology* 217 (1): 133–51.
39. Thompson RS, Khaskhely NM, Malhotra KR, Leggat DJ, Mosakowski J, Khuder S, McLean GR, Westerink JMA. 2012. Isolation and Characterization of Human Polyreactive Pneumococcal Polysaccharide Antibodies. *Open Journal of Immunology* 2: 98–110.
40. van Esch WJ, Reparon-Schuijt CC, Hamstra HJ, van Kooten C, Logtenberg T, Breedveld FC, Verweij CL. 2002. Polyreactivity of Human IgG Fc-Binding Phage Antibodies Constructed from Synovial Fluid CD38+ B-Cells of Patients with Rheumatoid Arthritis. *Journal of Autoimmunity* 19 (4): 241–50.
41. Koide S, Sidhu SS. 2009. The Importance of Being Tyrosine: Lessons in Molecular Recognition from Minimalist Synthetic Binding Proteins. *ACS Chemical Biology* 4 (5): 325–34.
42. Fellouse FA, Li B, Compaan DM, Peden AA, Hymowitz SG, Sidhu SS. 2005. Molecular Recognition by a Binary Code. *Journal of Molecular Biology* 348 (5): 1153–62.
43. Schöneich C. 2000. Mechanisms of Metal-Catalyzed Oxidation of Histidine to 2-Oxo-Histidine in Peptides and Proteins. *Journal of Pharmaceutical and Biomedical Analysis* 21 (6): 1093–97.

44. Ji JA, Zhang B, Cheng W, Wang YJ. 2009. Methionine, Tryptophan, and Histidine Oxidation in a Model Protein, PTH: Mechanisms and Stabilization. *Journal of Pharmaceutical Sciences* 98 (12): 4485–4500.

45. Liu M, Zhang Z, Cheetham J, Ren D, Zhou ZS. 2014. Discovery and Characterization of a Photo-Oxidative Histidine-Histidine Cross-Link in IgG1 Antibody Utilizing (18) O-Labeling and Mass Spectrometry. *Analytical Chemistry* 86 (10): 4940–48.

46. Dick LW Jr, Kim C, Qiu D, Cheng KC. 2007. Determination of the Origin of the N-Terminal Pyro-Glutamate Variation in Monoclonal Antibodies Using Model Peptides. *Biotechnology and Bioengineering* 97 (3): 544–53.

47. Tsubaki M, Terashima I, Kamata K, Koga A. 2013. C-Terminal Modification of Monoclonal Antibody Drugs: Amidated Species as a General Product-Related Substance. *International Journal of Biological Macromolecules* 52: 139–47.

48. Dick LW Jr, Qiu D, Mahon D, Adamo M, Cheng KC. 2008. C-Terminal Lysine Variants in Fully Human Monoclonal Antibodies: Investigation of Test Methods and Possible Causes. *Biotechnology and Bioengineering* 100 (6): 1132–43.

49. Brorson K, Krejci S, Lee K, Hamilton E, Stein K, Xu Y. 2003. Bracketed Generic Inactivation of Rodent Retroviruses by Low pH Treatment for Monoclonal Antibodies and Recombinant Proteins. *Biotechnology and Bioengineering* 82 (3): 321–29.

50. Sahin E, Grillo AO, Perkins MD, Roberts CJ. 2010. Comparative Effects of pH and Ionic Strength on Protein-Protein Interactions, Unfolding, and Aggregation for IgG1 Antibodies. *Journal of Pharmaceutical Sciences* 99 (12): 4830–48.

51. Arosio P, Barolo G, Müller-Späth T, Wu H, Morbidelli M. 2011. Aggregation Stability of a Monoclonal Antibody during Downstream Processing. *Pharmaceutical Research* 28 (8): 1884–94.

52. Arosio P, Barolo G, Müller-Späth T, Wu H, Morbidelli M. 2013. Aggregation Mechanism of an IgG2 and Two IgG1 Monoclonal Antibodies at Low pH: From Oligomers to Larger Aggregates. *Pharmaceutical Research* 30 (3): 641–54.

53. Ejima D, Tsumoto K, Fukada H, Yumioka R, Nagase K, Arakawa T, Philo JS. 2007. Effects of Acid Exposure on the Conformation, Stability, and Aggregation of Monoclonal Antibodies. *Proteins* 66 (4): 954–62.

54. Filipe V, Kükrer B, Hawe A, Jiskoot W. 2012. Transient Molten Globules and Metastable Aggregates Induced by Brief Exposure of a Monoclonal IgG to Low pH. *Journal of Pharmaceutical Sciences* 101 (7): 2327–39.

55. Brey RL, Cote SA, McGlasson DL, Triplett DA, Barna LK. 1994. Effects of Repeated Freeze-Thaw Cycles on Anticardiolipin Antibody Immunoreactivity. *American Journal of Clinical Pathology* 102 (5): 586–88.

56. Kiese S, Pappenberger A, Friess W, Mahler HC. 2008. Shaken, Not Stirred: Mechanical Stress Testing of an IgG1 Antibody. *Journal of Pharmaceutical Sciences* 97 (10): 4347–66.

57. González M, Murature DA, Fidelio GD. 1995. Thermal Stability of Human Immunoglobulins with Sorbitol: A Critical Evaluation. *Vox Sanguinis* 68 (1): 1–4.

58. He F, Hogan S, Latypov RF, Narhi LO, Razinkov VI. 2010. High Throughput Thermostability Screening of Monoclonal Antibody Formulations. *Journal of Pharmaceutical Sciences* 99 (4): 1707–20.

59. Goldberg DS, Bishop SM, Shah AU, Sathish HA. 2011. Formulation Development of Therapeutic Monoclonal Antibodies Using High-Throughput Fluorescence and Static Light Scattering Techniques: Role of Conformational and Colloidal Stability. *Journal of Pharmaceutical Sciences* 100 (4): 1306–15.

60. Ionescu RM, Vlasak J, Price C, Kirchmeier M. 2008. Contribution of Variable Domains to the Stability of Humanized IgG1 Monoclonal Antibodies. *Journal of Pharmaceutical Sciences* 97 (4): 1414–26.

61. King AC, Woods M, Liu W, Lu Z, Gill D, Krebs MR. 2011. High-Throughput Measurement, Correlation Analysis, and Machine-Learning Predictions for pH and Thermal Stabilities of Pfizer-Generated Antibodies. *Protein Science* 20 (9): 1546–57.
62. Jiskoot W, Beuvery EC, de Koning AA, Herron JN, Crommelin DJ. 1990. Analytical Approaches to the Study of Monoclonal Antibody Stability. *Pharmaceutical Research* 7 (12): 1234–41.
63. Shire SJ, Shahrokh Z, Liu J. 2004. Challenges in the Development of High Protein Concentration Formulations. *Journal of Pharmaceutical Sciences* 93 (6): 1390–402.
64. Das K. 2012. Protein Particulate Detection Issues in Biotherapeutics Development: Current Status. *AAPS PharmSciTech* 13 (2): 732–46.
65. Narhi LO, Jiang Y, Cao S, Benedek K, Shnek D. 2009. A Critical Review of Analytical Methods for Subvisible and Visible Particles. *Current Pharmaceutical Biotechnology* 10: 373–81.
66. Du W, Klibanov AM. 2011. Hydrophobic Salts Markedly Diminish Viscosity of Concentrated Protein Solutions. *Biotechnology and Bioengineering* 108 (3): 632–36.
67. Lavoisier A, Schlaeppi JM. 2015. Early Developability Screen of Therapeutic Antibody Candidates Using Taylor Dispersion Analysis and UV Area Imaging Detection. *mAb* 7(1): 77–83.
68. Wang W, Singh SK, Li N, Toler MR, King KR, Nema S. 2012. Immunogenicity of Protein Aggregates: Concerns and Realities. *International Journal of Pharmaceutics* 431 (1–2): 1–11.
69. Joubert MK, Hokom M, Eakin C, Zhou L, Deshpande M, Baker MP, Goletz TJ, Kerwin BA, Chirmule N, Narhi LO, Jawa V. 2012. Highly Aggregated Antibody Therapeutics Can Enhance the *In Vitro* Innate and Late-Stage T-Cell Immune Responses. *Journal of Biological Chemistry* 287 (30): 25266–79.
70. Joubert MK, Luo Q, Nashed-Samuel Y, Wypych J, Narhi LO. 2011. Classification and Characterization of Therapeutic Antibody Aggregates. *Journal of Biological Chemistry* 286 (28): 25118–33.
71. Liaw C, Tung CW, Ho SY. 2013. Prediction and Analysis of Antibody Amyloidogenesis from Sequences. *PLoS ONE* 8 (1): e53235.
72. Shankar G, Shores E, Wagner C, Mire-Sluis A. 2006. Scientific and Regulatory Considerations on the Immunogenicity of Biologics. *Trends in Biotechnology* 24 (6): 274–80.
73. United States Pharmacopeia (USP) 787. Subvisible Particulate Matter in Therapeutic Protein Injections.
74. Hong P, Koza S, Bouvier ES. 2012. Size-Exclusion Chromatography for the Analysis of Protein Biotherapeutics and Their Aggregates. *Journal of Liquid Chromatography and Related Technologies* 35 (20): 2923–50.
75. Philo JS. 2009. A Critical Review of Methods for Size Characterization of Non-Particulate Protein Aggregates. *Current Pharmaceutical Biotechnology* 10 (4): 359–72.
76. Carpenter JF, Randolph TW, Jiskoot W, Crommelin DJ, Middaugh CR, Winter G. 2010. Potential Inaccurate Quantitation and Sizing of Protein Aggregates by Size Exclusion Chromatography: Essential Need to Use Orthogonal Methods to Assure the Quality of Therapeutic Protein Products. *Journal of Pharmaceutical Sciences* 99 (5): 2200–8.
77. Arakawa T, Ejima D, Li T, Philo JS. 2010. The Critical Role of Mobile Phase Composition in Size Exclusion Chromatography of Protein Pharmaceuticals. *Journal of Pharmaceutical Sciences* 99 (4): 1674–92.
78. Stine WB Jr. 2013. Analysis of Monoclonal Antibodies by Sedimentation Velocity Analytical Ultracentrifugation. *Methods in Molecular Biology* 988: 227–40.
79. Esfandiary R, Hayes DB, Parupudi A, Casas-Finet J, Bai S, Samra HS, Shah AU, Sathish HA. 2013. A Systematic Multitechnique Approach for Detection and Characterization of Reversible Self-Association during Formulation Development of Therapeutic Antibodies. *Journal of Pharmaceutical Sciences* 102 (1): 62–72.

80. Brych SR, Gokarn YR, Hultgen H, Stevenson RJ, Rajan R, Matsumura M. 2010. Characterization of Antibody Aggregation: Role of Buried, Unpaired Cysteines in Particle Formation. *Journal of Pharmaceutical Sciences* 99 (2): 764–81.

81. Almagro JC, Raghunathan G, Beil E, Janecki DJ, Chen Q, Dinh T, LaCombe A, Connor J, Ware M, Kim PH, Swanson RV, Fransson J. 2012. Characterization of a High-Affinity Human Antibody with a Disulfide Bridge in the Third Complementarity-Determining Region of the Heavy Chain. *Journal of Molecular Recognition* 25 (3): 125–35.

82. Kimura S, Utsumi J, Yamazaki S, Shimizu H. 1988. Disulfide Bond Interchange in *Escherichia coli*-Derived Recombinant Human Interferon-Beta 1 under Denaturing Conditions. *Journal of Biochemistry* 104 (1): 44–47.

83. Glocker MO, Arbogast B, Deinzer ML. 1995. Characterization of Disulfide Linkages and Disulfide Bond Scrambling in Recombinant Human Macrophage Colony Stimulating Factor by Fast-Atom Bombardment Mass Spectrometry of Enzymatic Digests. *Journal of the American Society for Mass Spectrometry* 6 (8): 638–43.

84. Zhang W, Czupryn MJ. 2002. Free Sulfhydryl in Recombinant Monoclonal Antibodies. *Biotechnology Progress* 18 (3): 509–13.

85. Chaderjian WB, Chin ET, Harris RJ, Etcheverry TM. 2005. Effect of Copper Sulfate on Performance of a Serum-Free CHO Cell Culture Process and the Level of Free Thiol in the Recombinant Antibody Expressed. *Biotechnology Progress* 21 (2): 550–53.

86. Lee CV, Hymowitz SG, Wallweber HJ, Gordon NC, Billeci KL, Tsai SP, Compaan DM, Yin J, Gong Q, Kelley RF, DeForge LE, Martin F, Starovasnik MA, Fuh G. 2006. Synthetic Anti-BR3 Antibodies That Mimic BAFF Binding and Target Both Human and Murine B-Cells. *Blood* 108 (9): 3103–11.

87. Ellman G. 1959. Tissue Sulfhydryl Groups. *Archives of Biochemistry and Biophysics* 82 (1): 70–77.

88. Sears DW, Mohrer J, Beychok S. 1977. Relative Susceptibilities of the Interchain Disulfides of an Immunoglobulin G Molecule to Reduction by Dithiothreitol. *Biochemistry* 16 (9): 2031–35.

89. Liu H, May K. 2012. Disulfide Bond Structures of IgG Molecules: Structural Variations, Chemical Modifications and Possible Impacts to Stability and Biological Function. *mAbs* 4 (1): 17–23.

90. Wang X, Kumar S, Singh SK. 2011. Disulfide Scrambling in IgG2 Monoclonal Antibodies: Insights from Molecular Dynamics Simulations. *Pharmaceutical Research* 28 (12): 3128–44.

91. van der Neut Kolfschoten M, Schuurman J, Losen M, Bleeker WK, Martínez-Martínez P, Vermeulen E, den Bleker TH, Wiegman L, Vink T, Aarden LA, De Baets MH, van de Winkel JG, Aalberse RC, Parren PW. 2007. Anti-Inflammatory Activity of Human IgG4 Antibodies by Dynamic Fab Arm Exchange. *Science* 317 (5844): 1554–57.

92. Dunn-Walters D, Boursier L, Spencer J. 2000. Effect of Somatic Hypermutation on Potential N-Glycosylation Sites in Human Immunoglobulin Heavy Chain Variable Regions. *Molecular Immunology* 37 (3–4): 107–13.

93. Abel CA, Spiegelberg HL, Grey HM. 1968. The Carbohydrate Contents of Fragments and Polypeptide Chains of Human Gamma-G-Myeloma Proteins of Different Heavy-Chain Subclasses. *Biochemistry* 7 (4): 1271–78.

94. Zhu D, Ottensmeier CH, Du MQ, McCarthy H, Stevenson FK. 2003. Incidence of Potential Glycosylation Sites in Immunoglobulin Variable Regions Distinguishes between Subsets of Burkitt's Lymphoma and Mucosa-Associated Lymphoid Tissue Lymphoma. *British Journal of Haematology* 120 (2): 217–22.

95. Youings A, Chang SC, Dwek RA, Scragg IG. 1996. Site-Specific Glycosylation of Human Immunoglobulin G Is Altered in Four Rheumatoid Arthritis Patients. *Biochemical Journal* 314 (Pt. 2): 621–30.

96. Petrescu AJ, Milac AL, Petrescu SM, Dwek RA, Wormald MR. 2004. Statistical Analysis of the Protein Environment of N-Glycosylation Sites: Implications for Occupancy, Structure, and Folding. *Glycobiology* 14 (2): 103–14.

97. Wu SJ, Luo J, O'Neil KT, Kang J, Lacy ER, Canziani G, Baker A, Huang M, Tang QM, Raju TS, Jacobs SA, Teplyakov A, Gilliland GL, Feng Y. 2010. Structure-Based Engineering of a Monoclonal Antibody for Improved Solubility. *Protein Engineering, Design and Selection* 23 (8): 643–51.

98. Qian J, Liu T, Yang L, Daus A, Crowley R, Zhou Q. 2007. Structural Characterization of N-Linked Oligosaccharides on Monoclonal Antibody Cetuximab by the Combination of Orthogonal Matrix-Assisted Laser Desorption/Ionization Hybrid Quadrupole–Quadrupole Time-of-Flight Tandem Mass Spectrometry and Sequential Enzymatic Digestion. *Analytical Biochemistry* 364 (1): 8–18.

99. Shinkawa T, Nakamura K, Yamane N, Shoji-Hosaka E, Kanda Y, Sakurada M, Uchida K, Anazawa H, Satoh M, Yamasaki M, Hanai N, Shitara K. 2003. The Absence of Fucose but Not the Presence of Galactose or Bisecting N-Acetylglucosamine of Human IgG1 Complex-Type Oligosaccharides Shows the Critical Role of Enhancing Antibody-Dependent Cellular Cytotoxicity. *Journal of Biological Chemistry* 278 (5): 3466–73.

100. Raju TS. 2008. Terminal Sugars of Fc Glycans Influence Antibody Effector Functions of IgGs. *Current Opinion in Immunology* 20 (4): 471–78.

101. Kaneko Y, Nimmerjahn F, Ravetch JV. 2006. Anti-Inflammatory Activity of Immunoglobulin G Resulting from Fc Sialylation. *Science* 313 (5787): 670–73.

102. Bork K, Horstkorte R, Weidemann W. 2009. Increasing the Sialylation of Therapeutic Glycoproteins: The Potential of the Sialic Acid Biosynthetic Pathway. *Journal of Pharmaceutical Sciences* 98 (10): 3499–508.

103. Raju TS, Jordan RE. 2012. Galactosylation Variations in Marketed Therapeutic Antibodies. *mAbs* 4 (3): 385–91.

104. Goetze AM, Liu YD, Zhang Z, Shah B, Lee E, Bondarenko PV, Flynn GC. 2011. High-Mannose Glycans on the Fc Region of Therapeutic IgG Antibodies Increase Serum Clearance in Humans. *Glycobiology* 21 (7): 949–59.

105. Hossler P, Khattak SF, Li ZJ. 2009. Optimal and Consistent Protein Glycosylation in Mammalian Cell Culture. *Glycobiology* 19 (9): 936–49.

106. Nebija D, Urban E, Stessl M, Noe CR, Lachmann B. 2011. 2-DE and MALDI-TOF-MS Analysis of Therapeutic Fusion Protein Abatacept. *Electrophoresis* 32 (12): 1438–43.

107. Gupta R, Jung E, Gooley AA, Williams KL, Brunak S, Hansen J. 1999. Scanning the Available Dictyostelium Discoideum Proteome for O-Linked GlcNAc Glycosylation Sites Using Neural Networks. *Glycobiology* 9 (10): 1009–22.

108. Julenius K, Mølgaard A, Gupta R, Brunak S. 2005. Prediction, Conservation Analysis, and Structural Characterization of Mammalian Mucin-Type O-Glycosylation Sites. *Glycobiology* 15 (2): 153–64.

109. Zhang B, Yang Y, Yuk I, Pai R, McKay P, Eigenbrot C, Dennis M, Katta V, Francissen KC. 2008. Unveiling a Glycation Hot Spot in a Recombinant Humanized Monoclonal Antibody. *Analytical Chemistry* 80 (7): 2379–90.

110. Quan C, Alcala E, Petkovska I, Matthews D, Canova-Davis E, Taticek R, Ma S. 2008. A Study in Glycation of a Therapeutic Recombinant Humanized Monoclonal Antibody: Where It Is, How It Got There, and How It Affects Charge-Based Behavior. *Analytical Biochemistry* 373 (2): 179–91.

111. Huhn Cn, Selman MHJ, Ruhaak LR, Deelder AM, Wuhrer M. 2009. IgG Glycosylation Analysis. *Proteomics* 9 (4): 882–913.

112. Kelley BD, Switzer M, Bastek P, Kramarczyk JF, Molnar K, Yu T, Coffman J. 2008. High-Throughput Screening of Chromatographic Separations. IV. Ion-Exchange. *Biotechnology and Bioengineering* 100 (5): 950–63.

113. Urmann M, Graalfs H, Joehnck M, Jacob LR, Frech C. 2010. Cation-Exchange Chromatography of Monoclonal Antibodies. *mAbs* 2 (4): 395–404.
114. Shukla AA, Hubbard B, Tressel T, Guhan S, Low D. 2007. Downstream Processing of Monoclonal Antibodies: Application of Platform Approaches. *Journal of Chromatography B* 848 (1): 28–39.
115. Liu HF, Ma J, Winter C, Bayer R. 2010. Recovery and Purification Process Development for Monoclonal Antibody Production. *mAbs* 2 (5): 480–99.
116. Jefferis R. 2005. Glycosylation of Recombinant Antibody Therapeutics, *Biotechnology Progress* 21, 11–16.
117. Agrawal NJ, Kumar S, Wang X, Helk B, Singh SK, Trout BL. 2011. Aggregation in Protein-Based Biotherapeutics: Computational Studies and Tools to Identify Aggregation-Prone Regions. *Journal of Pharmaceutical Sciences* 100 (12): 5081–95.
118. Almagro JC, Teplyakov A, Luo J, Sweet RW, Kodangattil S, Hernandez-Guzman F, Gilliland GL. 2014. Second Antibody Modeling Assessment (AMA-II). *Proteins* 82 (8): 1553–62.
119. Almagro JC, Beavers MP, Hernandez-Guzman F, Maier J, Shaulsky J, Butenhof K, Labute P, Thorsteinson N, Kelly K, Teplyakov A, Luo J, Sweet R, Gilliland GL. 2011. Antibody Modeling Assessment. *Proteins* 79 (11): 3050–66.
120. Kuroda D, Shirai H, Kobori M, Nakamura H. 2008. Structural Classification of CDR-H3 Revisited: A Lesson in Antibody Modeling. *Proteins* 73 (3): 608–20.
121. Hamrang Z, Rattray NJ, Pluen A. 2013. Proteins Behaving Badly: Emerging Technologies in Profiling Biopharmaceutical Aggregation. *Trends in Biotechnology* 31 (8): 448–58.
122. Fernandez-Escamilla AM, Rousseau F, Schymkowitz J, Serrano L. 2004. Prediction of Sequence-Dependent and Mutational Effects on the Aggregation of Peptides and Proteins. *Nature Biotechnology* 22 (10): 1302–6.
123. Sánchez de Groot N, Pallarés I, Avilés FX, Vendrell J, Ventura S. 2005. Prediction of 'Hot Spots' of Aggregation in Disease-Linked Polypeptides. *BMC Structural Biology* 5: 18.
124. Frousios KK, Iconomidou VA, Karletidi CM, Hamodrakas SJ. 2009. Amyloidogenic Determinants Are Usually Not Buried. *BMC Structural Biology* 9: 44.
125. Tartaglia GG, Pawar AP, Campioni S, Dobson CM, Chiti F, Vendruscolo M. 2008. Prediction of Aggregation-Prone Regions in Structured Proteins. *Journal of Molecular Biology* 380 (2): 425–36.
126. Chennamsetty N, Helk B, Voynov V, Kayser V, Trout BL. 2009. Aggregation-Prone Motifs in Human Immunoglobulin G. *Journal of Molecular Biology* 391 (2): 404–13.
127. Chennamsetty N, Voynov V, Kayser V, Helk B, Trout BL. 2010. Prediction of Aggregation Prone Regions of Therapeutic Proteins. *Journal of Physical Chemistry B* 114 (19): 6614–24.
128. Lauer TM, Agrawal NJ, Chennamsetty N, Egodage K, Helk B, Trout BL. 2012. Developability Index: A Rapid *In Silico* Tool for the Screening of Antibody Aggregation Propensity. *Journal of Pharmaceutical Sciences* 101 (1): 102–15.
129. Conley GP, Viswanathan M, Hou Y, Rank DL, Lindberg AP, Cramer SM, Ladner RC, Nixon AE, Chen J. 2011. Evaluation of Protein Engineering and Process Optimization Approaches to Enhance Antibody Drug Manufacturability, *Biotechnology and Bioengineering* 108 (11): 2634–44.
130. Saxena V, Panicucci R, Joshi Y, Garad S. 2009. Developability Assessment in Pharmaceutical Industry: An Integrated Group Approach for Selecting Developable Candidates. *Journal of Pharmaceutical Sciences* 98 (6): 1962–79.
131. Yang X, Xu W, Dukleska S, Benchaar S, Mengisen S, Antochshuk V, Cheung J, Mann L, Babadjanova Z, Rowand J, Gunawan R, McCampbell A, Beaumont M, Meininger D, Richardson D, Ambrogelly A. 2013. Developability Studies before Initiation of Process Development: Improving Manufacturability of Monoclonal Antibodies. *mAbs* 5 (5): 787–94.

10 Developability Assessment Workflows to De-Risk Biopharmaceutical Development

Jesús Zurdo,* Andreas Arnell,† Olga Obrezanova,†
Noel Smith,† Thomas R. A. Gallagher,†
Ramón Gómez de la Cuesta,* and Ben Locwin‡

CONTENTS

* Research and Technology, Pharma & Biotech, Lonza Biologics plc., Granta Park, Great Abington, Cambridge, United Kingdom
† Applied Protein Services, Lonza Biologics plc., Granta Park, Great Abington, Cambridge, United Kingdom
‡ Lonza Biologics, Inc., Portsmouth, New Hampshire

10.1   INTRODUCTION: WHY DEVELOPABILITY?
WHAT ARE THE RISKS?

Recent estimates of the true cost of developing new drugs show that, on average, companies seem to spend between US$4 billion and US$11 billion for every new therapeutic treatment that is eventually commercialized.[*] What is the reason for this? Fundamentally, the extraordinarily high rate of failure observed in drug development. Approximately 90% of drug candidates will fail during clinical development—maybe over 99% if preclinical stages of development are also included.[†1] In this context there is a growing interest in maximizing the return on the investment made in the development of new therapeutic candidates, avoiding whenever possible late and expensive failures.

In this chapter we discuss how early risk assessment, using developability methodologies and computational methods in particular, can assist in reducing risks during development in a cost-effective way. We define specific areas of risk and how they can impact product quality in a broad sense, including essential aspects such as product efficacy and patient safety. Emerging industry practices around developability are introduced, including some specific examples of applications to biological products. Furthermore, we suggest workflow examples to illustrate how developability strategies can be introduced in practical terms during early drug development in order to mitigate risks, reduce drug attrition, and ultimately increase the robustness of the biopharmaceutical supply chain. We also discuss how the implementation of such methodologies could also accelerate the clinical access of new therapeutic treatments.

10.1.1   WHY DRUGS FAIL DURING DEVELOPMENT

Failure in drug development seems to be the rule rather than the exception. However, the true reasons behind drug failure during development remain a highly debated and elusive issue for many, primarily due to the lack of detailed and up-to-date data on the subject. In some cases, the specific information behind drug candidate discontinuation is not made public (often the case during preclinical development). In others, there is a combination of different elements playing a role in the demise of a particular drug candidate, making it difficult to identify specific contributing factors.

---

[*] Forbes, "The Truly Staggering Cost of Inventing New Drugs," http://goo.gl/C2KSB
[†] PhRMA, http://phrma.org/

A publication from Kola and Landis,[2] covering drug development between 1980 and 2000 and other analyses published since Arrowsmith and Miller[3] have shed some light on the subject, suggesting a collection of different causes behind drug attrition.

Obviously inadequate efficacy is perhaps the major single reason behind clinical failure, but other relevant causes of drug attrition include bioavailability and pharmacology shortcomings, safety and toxicology problems, or even stability and quality issues, the latter two being particularly challenging for biopharmaceuticals. Furthermore, strategic and commercial reasons for discontinuation can often hint at problems with the design, cost of goods, lack of competitive advantage over other products in the market or under development, and even insufficient funding. Whereas the failure of new drug candidates during late clinical development and registration is primarily linked to inadequate biological activity and efficacy, or pharmacology and dosage issues,[4] attrition during early clinical development fundamentally relates to problems with safety (immunogenic reactions in biopharmaceuticals) and pharmacology, the latter less common in biologic drugs. As indicated above, preclinical drug attrition is a very complex area to survey, but in our experience, manufacturing and quality issues related to product stability and even productivity are common problems observed. Equally important to note is that quality issues are one of the main reasons behind drug product recalls by regulatory agencies, primarily related to unacceptable impurity levels, instability, or patient safety concerns.[*]

10.1.2 Risk Awareness and Availability Bias: Why We Assume "It Won't Happen to Us"

One might think that our industry should be only too aware of the many risks that are ultimately behind the extremely high failure rates observed in drug development. While the pharmaceutical industry has developed highly sophisticated processes to develop new drugs, it is not immune to common bias in the perception of risk, likelihood of occurrence, and decision-making processes.[5] Chief among them is the availability bias that colors our perception of risk with the information available to us, through either direct experience or relative prominence because of external exposure, news, reports, and so forth. There is a large variability in the perception of risk across different functions in drug development, particularly in matters related to the assessment of the relative impact and prominence of different causes of drug failure. Perhaps this is motivated by the very complexity of drug development, the lack of a holistic view on the drug development life cycle, and the existing "divorce" between different stages of therapeutic development. It is also true that often first-hand knowledge of failure can truly help understanding in detail the consequences and costs associated with it. Below we include a few elements that contribute to this low preparedness for risk and their potential consequences and impact.

- **Only doing what is essential.** Developing new drugs is a long and costly process, and there is tremendous pressure to move along quickly and cost-effectively. However, not considering properly future risks during early

---

[*] http://www.fda.gov

stages of development can have enormous consequences in the form of lost time and additional costs down the line.

- **Resistance to change.** Adoption of new approaches or technologies is often contemplated with apprehension by the industry, particularly in aspects of development and manufacturing of new drugs. The default position is always to use tried and tested methodologies rather than risking potential complications with regulators. Paradoxically, it turns out that the introduction of risk mitigation approaches or more robust manufacturing platforms can be seen as high risk by potential adopters, which tend to consider the well-treaded path of current development praxis as the safer and simpler way forward.

- **False economies.** The implementation of de-risking efforts up front in development is too often regarded as a distraction, expensive, and so forth. This in turn is linked to a lack of awareness about the true cost of quality failures in later stages of development that we just mentioned. These failures include, for example, the cost of a failed batch, the cost of not being able to formulate a product for the desired route of administration, the cumulative costs incurred when withdrawing a product during clinical development because of quality or safety issues, and so forth. Juran was only too aware of this and developed a whole corpus dedicated to assess the cost of quality (CoQ), including, as some put it, the cost of poor quality.[6] Unfortunately, such future events are not always easy to spot or, worse, are completely invisible to scientists involved in early development, and hence the costs of poor quality (including drug failure) are very rarely factored into any major investment decision.

- **Highly fragmented (silo) development process.** The typical hierarchical architecture behind drug development very often implies a complete lack of knowledge of problems that might arise downstream, but also forces different product and process optimization activities to be done in complete isolation from one another—often, expecting that someone downstream will figure out how to put things right if needed. Processes are therefore optimized independently within each silo, not as a whole. Perversely, changes introduced to optimize one particular stage can have very negative consequences for another (i.e., titer optimization at the expense of impurity profile or formulation stability at the expense of patient compliance or administration costs).

- **Past experience or the "we're different" belief.** Past experiences tend to influence perception of risks. This is particularly acute in drug development, where drug candidates can take between 8 and 15 years of development to become a commercial product. This means that, at best, scientists only have a limited or partial exposure to the history of a given product and any potential issues related to its development.

10.1.3 POTENTIAL RISK FOR EVERYONE: THE COST OF FAILURE

As indicated above, the tragic consequence of the current fragmented approach to drug development is that design elements that are essential for the success of new therapeutics can be left out during discovery and early development of new drug candidates. This fundamental disconnect between process optimization and product optimization produces a gap that can cause subsequent significant problems that are often only discovered quite late in development. As a result, severe delays can ensue, requiring additional investments or, in the worse cases, triggering the discontinuation of an entire drug development program. Indeed, delays in development, reworkings, failed batches, or deviations are all frequent and costly issues observed during development and manufacturing of biological products and, on many occasions, can ultimately be traced back to poor design of the product candidate or manufacturing process.

Moreover, the financial impact of preclinical and clinical attrition is often overlooked. From a biopharmaceutical development perspective, a significant financial commitment is made for the development of a qualified manufacturing process well before the product has even been cleared for its assessment in clinical trials. In fact, fully commercially defined processes are usually developed for prototypes (drug candidates) that, in a majority of cases, will fail at some point during their development cycle.* Such investment is obviously at risk, subject to success at various preclinical and clinical development stages. In addition, potential manufacturing or safety concerns can also have a major financial impact in several other ways:

- Extend already long development timelines (reducing market exclusivity period)
- Require additional investment in process development, repeated work, or implementation of corrective measures
- Stop a program in its tracks, preventing it from entering or progressing toward later stages of clinical development
- Cause the failure of a program during clinical trials, requiring a repeat of the trials or preventing final commercial approval due to quality or safety concerns
- Require considerable investment in process redesign and adaptation, reformulation, or even product recalls

These issues can be worth many millions in lost opportunities or investments lacking a return. In this context, it is desirable to select or design a successful candidate early on by asking the right questions.

10.2   DEVELOPABILITY AND QUALITY BY DESIGN (QBD)

10.2.1   *DEVELOPABILITY: ANOTHER WORD FOR QUALITY*

When defining quality, Juran made a distinction between *small Q* and *big Q* for quality. The former is concerned primarily with manufacturing, manufacturing processes, and manufactured goods, whereas the latter, big Q, takes a more holistic view on quality, involving a much broader set of processes, stakeholders, and customer needs. Furthermore, big Q requires the introduction of what Juran calls the quality trilogy: quality planning (another name for quality by design), quality control, and quality improvement.[6] This broader definition of quality (in

---

big Q) and the emphasis on quality planning (designing quality) also has important connotations when defining the cost of quality (CoQ), particularly when understanding the true impact of noncompliance beyond manufacturing failed batches.[7]

ICH Q8(R2)* states that "the aim of pharmaceutical development is to *design a quality product* and its *manufacturing process* to consistently deliver the intended performance of the product" and that "*quality* should be built in *by design*" rather than *by testing*. Furthermore, it defines *quality* as "the suitability of either a drug substance or a drug product for its intended use," including attributes such as the identity, strength, and purity (ICH Q6A). However, as reviewed by Kozlowski and Swann,[8] ICH Q8(R2) and subsequent guidelines Q9, Q10, and Q11 are concerned primarily with manufacturing process understanding, but do not integrate product knowledge aspects, such as product design and product specifications for intended use.

Juran's take on *quality planning* or *quality design* includes, among others, establishing quality goals, identifying who the customers are (stakeholders), determining the needs of the customers (what the product should accomplish), developing product features that respond to customers' needs (product design), and developing processes able to generate the desired product (note these processes extend beyond manufacturing, like chain of custody, administration, reimbursement, etc.).[6,9] Therefore, it clearly delineates that at the core of quality design sits the design of product characteristics that meet specific customer requirements, which in the case of pharmaceuticals would include patients as well as other stakeholders (e.g., payers, healthcare providers).

Any new therapeutic candidate needs to answer the following questions: Can it be made (at the right cost)? Is it stable? Can it be formulated for the intended route of administration? Is it safe for patients? Can it access the desired tissue at the required dose and in the required length of time? Will it exert the desired biological activity and show adequate effectiveness? Even before a lead candidate is found, such requirements can be summarized in an intended performance profile, and from that profile, one can derive desired characteristics that will help ensure the development of a high-quality therapeutic candidate. In this context, developability concerns more than just purity or stability aspects of the manufactured product; it also provides a platform to incorporate early on a solid basis for product knowledge and defines, from the outset, a robust quality target product profile (QTPP)† that would guarantee a successful, safe, and efficacious drug product.

Developability looks at the different aspects that confer a given therapeutic candidate its desired characteristics in three main quality areas (see Table 10.1):

1. **Manufacturability:** Productivity (yield), stability of the product, degradation and product-related impurity profile, formulability
2. **Safety:** Immunogenicity, immunotoxicology, specificity, and so forth
3. **Pharmacology/biological activity:** Bioavailability, half-life, formulability for intended route of administration, immunomodulation, and so forth

---

* http://www.ich.org/
† See ICH Q8(R2) for a definition of QTPP

TABLE 10.1
Development Assessment in Biopharmaceuticals

Quality Areas/Pillars	Scope of Assessment	Type of Assessment	Questions Addressed
Manufacturability/ Processability	Yield/Productivity Stability (chemical/physical) Formulability Undesired modifications	*in silico, in vitro*	Can it be made? How much will it cost? Will it be active?
Safety & Toxicity	Immunogenicity Immunotoxicology/CRS Specificity	*in silico, in vitro, in vivo*	Will it be safe?
Pharmacology	Delivery/Bioavailability Formulability (for desired RoA) PK/PD Half-life	*in vitro, in vivo*	Will it be cost-effective? Will it be available? target organ/tissue required dose/time window
Biological Activity	Immunomodulation Dosing & patient segmentation Efficacy	*in vitro, in vivo*	Will it work?

Source: Adapted from Zurdo, J., *Pharm Bioprocess* 1: 29–50, 2013; Zurdo, J., *EBR* 195: 50–54, 2013.

Note: For simplicity, developability is divided into three quality areas or pillars: manufacturability/processability, safety/toxicity, and pharmacology/mode of action. The type of assessment refers to the type of tools required for a developability assessment to address each specific area of development. These include *in silico* or computational (predictive) approaches, *in vitro* surrogate analytics mimicking relevant clinical or biological parameters, and *in vivo* assays making use of animal models (particularly for pharmacology, toxicity, and mode-of-action studies). Abbreviations: RoA, route of administration; CRS, cytokine-release syndrome; PK/PD: pharmacokinetics/pharmacodynamics.

Interestingly, all these categories are interrelated. For example, low stability can cause aggregation and thus safety issues (immunogenicity), and the ability of a product to be formulated for a specific route of administration can indeed have an important impact on the bioavailability and pharmacology (and by extension, the efficacy) of a given candidate. For a more in-depth review of the subject of developability, please go to.[10, 11]

10.2.2 DEVELOPABILITY: A THREE-STAGE PROCESS TO IMPLEMENTING A HOLISTIC QbD

In the case of pharmaceutical products, a holistic understanding of QbD would address product understanding, (manufacturing) process knowledge, and the interdependence of the two. Unfortunately, the current level of implementation of QbD strategies in pharmaceutical products is limited. ICH Q8(R2), Q9, Q10, and Q11 primarily deal with process knowledge and its influence on the definition of critical quality attributes (CQAs) in the product, but do not address the possibility of managing quality by better understanding product requirements and design of the relevant characteristics in the product itself. Indeed, traditional implementation of QbD strategies in biopharmaceutical manufacturing has primarily focused on process validation and process robustness. This practice is definitely useful, but perhaps ignores the single most important aspect of designing product quality: the identification of required quality product attributes and the process of designing and optimizing such attributes within the molecular structure of a given therapeutic product. In fact, it has been recognized that the obvious way of designing quality into a drug is by designing an optimal molecular candidate with a defined QTPP.[12]

As indicated above, developability can be considered an extension of QbD guidance, providing a bridge between product knowledge and process understanding, addressing the influence of product characteristics in manufacturing and clinical outcomes, and helping expand the design space for a drug candidate. We show how developability can be applied to early de-risking and how it can be seamlessly integrated with both discovery and process development activities.

In essence, any developability assessment program consists of three different stages:

1. **Risk assessment.** The simplest and most cost-effective way of assessing risk is by implementing computational approaches able to predict specific developability features by using the sequence of the biopharmaceutical candidates as a single input. These methodologies can have an extraordinarily high throughput and are relatively simple to implement. Later in this chapter we describe a collection of different tools that could be utilized for risk assessment and decision-making purposes.

2. **Implementing a risk mitigation strategy: Select an alternative candidate, redesign the candidate, or modify the process.** Predicting potential problems serves no purpose if it is not accompanied by adequate corrective actions or mitigation strategies to address them. Depending on where in the process the risk assessment has been performed, there are different courses of action that can be considered. In the case where process development (i.e., cell line development) has not been initiated, two different routes can be explored: (a) selecting alternative candidates with a better risk profile and (b) redesigning a candidate to correct issues highlighted by the risk assessment.[13]

   Once the product has already been taken into process development, or in cases where reengineering is not an option, process-related interventions can help mitigate some of these potential problems. Strategies could potentially include the screening of larger numbers of clones during cell line development to identify those clones able to compensate for poor productivity, high aggregation, or stability problems. Other approaches could include the utilization of alternative downstream processes (i.e., different buffers, purification, or concentration steps) to minimize the occurrence of degradation or precipitation, or the introduction of more stringent purification protocols to reduce the presence of aggregated species.

   All these strategies are further described in detail later in this chapter, with information about how and when in the process they can be used, together with case studies that illustrate how risk mitigation can be applied to real-world examples.

3. **Validation of course of action.** The developability risk mitigation cycle is completed by introducing appropriate validation studies. For example, in the case of immunogenicity of biopharmaceuticals, candidates can be reengineered to eliminate the occurrence of specific T-cell epitopes in the sequence and then tested using relevant cell-based assays that make use of blood samples from human donors.

10.2.3 DEVELOPABILITY TOOLKIT: DIFFERENT METHODOLOGIES FOR EARLY DE-RISKING

There is no single method or pathway to implement developability during early biopharmaceutical development. Developability rather requires a combination of different tools, ranging from the use of computational tools to appropriate analytical methods and tests, as well as *in vivo* models and, whenever possible, adequate *in vitro* surrogates. Finally, as we will see in this chapter, decision-making methodologies are essential to compile and prioritize the information collected during a developability assessment and to extract meaningful conclusions from it.

We will describe some of these approaches, with a particular emphasis on the introduction of computational methodologies alongside specific *in vitro* assays. However, we first want to define what these different methodologies are useful for and what their limitations are.

10.2.3.1 Computational Methodologies

The use of computational tools in early development is experiencing growing attention due to their relative simplicity of implementation and flexibility, providing considerable benefits in terms of high throughput, low cost, and relatively short time of analysis. They can also be applied at any given point in time, given that they are usually not limited by material availability or assay constraints. These methodologies make it possible to begin building product understanding as soon as the sequence of a candidate is known. They offer a window into properties that would otherwise not be available until much later in the manufacturing or clinical development process, and can help build quality into the product by selecting or designing lead candidates with favorable characteristics. Currently, there are a number of computational methods available for the prediction of immunogenicity (safety) and physical and chemical stability (manufacturability) of biopharmaceuticals, among other properties. And we expect that in the near future, computational methods will also be able to assist in the design of purification protocols or formulation compatibility.[14, 15]

On the negative side, the efficacy of predictive computational tools is ultimately limited by the type, amount, and quality of the experimental information utilized in their development. This limitation is particularly important in instances where the intended output is determined by multiple contributing factors (e.g., immunogenicity) in a complex biological context (the human body) that also exhibits a great degree of variability across different populations (e.g., different genetic, phenotypic, or disease backgrounds across patients). In such circumstances, predictive tools often utilize a surrogate descriptor of the intended property as a predictive output of the algorithm (e.g., T-cell epitope content), which in turn might be defined by one or several contributing factors (e.g., major histocompatibility complex (MHC) binding, antigen processing, etc.). Therefore, it is important to understand that although computational tools offer extraordinary opportunities to help select and design optimal drug candidates, they also have intrinsic limitations that need to be counterbalanced by introducing adequate *in vitro* and *in vivo* assessments that can help to validate or disprove predictions and designs made using *in silico* tools. Computational methods are well suited to detecting potential problems; however, as we will see later, it is also important to qualify the identified risks (e.g., by taking structural considerations into account), as the methods can otherwise be overpredictive and flag an inappropriately high number of potential issues. This is where experimental *in vitro* or *in vivo* validation could be utilized to help quantify or qualify the actual risk impact. Therefore, as the development process progresses, computational tools take on a supporting role to traditional experimental methods, helping to guide when and where more expensive experimental resources can be most effectively applied. Table 10.2 summarizes the pros and cons associated with the application of computational tools in risk assessment.

10.2.3.2 Surrogate/Proxy Analytical Methodologies

One of the most important issues in the application of an effective and robust early risk assessment is the use of adequate analytical tools and assay methods that are fit for purpose. In this way, we cannot just transplant standard analytical practices and methodologies that are currently being used for late-product physicochemical characterization and stability studies. The reason for this is that although such practices provide high-content and high-precision information, they also typically require considerable amounts of product, resources, and time. Therefore, there is a drive toward developing methodologies that could potentially reduce material requirements by as much as 10^3- to 10^4-fold, as well as increasing sample throughput by 10^2- to 10^3-fold. Obviously, we cannot expect to collect the same level of information as that achieved by standard analytical technologies. However, at this early stage of development, qualitative information that can be obtained fast and cheaply, such as

---

TABLE 10.2

Advantages and Disadvantages Associated with the Application of Computational (*In Silico*) Predictive Tools in Early Risk Assessment

Advantages	Disadvantages
• Available throughout development process	• Can be potentially overpredictive
• Rapid	• Qualitative predictions in some cases
• Low cost	• *In vitro* validation/confirmation can be required
• High throughput	• As good as the data they have been built on
• Simpler, cheaper corrective measures (problem mitigation can be very costly and time-consuming later on in development)	

assessing whether a given molecule is better or worse than a reference or identifying the best molecule among a collection of different candidates, can be sufficient for decision-making purposes and to decide which lead moves forward in development.

In many cases this will mean that a surrogate or proxy assay is sufficient to define a given property for a product candidate. The analytical methods used in early-stage development are undergoing a rapid development toward miniaturization and high-throughput analysis,[16] and their integration with early, rapid, and low-cost analytical and computational methods lies at the heart of the concept of developability. For example, the assessment of a given candidate's general propensity to aggregate might not require highly sophisticated assessments (i.e., particle counting, light scattering, etc.), but could be sufficient with high-throughput capillary electrophoresis,[17] turbidity,[18] or indirect (proxy) immunoassays, such as the oligomer detection assay (ODA) that we have described elsewhere.[13] High-throughput strategies to assess stability at high concentrations have been explored by implementing a combination of machine learning approaches with high-throughput self-interaction chromatographic approaches to determine second virial coefficients (B22) as a surrogate for protein solubility.[19] Equally, a number of authors have proposed the use of the apparent hydrodynamic radius of latex beads of known size in a protein sample by means of dynamic light scattering (DLS) and photon correlation spectroscopy (PCS) or the diffusion interaction parameter (k_D) measured by DLS as surrogates for viscosity in biopharmaceutical preparations.[20–23]

Glycoprofiling is another important area of product quality where the combination of new analytical approaches and automation to increase sample throughput is opening the door to interesting new applications. These include increased efficiency in the selection of cell lines for biomanufacturing or more robust and faster process design and development. One can imagine that, in the near future, similar platforms will be routinely used as early de-risking approaches to identify potential quality flaws in specific product candidates.[24]

Ultimately, the assays we want to implement need to be fit for purpose in an early developability assessment context. Therefore, a given analytical method should provide the required amount of information (no less, no more), be adequate for early assessment, be economical in terms of low sample and resource requirements, be easily performed in a high-throughput format, and be fast to perform and simple to interpret.

10.3 MANUFACTURABILITY OF BIOLOGICAL DRUGS: AGGREGATION, CHEMICAL STABILITY, AND PRODUCTIVITY

As described in Table 10.1 under the three pillars for developability, key manufacturability parameters include productivity and process yield, chemical stability, physical stability (particularly protein aggregation), formulability (including solubility at high concentrations), and undesired (post-translational) modifications. Protein aggregation and chemical degradation are two particularly important issues that can appear at various stages of the manufacturing process, including fermentation, purification, formulation, and storage. They can impact not only the yield and economics of the manufacturing process, but also the target product profile, delivery, and ultimately, patient safety.[25, 26] From a process perspective, tackling aggregation and chemical instability through process development and manufacturing can be complex and lead to increased costs, lengthen development timelines, and result in limited and costly formulation and delivery options. From a regulatory and patient safety point of view, the incidence of aggregation in biopharmaceutical preparations can constitute a serious safety concern by itself,[27] or be a contributing factor toward immunogenicity of biopharmaceuticals.[28, 29] Furthermore, formulation and container–closure interactions with a product can be contributing factors to increased levels of aggregates, with potentially devastating effects in patients.[30–32] Furthermore, besides aggregation, the incidence of chemical degradation, or post-translational modifications (PTMs), can also have a negative impact on the immunogenicity of biological therapeutics.[33] Finally and equally important is the potential incidence of anaphylactic reactions to biopharmaceuticals that can be mediated by specific PTMs, such as abnormal (non-human) glycosylation patterns.[34]

10.3.1 PHYSICAL STABILITY AND AGGREGATION: COMPUTATIONAL APPROACHES

As mentioned earlier, protein aggregates can be present at various stages of development and comprise a heterogeneous group of structures that can be generated through several different mechanisms, reviewed by Philo and Arakawa.[35] These include:

1. Reversible association of native monomer
2. Aggregation of conformationally altered monomer
3. Aggregation of chemically modified product
4. Nucleation-controlled aggregation
5. Surface-induced aggregation

This section focuses on the first two mechanisms of aggregation (monomer mediated), while aggregation of chemically modified products is discussed later on in terms of chemical instabilities and post-translational modifications. Both nucleation-controlled and surface-induced (interface-induced) aggregation can be process-dependent or impurity-dependent mechanisms, and will not be addressed in this chapter. However, it is important to mention that while the various mechanisms of aggregation differ widely among themselves, they all share a common ultimate origin: the polypeptide chain and its intrinsic physicochemical and structural properties, as well as its chemical instabilities.

Over the years, a number of different models have been developed to predict the intrinsic aggregation propensity of proteins, and many of them have been reviewed elsewhere.[36–39] One of the earliest aggregation prediction methods was developed by Dobson and coworkers at Cambridge University, and was initially used to successfully link intrinsic aggregation propensity predictions to the aggregation rate of polypeptides in solution.[40] This model, which later became part of the AggreSolve™ platform, together with other tools, evolved over time to combine both intrinsic protein parameters and extrinsic environmental factors, such as pH and ionic strength, in order to account for the influence of formulation conditions.[41] This original method used the amino acid sequence of polypeptides as primary input, and was best placed to describe aggregation from a conformationally altered, unfolded, or partially folded state. However, subsequent evolution of the tool incorporated other descriptors for protein conformational stability, such as to describe the aggregation propensity from a fully folded state.[42] This methodology has been successfully implemented in the redesign of biopharmaceuticals to reduce their aggregation propensity.[13,43,44]

Other approaches, such as the spatial aggregation propensity (SAP) algorithm and developability index developed by Trout and coworkers, focus on the surface hydrophobicity of the protein, using the three-dimensional (3D) structure and conformational flexibility of the folded protein as inputs.[45] The SAP method was able to identify potential hydrophobic hotspots on the surface of an antibody by performing a highly detailed, fully atomistic molecular dynamics simulation of the entire antibody structure. The method has been further developed to predict the long-term stability prospects of antibodies by incorporating the protein's net charge.[46] Other researchers have focused on the charge contributions and claim that the introduction of negative charges in and around light-chain L-CDR1 is a generalizable method to reduce aggregation.[47] Another interesting development is in the area of high-concentration solutions and viscosity prediction, where coarse-grain molecular dynamics simulations were used to suggest a model of interaction for the observed difference in viscosity and behavior of two monoclonal antibodies.[48] The differences were thought to be the result of charge–charge and charge–dipole electrostatic interactions of the different Fabs.[21,49]

Aggregation prediction algorithms are generally useful when comparing the aggregation propensity of highly similar candidates (i.e., sequence variants of a parental molecule) and also for detecting and disrupting aggregation hotspots through protein engineering methods. However, it is still challenging to assess the aggregation risk of a given biotherapeutic in the absence of a reference protein of similar nature for which experimental aggregation properties are known. This is particularly true when assessing different classes of proteins with different native conformational stabilities, which are often linked to their biological function. Moreover, the environment to which proteins are exposed during their synthesis, purification, formulation, storage, distribution, and administration, primarily linked to specific constituents of the expression system, product concentration, pH, shear forces, or exposure to air–liquid and air–polymer interfaces, is often a crucial determining factor influencing their aggregation behavior. For example, a given molecule may be perfectly stable in its final product formulation or circulating in blood plasma *in vivo*, after administration, but have a high aggregation propensity at low pH, which could present specific challenges during manufacturing (i.e., viral inactivation).

In any case, many of the biopharmaceuticals currently approved for human therapeutic use are antibodies or antibody derivatives, and there is a large body of knowledge on their properties. Antibodies are extremely stable molecules when compared to other protein classes, but they can still occasionally exhibit very high levels of aggregation. In fact, the presence of aggregation-prone regions is not uncommon even in supposedly stable commercial antibodies.[25,50] The very nature of antibodies and the high variability required for their biological activity, usually extended by the use of novel architectures, random complementary determining region (CDR) libraries, or heterologous systems for display and selection of potential binders, contribute to the problem. Nature has selected molecules that can be efficiently assembled, expressed, and secreted by B-cells and optimized for half-life and biological requirements. However, most modern drug discovery platforms are not designed to select candidates on such requirements, but rather on binding affinity to their target. Antibody aggregation is often linked to the amino acid composition of the CDRs.[50, 51] Among those reported examples supporting this observation are the anti-IL-13 antibody CNTO607[52] and Dyax hMab-X antibody.[53] Of course, that is not the complete picture. In fact, antibody stability is also connected to the specific framework regions present in the variable domains of the molecule. Some germ line families in particular are commonly linked to better stability and higher expression levels.[54] For example, the V_H3 germ line family is known to be among the more stable of the heavy-chain frameworks and is most similar to the highly soluble camelid V_HH domains.[55] Moreover, it has been suggested that there is a network of complementary conserved interactions in the variable domain that significantly impact its biophysical properties.[56]

Our group has made similar observations in a number of antibody models, suggesting a link between amino acid composition in the variable regions (including both framework regions and CDR characteristics) and stability and productivity observed experimentally (see the following section on productivity). With this background, we have recently developed an antibody-specific algorithm to predict aggregation, based on experimental data obtained by expressing several hundred antibodies in a Chinese Hamster Ovarian-Glutamine Synthetase (CHO-GS) mammalian expression system, and then characterizing their differences in behavior.[57] The antibodies used in the experimental data set were based on public antibody sequences and selected to cover a wide physicochemical and structural space. Machine learning approaches were used to assign molecules to two different classes (low and high aggregation risk) using sequence and structural descriptors as input. The classification uses a predefined cutoff calibrated experimentally as an indicator of relative process risks linked to aggregation events during process development and manufacturing. The algorithm was further validated in a collection of 50 unrelated antibodies with good predictability results. Methods such as this are an important step in implementing high-throughput and inexpensive aggregation assessments that can be incorporated into a simple and actionable manufacturability risk.

10.3.2 Chemical Instability and Post-Translational Modifications

Modifications in the chemical composition of biopharmaceutical products, whether due to cellular processes, enzymatic, or chemical and degradation reactions, can result in a complex level of product microheterogeneity. It is estimated that up to 10^8 different species could be found in a single vial of a biopharmaceutical product.[8] Therefore, a computational method should not aim to answer the question "What modifications can happen to the product?" but rather "What modifications are likely to happen to the product that could compromise its performance?" The latter is

TABLE 10.3
Common Chemical Instabilities and PTMs Observed
in Biopharmaceutical Products

- Deamidation
- Isomerization
- Fragmentation
- Disulfide scrambling
- Glycosylation (N- and O-linked)
- Oxidation
- Pyroglutamate formation
- Glycation
- Citrullination*
- Phosphorylation*
- Sulfation*
- Racemization*
- C-terminal lysine cleavage

* Those modifications that are not generally found in therapeutic monoclonal antibodies.

aimed at addressing critical quality attributes (CQAs) that should be controlled and monitored.* Table 10.3 contains a list of modifications that should be considered during biopharmaceutical development, although not all of them are generally relevant to every biopharmaceutical product. For example, in the table we highlight those modifications that typically are not found in therapeutic monoclonal antibodies. An in-depth review of different types of chemical instabilities and PTMs can be found elsewhere.[58,59]

Antibodies are one of the most common biopharmaceutical product classes and, due to their intrinsic sequence variability, can experience a wide range of modifications, and therefore constitute a suitable topic for detailed discussion. One approach to assess potential risks associated with chemical degradation and undesired PTMs involves the use of computational tools to predict their occurrence in particularly sensitive sites, such as areas believed to be important for binding affinity or stability. For this purpose, primary sequence and structural constraints that might affect solvent accessibility, involved in subunit interfaces, or potentially interact with other solutes are important input elements to be considered. As we will see later in this chapter, this type of risk assessment will then be used to define a suitable mitigation plan.

The presence of free, solvent-exposed thiol groups from cysteine residues constitutes one of the chemical instabilities with highest risk for biopharmaceutical products. The reactivity of these groups can potentially promote protein misfolding and aggregation, as well as increase the risk of immunogenic reactions. This is why, unless required for a functional purpose, such as the provision of a conjugation site in the molecule (i.e., to incorporate polyethylene glycol or a payload), free cysteines should preferably be removed prior to entering development. There are several

* ICH Q8(R2)

examples in the literature where the removal of such high-risk thiol groups had a beneficial effect on product quality. For example, the elimination of a free cysteine in a variable domain framework of the anti-angiopoietin 2 mAb MEDI-3617 resulted in increased product homogeneity, a 26-fold increase in productivity, and a 11°C increase in the melting temperature.[60]

The potential impact of chemical instability on product performance is not always easy to determine. Consider, for example, asparagine deamidation and aspartate isomerization, which are two of the most commonly found chemical instabilities in antibodies. Depending where these modifications occur in the sequence, they can either have very little impact on stability and functionality of the molecule or, in severe cases, potentially cause loss of activity, high product heterogeneity, and promote aggregation and fragmentation. The reaction rate of these modifications is influenced by pH, temperature, sequence (primarily the size and hydrogen bonding capability of the succeeding residue), and solvent accessibility, with isomerization occurring at a significantly slower rate than deamidation.[61] These types of instabilities can potentially be managed by process control and formulation,[62, 63] but they may also, in some instances, require protein engineering due to their high impact on product quality.[64]

Methionine and tryptophan are the two amino acids most susceptible to oxidation independently of the sequence context in which they are located. However, the degree of sensitivity of these two residues is determined in part by the solvent exposure of the side chain. Buried residues are less sensitive or take longer to react with oxygen.[59] Therefore, to properly assess the risk of oxidation for these two amino acids, structural information has to be taken into account, including the relative location of the affected residue in the molecule. Oxidation events in Fc and CDRs of antibodies can potentially affect a number of CQAs (aggregation, purity, identity), but can normally be managed by process control.[65]

Product heterogeneity and instability can also be the result of cellular processes such as glycosylation. Not only is proper glycosylation important to confer specific biological characteristics to a given biopharmaceutical, including its potency and pharmacological properties, but also it can be a determinant factor in the adequate folding and assembly of a product, and often defines other key attributes, such as stability, solubility, and immunogenicity.[34, 66] However, the incorporation of unintended glycan structures in or near binding interfaces such as the CDRs may occlude the binding region of the product or introduce steric hindrances that could negatively impact binding affinity. Glycan structure can vary in branching and composition, thereby introducing further heterogeneity that could potentially impact product identity and other CQAs. Furthermore, it is also important to mention that the presence of non-human glycans in a product is a known risk for hypersensitivity and anaphylactic reactions to biopharmaceutical products.[67] While a glycoengineering approach can be useful in order to increase stability or modulate the activity of a biotherapeutic,[52, 68] complementary studies could be required to determine the level of influence of process conditions (including host) in any given glycosylation site. Finally, chemical glycation can also occur in a product due to reaction with sugars (normally present in media) during bioprocessing, and can potentially affect identity, stability, and immunogenicity CQAs. Potential susceptible sites for glycation (i.e., solvent-exposed Lys) can be predicted. However, forced glycation studies could be more useful not only in confirming susceptible positions, but also in helping to define the magnitude of the problem, as well as determine the conditions that promote or prevent its occurrence.

10.3.3 PRODUCTIVITY AND YIELD

One important aspect not often recognized in biomanufacturing is the relationship between productivity or yield, on one hand, and product stability, primarily protein aggregation, on the other. As we have described before,[10, 69] protein aggregation and stability do, in fact, have many different faces. Aggregation not only manifests as turbidity, precipitates, or product loss during bioprocessing, but can also appear in the form of intracellular inclusions, low cell/culture viability, or low levels of productivity. Why is this so? Biological systems have evolved to deal effectively with protein synthesis, folding, assembly, trafficking, and secretion. To do so, they have developed an array of tools and systems specially tailored to prevent misfolding and aggregation, among other things. Unfortunately, we have not yet been able to fully adapt the biological platforms used in biomanufacturing to cope with the always-growing demands from the industry and its more challenging applications. Aggregates are often toxic to cells, probably by different mechanisms, and it is long known that exposure of cells to protein aggregates, or expression of aggregating proteins, can significantly affect cell viability. Prokaryotic cells have the ability of forming inclusions to isolate the problem, whereas such responses are less common in eukaryotic cells. Furthermore, secretory cells, such as the mammalian systems commonly used in biopharmaceutical manufacturing, have a very effective quality control system that prevents misfolded or misassembled proteins from being secreted. When abnormal events do occur, usually proteins get stopped in the secretory pathway and pushed toward either a refolding or degradation pathway. As a result, products that tend to misfold would naturally have a lower chance to be secreted by the cell. We have observed this pattern in multiple products with varying degrees of stability and aggregation. This is particularly evident in mammalian systems, and usually shows that high-aggregation propensity tends to go hand in hand with low productivity. Of course, we also observe exceptions, which tend to be associated with the fact that clonal selection can, occasionally, offset the intrinsic challenges contained within the polypeptide chain to be expressed. Leaving exceptions aside, we usually see a high degree of correlation between observed productivity and observed aggregation propensity. Figure 10.1 shows a systematic analysis performed in three different monoclonal antibody families that were built by incorporating single and double mutations intro three different parental monoclonal antibodies. The graph shows that there is a high level of correlation between the productivity and aggregation levels observed within each of the families. We have also observed a similar pattern when reengineering antibodies to reduce their aggregation levels (see case studies later in this chapter). In all these cases, we consistently see a substantial increase in product titer together with lower aggregation levels. Therefore, aggregation prediction could be potentially utilized as a surrogate for productivity levels, particularly in biopharmaceuticals expressed in mammalian systems.

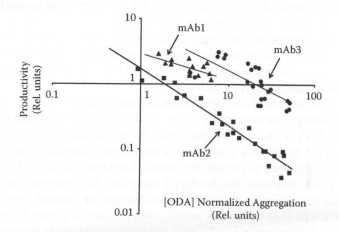

FIGURE 10.1 Analysis of the relationship between antibody productivity and aggregation in three different antibody families: mAb1, mAb2, and mAb3. Data points correspond to collections of antibody variants (single and double mutations) derived from three different parental antibody molecules. All different antibodies were expressed transiently under identical conditions to minimize any clonal variability in expression. Relative aggregation was assessed using Lonza's oligomer detection assay (ODA).[13] The plot suggests some degree of correlation between the observed aggregation and productivity levels within each antibody family. (Adapted from Zurdo J. et al. *Biomed Res Int*, Vol. 2015, Article ID 605 427, 2015.)

Furthermore, in the case of therapeutic antibodies, we and others have found a linkage between the amino acid composition of specific areas of the antibody molecule and the productivity observed in mammalian systems. These observations open the door to the development of predictive platforms to assess product expression by means of computational tools.[70,71]

10.3.4 IMPORTANCE OF FORMULATION: FORMULABILITY ASSESSMENT

Biotherapeutic delivery is becoming an area of increasing importance, not only because of its potential impact in the pharmacology of the product, and hence its efficacy, but also because of its impact on other important product attributes that are linked to patient compliance and even costs associated with a given treatment. For example, it has been estimated that, in some extreme cases, the costs associated with the infusion of a biopharmaceutical product, in a hospital and under specialized supervision, could surpass the cost of the product dose itself.[72] This brings pressure from payers to reduce the total cost of treatment,[73] but also to facilitate patient compliance by developing formulations and delivery methods that could potentially facilitate self-administration.[74]

Subcutaneous delivery of biotherapeutics presents a number of advantages compared to traditional infusion approaches. It simplifies the administration of the drug, is less invasive, reduces patients' discomfort, and can modulate the product pharmacology by facilitating a gradual/sustained release of the product. Furthermore, it is compatible with autoinjector devices, facilitating patient self-administration.

However, subcutaneous doses are limited to a 1.5 ml volume and typically involve product concentrations ranging between 100 and 200 mg/ml, in order to deliver a sufficient dose. The use of such high-concentration formulations introduces new challenges to product quality, such as solubility constraints, high viscosity, the presence of aggregates, and phase separation that could block the use of a subcutaneous route of administration (RoA) for delivery.[16, 75]

This is one of the reasons why a good understanding of formulation compatibility or formulability—the assessment of the suitability of a given product candidate for a desired RoA or delivery method—can be crucial during product development. However, formulability is also a very important parameter in other areas of development, particularly in process development and manufacturing. For example, most development processes are developed using standard methods and solutions (including chromatographic purification, diafiltration, or product concentration), which might not be at all compatible with a given product candidate. Product losses are unfortunately common in cases where products are not stable in a given buffer or do not tolerate a specific pH range. This is also the case where products need to be concentrated during the manufacturing process. For example, during downstream processes, drug substance storage can require concentrations ranging from 25 to 200 g/L, because of limitations in volumes that a plant can store at a given time. This problem has been exacerbated by the increase in product titer that can be achieved in today's manufacturing platforms.[76]

We still lack simple formulability platforms to assess the suitability of a given product candidate to be formulated at high concentrations (i.e., for subcutaneous administration) or compatibility with basic solution and process conditions, in order to maximize product stability and recovery during manufacturing processes. However, a number of promising approaches, described below, are being developed in this space, and hopefully some of them will produce new developability tools in the near future.

Formulation screening can be informed by computational methods in terms of aggregation propensity (see earlier), long-term stability,[46] and selection of excipients.[77] However, formulation design, as it stands today, is fundamentally an analytical discipline where the results of multiple orthogonal methods usually have to be taken into account, as well as the changing conditions as the process is transferred to a manufacturing scale.[75] Nevertheless, the utilization of high-throughput methods is key in order to screen a wide formulation space, whereas computational support can be utilized to support the analysis.

A number of high-throughput strategies have been explored to assess protein stability at high concentrations. One approach relies on the use of the second virial coefficient as a surrogate for protein solubility through the use of dynamic light scattering (DLS)[78] Furthermore, a number of authors have proposed the use of surrogate parameters to assess the viscosity of biopharmaceutical preparations by harnessing the Stokes–Einstein equation. These involve the measurement of the apparent hydrodynamic radius or the diffusion interaction parameter (k_D) of latex microbeads with known size in a protein sample, from which the solution viscosity can be derived. The methods proposed include DLS and photon correlation spectroscopy (PCS).[20–23]

Even with relatively high-throughput methods as those described above, the number of potential formulation conditions to be explored can be overwhelming. In such circumstances, the use of computational methods can help make the screening process more manageable. One interesting approach involves a combination of machine learning computational tools with high-throughput self-interaction chromatography (SIC) to estimate the second virial coefficient of protein solutions covering a very wide formulation space (up to several thousand different conditions), allowing the design of biopharmaceutical formulations with very limited product availability and early on in the development process.[19] Furthermore, computational methods are also useful in integrating measurements from different orthogonal analytical methods, potentially allowing the analysis of large data sets. Examples of this type of approach include Chernoff faces, star charts, and empirical phase diagrams.[79]

10.4 SAFETY IN BIOPHARMACEUTICALS: IMMUNOGENICITY AND IMMUNOTOXICOLOGY

Biopharmaceuticals are generally considered to be safer for patients than small-molecule pharmaceuticals. However, their administration to patients can cause a number of undesirable side effects, usually related to pharmacology issues, mechanism of action, or more commonly, immunogenic reactions.[80, 81] As a result, immunogenicity is often considered to be one of the principal safety concerns for biotherapeutics and one of the primary causes for attrition during early clinical development.

Current clinical data suggest that the majority of therapeutic proteins are to a variable extent immunogenic.[82] The generation of an unwanted immune response can negatively influence both the efficacy and safety of the therapeutic protein. Therefore, the incorporation of an immunogenicity assessment early on during preclinical drug development can significantly reduce the risk of generating an unwanted immune response in the clinic, which could potentially modify the pharmacology of the product or render it completely inefficacious. In extreme circumstances, biopharmaceuticals can also cause severe hypersensitivity and anaphylactic or immunotoxicology reactions that can put patient's life at risk.[34, 83–85]

Early identification of immunogenicity risk factors can also allow risk mitigation strategies to be employed (e.g., deimmunization), and hence increase the chances of success for a given product. In the following pages we will describe a number of preclinical methodologies that can be utilized to assess the immunogenic risk of biopharmaceutical candidates in development and their specific applications in various process and product contexts.

10.4.1 NEED FOR PRECLINICAL IMMUNOGENICITY ASSESSMENT

Immunogenicity is the ability of a substance to provoke an immune response in the body of a human or animal and can include both humoral and cell-mediated responses. Immunogenicity can be both wanted (e.g., raising an immune response against a virus or bacterium during vaccination) and unwanted (immune responses to therapeutic proteins that reduce the efficacy and safety of the treatment).

Once a biotherapeutic protein is administered, it encounters many different cells in the body, including antigen-presenting cells (Figure 10.2). These antigen-presenting

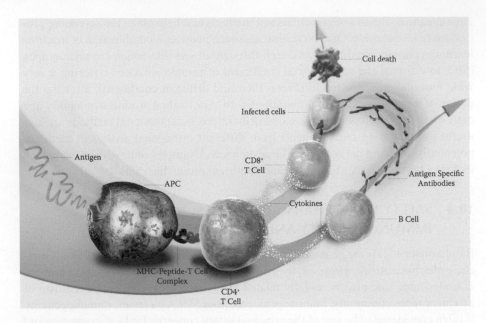

FIGURE 10.2 Schematic showing the paths that trigger immune responses to antigens. (Copyright Lonza Biologics plc.)

cells (including dendritic cells and B-cells) take up and process the protein internally. Some peptides derived from the protein can bind human leukocyte antigen (HLA) class II molecules (also known as major histocompatibility complex [MHC] class II) and be displayed on the surface of the antigen-presenting cells. These HLA–peptide complexes can then interact with the T-cell receptors on the surface of the T-cell and activate a T-cell response. The T-cells activated by specific T-cell epitopes from the biotherapeutic protein can then interact with B-cells that have also encountered the biotherapeutic protein (via binding to the B-cell receptor) and provide help to generate specific antibodies against the biotherapeutic protein, thus generating the unwanted response against the protein.

The immune response to therapeutic proteins is highly complex, with both intrinsic and extrinsic factors contributing to the observed immunogenicity. Intrinsic factors include the presence of B-cell and T-cell epitopes and non-human post-translational modifications (e.g., glycosylation) in the protein. Extrinsic factors include formulation excipients, aggregation, degradation, and any contaminants and impurities present in the final product. In addition, treatment factors, such as route of administration, dose regime, length of treatment, and characteristics of individual patients (genetic status, disease indication, and concomitant medication), can also affect the immunogenicity of a therapeutic protein.[86]

In general, immune responses to therapeutic proteins are assessed in the clinic by monitoring the generation of antibodies raised against the protein. These humoral (i.e., not cell-mediated) responses can be both T-cell dependent or independent. T-cell-independent antibody responses are generated when B-cells are able to recognize and bind to epitopes in the protein, but in the absence of T-cell help, these are generally low-affinity, transient IgM antibodies. When a T-cell response is also induced by the therapeutic protein, then the antibody response can lead to high-affinity, long-lived IgG antibodies that are much more likely to affect the safety and efficacy of the therapeutic protein in the clinic. Due to the importance of the T-cell response in the development of long-lived, high-affinity antibodies, there is much focus on the identification and removal of T-cell epitopes during the development of therapeutic proteins to reduce their potential immunogenicity risk.

Interestingly, regulatory bodies encourage innovators to explore the use of preclinical methodologies that could give an early indication of immunogenicity risks to patients, including both *in silico* and *in vitro* methodologies.[87–89] In the following sections we will cover some of the approaches that can be used to this aim.

10.4.2 ALTERNATIVES TO ANIMAL TESTS FOR PRECLINICAL IMMUNOGENICITY

Current regulatory guidelines require a preclinical immunogenicity assessment to be carried out in animal models (e.g., rodent, non-human primates). Although these models can be useful in the design and development of immunogenicity assays, it is widely accepted that these non-human models are not predictive of the levels of immunogenicity that can be expected in the clinic. There is also a drive to reduce the use of animals in pharmaceutical research (e.g., the National Centre for the Replacement, Refinement and Reduction of Animals in Research in the UK) from both an ethical and a cost perspective.* The use of fully human preclinical technologics to complement the required animal testing is also encouraged by regulatory bodies, especially in high-risk cases.[90]

10.4.3 *IN SILICO* PREDICTION OF IMMUNOGENICITY

As we have mentioned earlier, the immunogenicity of therapeutic proteins can be determined by many different factors. These can be intrinsic (T-cell epitopes or abnormal glycosylation) or extrinsic (aggregates, degradation products, contaminants, excipients, etc.), as well as related to the treatment itself (route of administration, dosing regime, or characteristics of individual patients). Because of the complexity of immune responses, it would be extremely challenging to develop a predictive computational model able to integrate all these variables. Instead, the presence of T-cell epitopes is commonly used as a surrogate predictor of immunogenicity. The rationale behind this approach is that the generation of antibodies by B-cells requires antigen-presenting cells (APCs) to present nonself T-cell epitopes. Therefore, the presence of sequences that could be potentially presented as nonself by HLA receptors is a prerequisite for many immunogenic reactions. *In silico* predictive tools therefore focus on the identification of potential T-cell epitopes present in proteins. In spite of its limitations, *in silico* screening offers a high-throughput and inexpensive approach to immunogenicity assessment during drug design and lead candidate selection stages, even before the molecules have been expressed and characterized.

---

* http://www.nc3rs.org.uk/

During the last two decades a number of computational methodologies have been developed for the prediction of immunogenicity. Most of these tools assess the T-cell epitope content in proteins by predicting the binding specificities of peptide fragments from the protein of interest to HLA class II receptors. Such tools are reviewed elsewhere.[91, 92]

We will describe here the Epibase™ in-house computational platform for immunogenicity screening, and exemplify its utilization in the early stages of drug development. The immunogenicity risk platform identifies potential T-cell epitopes in a protein sequence by predicting the binding of peptide sequences to HLA class II receptors. The tool is built using statistical methods and is based on structural characteristics of HLA molecules, sequence-based peptide features, and experimentally measured peptide/HLA binding affinities. During the assessment, a protein sequence is divided into 10-mer overlapping peptides, the binding affinity of each peptide is predicted for HLA class II allotypes, and peptides are classified into binders and nonbinders. Later in this chapter, we also discuss a HLA class I tool, which has been developed to assess immunogenicity of epitopes to support the design of vaccines.

HLA receptors are highly polymorphic; hence, the assessment of the potential immunogenicity risk of a biotherapeutic drug in a given population requires the analysis of the binding specificities of each constituting peptide to a great number of HLA allotypes. Some allotypes have a higher prevalence in a given population than others. Therefore, knowledge of population frequencies for different allotypes is important for the selection of relevant HLA types and for the proper assessment of immunogenicity risk in a specific population. For the purpose of assessing the relative immunogenicity risk of a protein, the Epibase platform utilizes HLA class II allotype frequencies in eight major population groups and a global population frequency distribution. The relevant allotypes for a given population are selected based on their frequency of occurrence. The global allotype set provides 99% global population coverage.

In silico T-cell epitope profiling tools can be efficiently applied during the lead selection and optimization stages in three ways: (1) to rank protein leads based on their relative immunogenicity risk, (2) to identify specific peptides within a protein sequence with high immunogenicity risk, and (3) to guide protein reengineering by helping remove T-cell epitopes, a process known as deimmunization.

In order to rank protein leads as a function of their relative immunogenicity risk, several criteria need to be taken into account. These include the number of potential epitopes in a protein and the promiscuity of binding, among others. HLA-DRB1 allotypes would typically be one of the primary elements of focus for an immunogenicity profiling, as DRB1 receptors are expressed at much higher levels than other allotypes. Also, self-peptide filters may be employed to exclude the contribution of false positives from sequence motifs present in endogenous proteins; different filters can be applied, depending on the type of biotherapeutic being analyzed. Lonza's Epibase platform incorporates in a single score a number of different contributing factors to rank different biotherapeutic lead candidates according to their respective potential immunogenicity risk. Figure 10.3 shows the relative immunogenicity risk predicted for a selection of commercial therapeutic antibodies. As expected, the

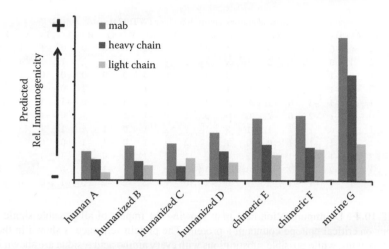

FIGURE 10.3 Predicted relative immunogenicity for a selection of marketed therapeutic antibodies.

predicted risk of immunogenicity is higher for murine and chimeric antibodies and smaller for human and humanized antibodies.

In cases where there is a high risk of immunogenicity, computational tools can help to identify potentially immunogenic peptides and guide the removal of T-cell epitopes by means of protein reengineering. *In silico* mapping of T-cell epitopes considerably reduces the work and time required by facilitating the selection of high-risk peptides in the molecule, rather than having to screen every single peptide making up the protein sequence. Furthermore, *in silico* tools can also be used to evaluate, in a very high-throughput manner, the relative impact of potential sequence modifications on the predicted immunogenicity. This facilitates the rapid selection of optimal modifications that can effectively reduce the immunogenicity risk of a biotherapeutic.

The Epibase platform uses the same principles to rank individual epitopes and whole proteins; the higher the score of a peptide, the higher is the risk of immunogenicity associated with this peptide. If high-risk epitopes need to be removed and protein engineering is required, a deimmunization heat map, as shown in Figure 10.4, provides an invaluable visual tool to enable quick identification of potentially suitable mutations to reduce the immunogenicity potential of a protein, in combination with structural modeling of the product.

10.4.4 *In Vitro* and *Ex Vivo* Cell-Based Assays

In vitro and *ex vivo* cell-based assay platforms have the advantage of being able to evaluate and characterize the immune response to a therapeutic protein in a fully human system, thus providing important information on the safety of the protein prior to first-in-human trials. Human *ex vivo* cell-based assay platforms have the additional advantage of being able to assess much more than just the potential T-cell epitope content of the primary amino acid sequence. These assays can also

Critical epitope count difference (WT: 124)

Pos Res	A	C	D	E	F	G	H	I	K	L	M	N	P	Q	R	S	T	V	W	Y
8 G	1	0	0	0	1	0	0	1	0	1	1	1	0	0	0	0	1	1	1	1
9 P	2	1	0	0	2	2	1	3	2	2	3	1	0	2	2	1	1	4	1	3
10 G	1	0	-1	-1	0	0	0	0	0	0	0	0	0	0	0	0	0	1	0	0
11 L	0	-1	-1	-1	2	0	0	1	1	0	1	1	-1	0	2	0	-1	1	1	2
12 V	-1	-3	-4	-3	0	-1	-3	0	-2	0	0	-2	-3	-2	0	-1	-1	0	-1	0
13 A	0	-1	-1	-2	0	-2	0	0	0	1	0	-1	-1	-2	1	0	-1	0	0	0
14 P	1	0	0	1	2	1	-1	2	1	2	2	1	0	1	1	1	1	2	2	2
15 S	1	-1	-1	-1	1	0	0	1	1	2	1	-1	0	-1	1	0	0	1	1	1
16 Q	2	1	-1	-1	1	1	2	1	2	1	1	1	0	2	1	1	2	1	1	1
17 S	2	0	-2	-2	2	0	0	3	1	3	3	2	1	-1	1	0	0	3	1	3
18 L	-1	-1	-1	-1	-1	-2	-2	0	-1	0	0	-2	-2	-2	0	-1	-2	0	-1	-1

FIGURE 10.4 Deimmunization heat map showing the impact of all possible single point mutations on critical epitope counts in a protein. The protein sequence is shown in the two leftmost columns, while possible substitutions with every amino acid residue are shown in the top row. Cells with negative values (green) correspond to mutations that reduce the protein epitope count and therefore reduce potential immunogenicity. Cells with positive values (red) correspond to mutations that increase the protein critical epitope count.

include the analysis of any conformational epitopes (e.g., B-cell epitopes), impurities (e.g., aggregates or particles), and contaminants (e.g., host cell protein, endotoxins) present in the protein sample.

The source and quality of the human primary cells used for the *ex vivo* assays are of critical importance. Obtaining highly viable and highly functional peripheral blood mononuclear cells (PBMCs) will ensure that optimal assay sensitivity and robustness are maintained at all times. In our experience, dedicated blood collection from human donors and immediate, optimized isolation and cryopreservation of PBMCs are essential in ensuring that cells produce consistent results and that the different *ex vivo* assay platforms in which the cells are used have the required level of robustness. The donors used in these *ex vivo* assays should be selected to match the intended target population (e.g., a global population that would closely represent a phase I clinical trial). In general, the PBMC fraction isolated from healthy donors is suitable for the assessment of both T-cell and B-cell responses to therapeutic proteins. However, it is also important to note that PBMCs can be sourced from patients suffering from a specific disease indication or with a given genetic or ethnic background that could be relevant for the therapeutic agent being developed. For example, PBMCs can be sourced from patients suffering from rheumatoid arthritis to assess their response to a therapeutic protein being developed to treat this condition, thus taking into account both the immune status and genetic background (i.e., HLA allotype makeup) of the intended patient population. The use of PBMCs taken from patients with the targeted disease indication may ultimately be more representative of the type of immune responses that could be observed in subsequent clinical trials.

A number of fully human *ex vivo* assay platforms are currently being used to assess immunogenicity risk, and some are discussed briefly below for completeness; however, *in vitro/ex vivo* assay platforms are discussed in more detail elsewhere in this book.

10.4.4.1 MHC-Associated Peptide Proteomics

While *in silico* tools are able to predict the potential T-cell epitopes that may be present in a protein sequence, they are currently unable to accurately predict the exact way in which the therapeutic protein will be processed within the endosome after uptake into APCs. The MHC-associated peptide proteomics (MAPPs) assay is used to identify naturally processed HLA binding peptide sequences from therapeutic proteins.[89, 92–95] During the MAPPs assay, CD14+ monocytes are isolated from PBMC samples and differentiated with GM-CSF and IL-4 into immature dendritic cells (iDCs). iDCs are then harvested and loaded with the therapeutic protein before being activated with a cytokine cocktail (typically including lipopolysaccharide (LPS) or tumor necrosis factor-alpha (TNFα)/IL-1β) to produce mature dendritic cells (mDCs). iDCs are very efficient at protein uptake, whereas mDCs are geared toward protein processing and presentation via MHC class II on the cell surface. The mDCs also express high levels of T-cell co-stimulatory molecules on the cell surface and are hence primed to interact with and activate CD4+ T-cells. These primed mDCs express large amounts of MHC class II molecules loaded with peptides from both endogenous cellular proteins and the therapeutic protein that the DCs were loaded with. These mDCs can then be harvested, lysed, and the MHC class II (typically DRB1) molecules purified by affinity chromatography. The bound peptide can subsequently be eluted from the complex and identified using mass spectrometry. The identified peptides can then be compared to the therapeutic protein sequence to identify the binding peptides that will be presented to CD4+ T-cells. The MAPPs assay more closely represents the HLA binding that will occur *in vivo* and will reduce the false positive rates compared to *in silico* predictions, but this assay does not offer the throughput and cost-effective benefits that *in silico* tools offer. It is also important to note that both the *in silico* and MAPPs tools only predict potential T-cell epitopes in the therapeutic protein, and subsequent assessment of whether these identified epitopes are able to bind to the T-cell receptor (TCR) and activate CD4+ T-cells is required.

10.4.4.2 T-Cell Activation Assays

A wide variety of biological outcomes for T-cell activation can be measured *in vitro*, including intracellular cytokine expression or cytokine secretion, as well as cell surface activation marker and proliferation.[96] Intracellular cytokines and T-cell proliferation can be assessed by flow cytometry, which has the advantage of being able to label the cells for the expression of other surface markers (e.g., characterizing T-cell phenotypes, including Th1, Th2, and Th17). The levels of cytokines secreted into the supernatant during an *in vitro* T-cell assay can be assessed to determine both the magnitude and quality of the immune response. Secreted cytokines can be assessed by multiple methods, including enzyme-linked immunosorbent assay (ELISA), Luminex, or ELISpot.[97]

The format of the *ex vivo* T-cell assay is also very important, and a number of product-related factors should be considered when selecting the most suitable assay. These product-related factors include the nature of the protein (e.g., peptides, antibodies, antibody fragments, novel protein scaffolds, fusion proteins, and recombinant proteins), mode of action of the protein (e.g., toxic or immunomodulatory proteins can interfere with some assays), and purity of the protein (e.g., some assay formats are more sensitive to endotoxins and aggregates). Often an optimization of the intended assay format is required to ensure that the most suitable assay format is being used for the therapeutic protein. Optimization parameters often include protein dose, kinetics of the assay, and interference in the assay (e.g., co-culture with a positive control to identify any inhibitory effects of the test protein).

PBMC-based *ex vivo* assays can be used to assess T-cell responses to short peptides (often used to identify specific T-cell epitopes in a protein sequence) or whole proteins (to provide an assessment of the immunogenicity risk of the intact protein). Typically, a screening project incorporates PBMC samples from at least 50 donors representative of the target population (often healthy donors, but can also be samples from diseased patients). Whole PBMC assays simply co-culture the PBMC with the test protein alongside suitable controls/reference proteins, and after 5–7 days assess the activation/proliferation of T-cells by flow cytometry or ELISpot. The sensitivity of whole PBMC assays can be increased by depleting the CD8+ T-cells, hence enriching the CD4+ T-cell population. Whole PBMC assays are often unsuitable for products that modulate the immune system (e.g., T-cell targeting therapeutics), with many of these proteins inhibiting T-cell proliferation and masking the impact of any intrinsic T-cell epitopes. For immunomodulatory proteins, dendritic cell–based assays are usually employed. In these assays, monocytes are isolated from PBMC and differentiated into dendritic cells in much the same way as for the MAPPs assay (see above section). The DCs are loaded with the therapeutic protein and activated, so that any potential T-cell epitopes in the protein sequence are expressed on the cell surface as HLA–peptide complexes. The activated DCs also express high levels of T-cell co-stimulatory molecules and are hence very potent activators of any T-cells that are able to recognize the HLA–peptide complex. These activated DCs are then co-cultured with autologous CD4+ T-cells, and T-cell activation/proliferation is monitored typically by flow cytometry. This assay is highly sensitive due to the potent antigen-presenting cells used.

10.4.4.3 Artificial Lymph Nodes

A number of platforms are being developed to represent more closely the human immune system *in vitro* and attempt to replicate the natural environment of the lymph node, including 3D structures and the ratios of antigen-presenting cells, T-cells, and B-cells. Some of these models also include the addition of endothelial cells and flow-through systems to promote assembly of lymph node-like structures. These artificial lymph node systems have already generated data on the immunogenicity of protein-based vaccines and could also be applied to predicting the immunogenicity of therapeutic proteins.[98]

10.4.4.4 B-Cell Assays

Although the detection of primary B-cell responses *in vitro* is highly challenging, detecting preexisting B-cell responses is possible. There is increasing concern about the prevalence of preexisting antibodies to many of the novel protein therapeutics that are currently being developed. Many novel protein scaffolds and small antibody fragments are being modified to extend the half-life of the molecules. One such half-life extension technology is PEGylation, and there are recent reports showing that up to 20% of the healthy general population has detectable preexisting antibodies to PEG groups. Some novel antibody scaffolds have also reported problems with pre-existing antibodies in the clinic,[99] leading to significant delays and increased costs associated with identifying B-cell epitopes and reengineering the molecule. *In vitro* B-cell assays could potentially be used as a screening tool during preclinical development to provide an assessment of the prevalence of preexisting antibodies to the therapeutic protein.

To assess the prevalence of preexisting B-cell responses to a therapeutic protein, PBMC samples from healthy individuals (or samples from patients with the disease indication) undergo polyclonal stimulation to differentiate the memory B-cells into plasma cells. These plasma cells can then be assessed by B-cell ELISpot to determine if they are producing antibodies that cross-react with the therapeutic protein being assessed.[100] Screening for preexisting B-cell responses could be applied during lead selection to select candidates with the lowest risk of clinical immunogenicity.

It is also known that adverse reactions to some biopharmaceuticals could be linked to preexisting antibodies (IgA, IgM, IgG, or IgE) that recognize non-human epitopes present in the product, such as non-human glycol epitopes. Interestingly, it has been proposed that at least some cases of hypersensitivity to biotherapeutics could be associated with the presence, before the start of the therapy, of IgE antibodies able to react with the product.[67] All this suggests that this type of assay could perhaps help avert hypersensitivity or anaphylactic reactions before entering the clinic.

10.4.5 AGGREGATION AND IMMUNOGENICITY

As discussed earlier in this chapter, aggregation is an important product quality attribute to consider during the developability assessment of a therapeutic protein to ensure that the protein is manufacturable. There is also a well-documented link between aggregation and immunogenicity, with aggregated forms of a protein inducing higher immunogenicity rates than the monomeric form of the protein.[29] Thus, many of the concepts discussed earlier in this chapter to reduce aggregation will also reduce the risk of immunogenicity. However, the majority of therapeutic proteins contain at least a low level of aggregates, and it is not currently known what type and amount of aggregation can pose a risk for increased immunogenicity.[101] There are some examples from the clinic where increased levels of aggregation are linked to immunogenicity. Eprex® is a human erythropoietin (EPO) that underwent a formulation change that was subsequently linked to increased antibody formation to the endogenous form of EPO. This increased immunogenicity was associated with pure red cell aplasia (PRCA). Studies have subsequently shown that this formulation change could encourage the formation of micelles to which EPO molecules associate and display multimeric epitopes.[102] Another example is IFNβ1a, where there are two products available for the treatment of multiple sclerosis (MS) in the clinic: Avonex® and Rebif®. Avonex seems to induce low levels of immunogenicity (~2%), whereas Rebif is highly immunogenic (~25%) in MS patients. In this particular case, the observed rates of immunogenicity can be linked to aggregate levels found in each of the products, with Rebif exhibiting higher levels of aggregation than Avonex.[103] There are also recent data to suggest that aggregation of monoclonal antibodies can lead to a significant change in the potential T-cell epitopes that are presented via HLA class II molecules on the surface of dendritic cells *in vitro*.[95]

The factors that drive both immunogenicity and aggregation are often located in the same region of the therapeutic protein (e.g., in and around the CDRs of monoclonal antibodies). It is important to consider the impact on multiple factors when selecting lead candidates or engineering molecules for lead optimization. Interestingly, some recent studies suggest that, at least in some biotherapeutics, there might be some co-localization of T-cell epitopes with aggregation-prone regions, the latter primarily influenced by the presence of a relatively high number of hydrophobic residues.[104, 105]

10.4.6 PRECLINICAL IMMUNOTOXICOLOGY

The current trend for developing therapeutic proteins targeting components of the immune system (e.g., immune cells or cytokines) also raises the risk of overstimulating the immune system and contributing to an immunotoxic response to the therapeutic protein. This was clearly seen during first-in-human trials for the anti-CD28 monoclonal antibody TGN1412, where a severe inflammatory response was induced, which included cytokine release syndrome (CRS) ("cytokine storm") and multiple organ failure. Subsequent studies have indicated that it was the CD28 agonist activity rather than any sample contamination or errors in the manufacturing, formulation, dilution, or administration of TGN1412 that led to the CRS response. In this particular case, preclinical studies both *in vitro* and *in vivo* failed to predict the induction of CRS, mainly due to suboptimal conditions using human PBMC *in vitro* and differences in the immune system between humans and primates *in vivo*.[106] More recently a number of new *in vitro/ex vivo* assays have currently been developed that seem to be able to detect CRS responses, and therefore have been proposed as a new tool to assess this type of risk during preclinical development of therapeutic proteins.[107]

10.4.7 IMMUNOGENICITY AND VACCINES

In the development of subunit vaccines, immunotherapies, and diagnostics, the initiation of an immune response is a wanted and required product attribute. The aim in these cases is to identify regions of viral proteins or human tumor antigens that could induce protective immunity against viruses or cancer cells by activation of cytotoxic T-lymphocytes (CTLs) and T-helper (T_h) cells. CTLs recognize peptides presented by HLA class I molecules, and T_h cells recognize peptides presented by HLA class II molecules. To this aim, computational tools can be used to identify potential CTL epitopes and T_h epitopes in a pathogen or target protein sequence by predicting peptide binding to both HLA class I and II molecules. The discovery of suitable immunogenic peptides may involve the screening of the whole pathogen proteome, which could make experimental studies unfeasible in terms of time and costs. In such cases, the only viable alternative is the application of *in silico* tools, which can screen large numbers of sequences in a short time span, to select peptides predicted as potential epitopes. Such epitopes can then be confirmed in cell-based *in vitro* studies. The Epibase platform can predict binding affinities to HLA class I allotypes, providing coverage for up to 91% of the global population. Similarly to the HLA class II tool, the Epibase class I platform also combines a number of contributing parameters (epitope binding affinities of epitopes, affected allotypes, etc.) into a single score, which can then be used to rank peptides based on their predicted relative immunogenicity and select those epitopes with the highest immunogenic potential.

10.5 DEVELOPABILITY AND DECISION-MAKING TOOLS

Developability assessment can be considered a truly integrative approach to quality by design (QbD) methodologies. As we have seen in previous sections, developability is not about listing every single type of problem that can negatively impact a given therapeutic candidate. Neither is it about measuring every single property that can define the identity of a given product or to characterize every potential impurity, its stability, and its evolution over time under every imaginable condition. Developability is about assessing and managing risk by identifying the most significant potential causes of failure for a given product based on its required properties and intended utilization in patients, and then implementing adequate control and corrective measures for those key risk parameters.

10.5.1 QUALITY RISK MANAGEMENT: DECISION MAKING UNDER UNCERTAINTY

Two of the key questions asked by many scientists when presented with developability methodologies is "What should we measure?" and "How can we combine multiple pieces of information into a single output or decision?" In this context, it is worth mentioning that mining large amounts of available information (also known as death by testing) can often hinder the implementation of a robust risk assessment decision-making process. This could be a clear case of data-rich/knowledge-poor syndrome. We earlier introduced a new paradigm for early developability risk assessment that leverages the use of computational methodologies and surrogate or proxy analytics to offer information on specific risk parameters on a relatively large scale. For this type of assessment, it is not important that the available information is extremely precise. However, it is essential that the required data are easily accessible (in terms of time, material, and cost), and that they provide adequate information about specific areas of concern. In other words, the data provide sufficient depth and context to enable a decision to be made.

Quality risk management requires a structured approach to risk management during development to facilitate decision making.[108] This can sound simple, but it is actually quite complex to implement, particularly when its application is in the context of a multidisciplinary and convoluted environment such as pharmaceutical drug development. We are not pretending we will solve such a conundrum in just a few pages; however, in the sections below we give some examples of the type of thinking and decision-making tools that can be used in the application of early developability assessment strategies. We hope this will spur additional discussions in the future and, hopefully, the development of initiatives aimed to provide guiding pathways toward simpler and more effective biopharmaceutical development.

10.5.2 UNDERSTANDING CAUSALITY

Even though developments in quantum mechanics and relativity usurped the presumed exactness of Newtonian mechanics in the early part of the twentieth century, for complicated elements we operate under a strong presumption of causality; that is, for every effect, there is a (or are) definite cause(s). As a result of thousands of successes in problem solving, designed experiments, and process optimizations, we categorically use causality-based structured problem-solving mechanisms. Any outcome (expected or unexpected) occurs as a result of the initial conditions that eventually lead to the outcome.

10.5.2.1 Understanding Causality in a Multifactorial Context

The search for causality in complex processes that involve many contributory factors introduces greater probabilities of factor interactions, and therefore much less ability to use anecdote, opinion, or heuristics to guess at an outcome. Even though these complicated processes may evade guesses at their outcomes, they cannot evade empirical science. This is the basis for the switch to design of experiments (DoE) over one factor at a time (OFAT). DoE allows for the elucidation of several simultaneous factor effects on a singular or multiple outcomes. Interactive factors will be captured as such, and insignificant factors in the models will be eliminated via the principle of parsimony.[109]

The tendency to make decisions biased toward personal heuristics (and away from evidence and data) is ubiquitous in the industry and our lives in general. Decision making should favor circumstances in which risk is reduced to practically achievable levels, and if it is possible to make several choices, decision quality is highest in those paths, with the lowest risk (assuming end results are identical in each path).

The optimal way we present here for decision making is a structured thinking process by which the solution derivation itself is done in a formal and repeatable way, rather than a trial-and-error process that ends up being significantly more costly and risky in the long run. From the combination of basal intelligence and experiential subject matter expertise we make decisions influencing multifactorial systems—and these heuristic decisions have virtually no repeatability to them, because each decision maker is doing so with his or her own preset biases. Because of this significant potential energy well to using proper risk management tools, individuals often default to their built-in systematic biases. The magnitude of problems this poses includes enormous financial, temporal, and labor-related costs. It is so important that we offer the following list of risk assessment and problem-solving tools in the sections below, along with the principles of their use.

10.5.2.2 Cause-and-Effect Models: Fishbone Diagram

So named for its morphology that resembles a fish skeleton, the fishbone diagram is ubiquitous in most robust problem-solving and quality programs. It represents one of the seven basic tools of quality, and is so foundational that quality improvement guru Kaoru Ishikawa, the inventor of the fishbone diagram for problem solving, said

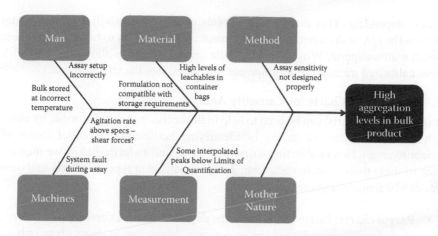

FIGURE 10.5 Fishbone diagram detailing the problem statement (high aggregation levels found in bulk product), as well as the major causal categories and subelements potentially contributing to the observed failure.

that the skillful use of seven quality control tools will resolve 95% of workplace problems.[110] In Figure 10.5, we show a typical fishbone diagram corresponding to the problem of high aggregation levels in bulk product. In this diagram, each rib of the fish represents a major category of potential causes, leading to the ultimate effect under investigation. With sufficient data gathered, probabilities can also be ascribed to the ribs of the fish to represent the relative likelihood of one causal factor versus another. In this manner, a problem (head of the fish) can be visually deconstructed into its causal factors to allow the highlighting of those needing further investigation and amelioration.

10.5.2.3 Five Whys

If disparate items can be captured on a fishbone diagram, the five whys is a recursive tool to delve deeply within one particular failure mode.[108] For example, five whys questioning may go as follows:

- **Problem:** pH of sample setup repeatedly not in range
- **Why?** Samples not prepared similarly from one analyst to the next
- **Why?** Analysts use different pipettors for the same preparation procedure
- **Why?** Standard operating procedure (SOP) doesn't specify one pipettor versus another, allowing measurement variation to creep in
- **Why?** Standard operating procedures not written with area subject matter experts

In this particular case, four whys were sufficient to derive specificity in causal factors, including or close to the root causes. The reference to five whys is a legacy from George Miller's "The Magical Number Seven, Plus or Minus Two"[111] and does not have to be taken literally; it just refers to not stopping after the first why. Sometimes it may take many more to determine the root causes, as it is entirely process dependent. This example also points to other potentially systemic issues (Why is the QA or documentation department allowing SOPs to be written without subject matter experts? Why is the lab or the calibration group allowed to bring different calibrated measurement equipment [pipettors] into the testing space?).

10.5.2.4 Use of Charts for Causality Analysis

Graphical tools (charts) can be used to help in the decision-making process, by either highlighting deviations in quality or identifying bottlenecks, potential causes of variability, main factors affecting a desired output, and so forth. Below we mention some of these tools as an introduction; details on different types of charts and their uses can be found elsewhere.[9,112]

- **Pareto charts:** Pareto charts show the degree to which a certain proportion of causes differentially lead to an outcome. Many Pareto charts show only the few factors that account for the preponderance of the effects in order to highlight those that are more important. Some call it the 80–20 rule.
- **Flowcharts:** Flowcharts depict how a process should run ideally. Flowcharts can be used to highlight deviations from an ideal process flow and fix them quickly.
- **Histograms:** Histograms are used to assess the distribution of observed frequencies for different features (bins). The shape, center, and spread of the distribution can help identify anomalies and support further investigation into potential causes or failure modes.
- **Run charts:** Run charts are linear plots showing the change of a single variable over time and are used to quickly identify deviations (preferably in real time) that could facilitate adoption of a corrective action.
- **Control charts:** Control charts are a variation of run charts that are commonly used in process control. Generally, control charts include a centerline (the mean) and upper and lower control limits. Control limits are calculated based on historical data and define acceptable levels for a given parameter or observable. They are often used to define when a process runs within or outside specifications and identify deviations that would require additional investigation.

10.5.3 MULTIPARAMETER RANKING METHODOLOGIES ENABLING DECISION MAKING

As we have mentioned earlier in this chapter, a successful biological drug needs to fulfill multiple criteria. Of course, it needs to show adequate biological activity and potency, but it also has to comply with other requirements, such as its suitability for being manufactured at acceptable yields and costs, or formulated for its intended route of administration. Furthermore, it should be sufficiently stable during storage and administration and safe for patients.

In this context, perhaps the most important task that drug developers have to face would be the selection and optimization of candidates to incorporate optimal properties that are ultimately responsible for their success as a therapeutic treatment. This requires the implementation of assessment- and decision-making tools that would enable ranking and optimization of therapeutic candidates according to multiple criteria.

In the field of small-molecule drug discovery, a number of computational tools and approaches have been developed to rank and optimize compounds based on multiple properties (see reviews by).[113–115] However, in the case of biological drug development, the use of similar tools to drive the decision-making process is still in its infancy or, in many instances, largely unheard of. Perhaps some of the reasons for such differences lie in the lack of robust computational predictive tools for many biological drug properties that could be relevant for their success as therapeutic agents, as well as the relatively low uptake of computational tools in biopharmaceutical development, compared to small molecules. For example, quantitative structural–activity relationship (QSAR) methods are standard practice in small-molecule drug discovery and development, whereas their application to biotherapeutics is, at best, limited.

There are a number of elements that can have an important impact on multiparameter ranking and optimization approaches:

- Relative importance/weight of different properties in the required outcome
- Desired data ranges (or thresholds) for the different properties to be considered
- Uncertainty or reliability of available data: confidence interval for model, assay accuracy, or prior knowledge
- Multiple values to describe specific properties, for example, *in silico* and *in vitro* data

Below we describe some common strategies used in multiparameter optimization and how these different elements can pose specific challenges affecting the quality of the decisions made.

10.5.3.1 Consecutive Filtering

One of the most common methods employed for decision making and candidate selection is a *consecutive filtering* approach. The idea behind this approach is that only one property is considered and optimized at a time. A suitable threshold for a property or a desired range of values is decided upon, and then lead molecules satisfying this criterion are progressed into subsequent stages. An example of such an approach is the early screening and selection of drug candidates based on their activity. A common problem with the consecutive filtering strategy is that the optimization of one specific property (e.g., binding affinity) may make other desirable properties (e.g., yield, immunogenicity, or aggregation) worse. As indicated before, there is a need for more advanced strategies that would allow multiparameter classification and optimization early in the drug discovery process. The use of *in silico* tools enabling profiling for multiple properties will help to resolve this problem.

Another factor contributing to disadvantages of the filtering approach is that often the uncertainty present in data is not considered while employing hard cutoffs.

Uncertainty is present in all types of data: laboratory measurements suffer from experimental error, while model accuracy will influence confidence intervals for *in silico* predictions.

10.5.3.2 Weighted Sums

Another common approach to multiparameter optimization is to use a weighted sum of ranking values corresponding to different properties to obtain a balanced or weighted ranking of lead candidates. For example, different developability properties could be combined in an objective developability ranking function for a given molecule n (D_n) that could take the following form:

$$D_n = -w^t * T_n + w^a * A_n + w^i * I_n \qquad (10.1)$$

where w^t, w^a, and w^i are weight factors reflecting the relative importance of the different developability properties, and T_n, A_n, and I_n are, respectively, the scaled values (measured or predicted) for the corresponding properties of titer, aggregation, and immunogenicity (standard deviations of property values across considered data set are often taken as scaling factors). Optimal molecules across these three properties would have the smallest rank; the negative weight for titer property reflects the fact that we would like to maximize a molecule's titer while reducing aggregation and immunogenicity.

This method requires *a priori* assignment of weights—relative importance for different properties. The ranking results also can be very sensitive to the values assigned to the weight factors. A slight change of weights may lead to different ranking order and a different optimal solution. Furthermore, the task of assigning relative importance or weights to different properties requires human intervention, and as such, it can be very subjective. Finally, this type of approach does not take into account uncertainties in data.

10.5.3.3 Pareto Optimization

The Pareto optimization method can help to explore multiparameter situations when optimizing one given property is likely to make a different one worse. This method can identify Pareto-optimal leads with best trade-offs across multiple properties. Figure 10.6 demonstrates the concept in two dimensions. Ideally, the aim would be to maximize desired values for both properties (i.e., the ideal molecule is in the top-right corner). The real situation, though, is that by increasing property 1, there is a decrease in property 2. The highlighted points on the Pareto front (solid circles) represent those conditions with superior combined properties, when compared to the rest of the conditions (empty circles). These Pareto-optimal points are better in at least one of the two properties analyzed, while they are no worse in the other property. This method usually provides a number of Pareto-optimal conditions from which the best condition can be selected by a development scientist.

The Pareto optimization method is most suitable where only a small number of properties need to be optimized. The number of possible points to consider during optimization increases dramatically with the number of properties. Of course, in the case where only a limited number of molecules needs to be assessed, this may not be a

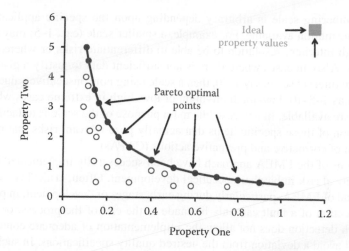

FIGURE 10.6 Example of Pareto optimization for two different properties. Pareto-optimal points (Pareto front) are shown as solid circles, and non-optimal points as empty circles. The green square denotes ideal property values.

problem. Also, Pareto optimization does not require any *a priori* assignment of importance weights to properties. Pareto optimization has been applied in drug development to guide the design and selection of small molecules with optimal absorption, distribution, metabolism, excretion, and toxicity (ADMET) properties.[116]

10.5.3.4 Incorporating Uncertainty in Decision Making: FMEA and Criticality

None of the methods for multiparameter ranking and optimization described above take into account the uncertainty present in all types of data. To address this, probabilistic scoring methods incorporating experimental and computational data and taking into account uncertainty in the data and the relative importance of individual properties have been developed. This type of approach has been used in small-molecule drug discovery to optimize ADMET drug profiles.[117]

Failure modes and effect analysis (FMEA) forms the basis of a good deal of the quantitative and semiquantitative risk assessment principles in use within QbD and quality risk management (QRM). FMEA measures, by way of multiplicative product, the assumed risk of a process's subelements by scoring each item by severity of risk (S), probability of occurrence of each risk element (O), and (lack of) detectability for each element (D). The product of these factors is often termed risk priority number, or RPN for simplicity. RPNs for different risk factors are then often rank-ordered from greatest overall risk to least overall risk in order to determine their priority order and define risk mitigation strategies accordingly. As a result, application of effort and resources can be made a priority to address those elements that pose the highest risks, whereas reduced efforts and attention are spent on those elements that present lower risks to the process or system.

$$RPN = S \times O \times D \qquad (10.2)$$

The numbering scale is arbitrary depending upon the specific application and level of information available. For example, a smaller scale (e.g., 1–5) may not provide enough interitem resolution to be able to differentiate risks or where to apply resolutions. Also, in cases where there is not sufficient data to justify a given linear consecutive interval scale, say 1–10, then a scale using nonconsecutive values can be used, such as 1–5–10, low–medium–high, or low–high in extreme cases when very little data are available. In any case, the main purpose of the scale is to facilitate the identification of those specific items that actually pose relevant risks that merit the application of corrective and preventive actions (CAPAs).

Variations of the FMEA approach have been successfully implemented in determining critical risk in biopharmaceutical development. Often, criticality ranking is used instead of FMEA, particularly during early stages of development, in processes where detection of a fault can only be made at the end of the process, or in cases where such detection does not allow the implementation of adequate control loops that would avoid a deviation from the desired quality specifications. In such events, risk ranking is often done without a detectability parameter.

In biopharmaceutical development, the application of QbD provides an example of how decision-making strategies can integrate various types of data and the uncertainty behind that data, particularly in the implementation of risk assessments. The A-Mab case exemplifies the use of two different risk assessment tools in biopharmaceutical process development for a humanized monoclonal antibody, A-Mab.[*][118] The tools provide an assessment of the criticality of several quality attributes (QAs), including aggregation, glycosylation, leached protein A, oxidation, and deamidation, among others, and assess the impact of such QAs on efficacy, PK/PD, immunogenicity, and safety. For the completion of the criticality assessment, the tools utilize various types of available information, including process experimental data, prior knowledge on similar products, or clinical data.

> **Tool 1: Risk ranking tool.** One of the tools defines criticality of an attribute in terms of its impact and uncertainty, representing relevance of information (e.g., literature data or clinical studies), which is used to assess the impact:

$$\text{Criticality (Risk Score)} = \text{Impact} \times \text{Uncertainty} \qquad (10.3)$$

> The impact reflects known or potential consequences on safety, pharmacology, and efficacy, whereas the uncertainty reflects the relevance of the information used to assign the impact ranking. The risk is broken into contributing factors, a score for an attribute is evaluated separately for each individual factor (immunogenicity, efficacy, etc.), and then the maximum score is taken as an overall criticality score for the attribute.
> **Tool 2: QA criticality ranking tool.** The second tool defines criticality in terms of severity and likelihood scores:

$$\text{Criticality (Risk Priority Number [RPN])} = \text{Severity} \times \text{Likelihood} \quad (10.4)$$

Severity is similar to impact but based mostly on prior knowledge and product-specific information, and it considers risks associated with patient safety, pharmacology, and efficacy. Likelihood estimates the probability that an adverse event caused by a quality attribute impacts safety or efficacy and is based on clinical, nonclinical, and relevant literature information.

The criticality scores resulting from the risk assessment are used to define critical quality attributes for the process development for A-Mab.

Although the A-Mab case study is not an approach for the ranking of lead candidates across multiple properties, some of the ideas of the case study on using uncertainties, the potential impacts of attributes, and the relative importance of different attributes could be extended to multiple-property ranking tools for lead selection, as we will see later.

10.5.4 APPROACHING DEVELOPABILITY FROM A MULTIPARAMETER RISK ANALYSIS PERSPECTIVE

Earlier in this chapter we illustrated how a developability assessment can be introduced at various stages of development by making use of different tools (computational and analytical). Below, we offer a brief illustration about how such risk assessment could be used and how mitigation strategies could be implemented. We want to emphasize that there is no such thing as a single developability risk management tool; rather, different approaches from those described above can be combined in a risk management process that should be adapted to specific product requirements or stage of development.

For example, control charts might highlight a relatively higher incidence of a specific chemical degradation event (i.e., deamidation) in a given part of the manufacturing process. This could trigger a red flag during an FMEA or criticality analysis if this event is susceptible to impact product efficacy or safety. Following the rationale described above, the probability of occurrence (O) of such a degradation event must be balanced by incorporating an assessment of its potential impact or severity (S) (i.e., manufacturing, biological, or clinical consequences derived from this event). As described above, the combination of the two can provide a measure of the criticality of such a particular event, and therefore suggest whether it is advisable to introduce a corrective course of action (mitigation strategy) for this event. This type of approach fits well with traditional QRM tools utilized in QbD strategies.[12]

$$\text{Criticality } (RPN) = \text{Probability of Occurrence } (O) \times \text{Severity } (S) \quad (10.5)$$

The example in Figure 10.7 shows how different parameters or events can be assessed based on their probability of occurrence and their relative impact and either used as part of a risk index for the entire molecule or assessed individually to define a mitigation plan (mitigation actions required).

Risk	Potential Effects	Severity (S)	Occurrence (O)	Criticality	Risk	Risk Mitigation Action Required
Deamidation V_H CDR3	Loss of activity, degradation, aggregation	10	5	50	High	• Re-engineer product • Additional tests to detect / mitigate excursions that could affect activity. • Etc.
Deamidation C_H2	Different charged species, potential degradation	1	5	5	Low	
Oxidation V_L FW2	Loss of activity, aggregation	5	5	25	Low	
Aggregation	Loss of activity, process yields, failing specs, safety / immunogenicity	10	10	100	High	• Re-engineer product • Screen higher No. of clones during Cell Line Development • Additional purification steps to remove aggregates • Etc.
...	...	...	...	...	...	...

FIGURE 10.7 Example of a criticality assessment for a monoclonal antibody. Different events are categorized and assessed based on their relative severity (*S*) and probability of occurrence (*O*). To simplify the assessment, both severity and occurrence are assigned the following risk parameters: 1 = low risk, 5 = intermediate risk, 10 = high risk. In this particular case, occurrence corresponds to relative propensities of occurrence as determined by *in silico* methods.

Alternatively, different risk parameters can be combined using weighted functions or Pareto optimization, to produce a more relevant assessment of a given risk. One possible approach could involve the modification of the weighted sum model proposed earlier by defining each single risk scoring factor as a function contributed by various other elements. For example, the immunogenicity score I_n defined in Equation 10.1 could be built as a function of multiple contributing factors, such as T-cell epitope content, aggregation propensity, and chemical degradation propensity.

10.5.5 Decision Making Can Be Product Specific

One aspect we want to emphasize in this section is that there is no such thing as one size fits all for a developability risk assessment. As we have discussed earlier, the intended use of the therapeutic agent, its target population, and even the available reimbursement regime can place specific restrictions and requirements on how the product is designed and manufactured. In this way, different quality attributes will be linked to specific risk profiles and risk management strategies. It is therefore important to understand, right at the outset, what exactly those requirements are in order to define a suitable risk management approach.

In the following sections, we will illustrate how different considerations could be integrated when defining a developability program for a specific product with a particular set of requirements defined by disease condition, target patient population, commercial requirements, desired clinical outcomes, platform utilized, and company strategy, among others.

10.6 DEVELOPABILITY APPLICATIONS TO BIOPHARMACEUTICALS: SOME EXAMPLES

10.6.1 WHEN AND HOW TO USE DEVELOPABILITY RISK ASSESSMENTS

As we have indicated above, developability methodologies do have a strong beneficial potential if they are applied from the early stages of discovery, ensuring that the right quality attributes are designed into the chosen candidates for development. In the case of monoclonal antibodies, selecting an optimal candidate could begin, for example, with developing a suitable display library that is enriched with favorable biophysical properties (i.e., with chemically instable sites and regions prone to trigger aggregation removed or with T-cell epitope content reduced) and that can be easily converted into a full antibody format.[119–121] Also, there are a number of novel antibody platforms making use of humanized rodents[122] or human B-cell libraries or even antibodies isolated directly from patients.

10.6.1.1 Lead Discovery

What the discovery sources above have in common is that they generate a lot of candidates and a lot of data. The most effective developability strategies to be used at this stage are computational methods that automatically categorize or flag high-risk candidates and help guide the selection of more favorable molecules at an early stage. For example, the antibody aggregation prediction tool described earlier, used in combination with the detection of potential chemical instabilities and Epibase immunogenicity (T-cell epitope-based) scores, is one potential way to integrate early on both physical and chemical stability with immunogenicity risks in a simple manner without having to perform any additional experiments. In fact, while the amount of *in vitro* and *in vivo* data continue to build up as the product candidates progress into development, most of the *in silico* data and predictions can already be available during the early lead discovery phase (see Figure 10.8 for a schematic representation). The two sources of data (*in silico*, *in vitro*) are complementary, but as the project moves further into preclinical development, computational information should be complemented or transitioned toward more experimentally verified data, facilitated by the lower throughput required (i.e., lower number of candidates that have made it to that point).

10.6.1.2 Lead Selection

Lead selection from, for example, up to 10 antibody candidates should be performed on both an experimental and a computational level. A developability-driven approach to lead selection is to begin by reviewing the sought QTPP/target profile and determining what the critical or limiting characteristics are. With a list of the key/critical characteristics at hand, a suitable set of evaluation techniques can then be selected. Indeed, as we discussed earlier in this chapter (Section 10.2), it is important to adapt the analytical methodologies utilized to the scope (information detail and throughput required), as well as resource and material availability at this stage. This is one of the main motivations for the use of computational methods, which can be of great help early on in discovery and development and why surrogate or proxy analytics would be required.

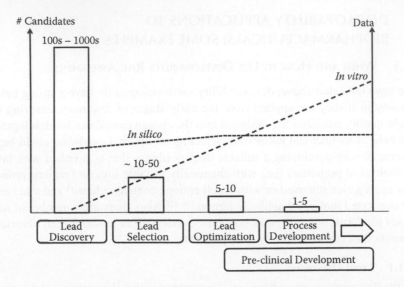

FIGURE 10.8 Implementation of developability methodologies in different stages of discovery and development and their relationship with the number of lead candidates and available data. As the number of potential candidates converges into a smaller number (could be a single lead), the amount of experimental data available increases steadily. *In silico* computational methods can, by comparison, yield a lot of information at an early stage. However, the amount of information does not grow to the same extent as with the introduction of *in vitro* analytics. Although more in-depth *in silico* analysis can also render further information at later stages, particularly by integrating experimental data, its primary benefit is realized during earlier stages of discovery/development, when limited or no experimental information is available. (Adapted from Zurdo J. et al. *Biomed Res Int,* Vol. 2015, Article ID 605 427, 2015. doi:10.1155/2015/605427)

Unless a protein structure has been determined or a structural homology model generated, the actual amount of computational information has not changed. However, with fewer candidates to compare, it is viable to go into more detail. At early screening stages, a number of factors, such as immunogenicity potential, can be estimated by a representative score (see Section 10.4). Similarly, details on potential chemical instability or degradation risks can be included in the risk assessment. Whereas in the lead discovery the presence or number of potential issues could be sufficient, at the lead selection/optimization stage, it is feasible to consider, for example, the type of risk, its location within the protein sequence–structure, its probability of occurrence, and of course, its potential impact on activity and safety as a measure of its criticality (see Section 10.5). As we show in the example in the previous section, the detection of potential chemical instabilities that impact CQAs in the final product is one of the most common uses of computational methods in the industry today. Unfortunately, their utilization is not yet as widely spread as one might like, and their scope remains still somewhat limited.

A wide range of analytical techniques are available for biophysical characterization of the candidates, and a list of common biophysical methods is given in Table 10.4. However, their suitability for small-scale/high-throughput analysis varies, as we have discussed in Section 10.3, and it is important to understand the balance between the level

TABLE 10.4

Common Biophysical Methods for Biophysical Characterization of Biopharmaceuticals

Differential scanning calorimetry (DSC)

Fluorescence spectroscopy

Absorption spectroscopy

Circular dichorism (CD)

Fourier transform infrared spectroscopy (FTIR)

Raman spectroscopy

Static light scattering (SLS)

Dynamic light scattering (DLS)

Analytical ultracentrifugation (AUC)

Reverse-phase HPLC

Bioanalyzer

LC/MS

Micro-HPLC

Rheometry

Size exclusion chromatography

Sodium dodecyl sulfate polyacrylamide gel electrophoresis (SDS-PAGE)/capillary electrophoresis (CE)

Field flow fractionation (FFF)

Surface plasmon resonance (SPR)

of information provided, on one hand, and the throughput, time, and sample requirements, on the other hand. For example, analytical ultracentrifugation (AUC) is an excellent tool to characterize protein–protein interactions and types and shapes of protein oligomers formed in solution, clearly information-rich. However, despite advances in recent years, it still remains a low-throughput methodology that requires substantial amounts of material and time of analysis. Therefore, AUC is not suitable for the selection of optimal candidates or the exploration of formulability or stability in a wide range of conditions. This is why it is important to select the right tool that is fit for purpose (see Section 10.3). In addition to biophysical characterization of lead candidates, a safety risk profile can be improved by the introduction of preclinical immunogenicity assessments as defined in Section 10.4. Finally, as discussed above, it is possible to evaluate the candidate's relative yields and suitability to be used on a given production system.

At this stage, candidates should ideally be compared using the final format and host cell type, even if standard buffers and solutions are used. If the final format has not yet been determined (e.g., both IgG1 and IgG4 isotypes are being considered suitable scaffolds), it is important to remember that candidates can perform better in one format than in the other, and this is not always predictable. While potential reengineering is discussed under the next stage, it is also available at this point in time, if needed, to address a given problem.

10.6.1.3   Lead Optimization and Potential Reengineering

A lead optimization stage is common in small-molecule development, but often overlooked in biologic development. Modifying the primary sequence of the lead candidate is, despite the title, not done lightly. However, given the jump in resources and regulation that initiating cell line construction and process development entails, we recommend conducting at least a risk assessment of the lead candidate based on computational methods before proceeding. The risk assessment should determine whether:

- Process development (PD) can proceed utilizing the platform processes and standard optimization
- A risk can be mitigated in PD by an enhanced QbD approach or control strategy to one or more areas/processes
- Further experimental quantification and validation of a risk is necessary
- Protein engineering should be attempted to remove or reduce a potential problem

Figure 10.9 shows a schematic workflow of this approach. Computational predictions should turn into experimentally verified information in this stage, and the use of computational methods transitions from screening and selection to risk assessment and protein engineering. Even if there is no scope for additional experiments, the computational risk assessment can highlight potential manufacturability problems and a mitigation plan or enhanced control strategy can be implemented. The platform process may already be informed by a QbD approach, in which case the risk assessment and a proactively managed control strategy is par for the course.

Two recent studies showed practical examples of developability methodologies to assess and mitigate manufacturability risks in monoclonal antibodies.[64, 65] In one of these studies, different mitigation strategies were devised to address the issues identified, involving the implementation of a process control strategy or, in one particular case, requiring protein reengineering.[64] These studies utilize analytical tools, rather than computational, as the main evaluation criteria. However, as we will see later on, one could envisage circumstances where the use of *in silico* methodologies could be useful in both the assessment of a given risk and the design of alternative variants with improved performance.

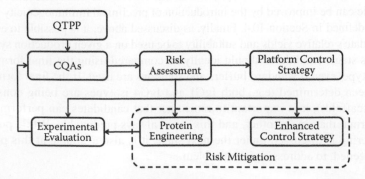

FIGURE 10.9 Preprocess development risk assessment and lead optimization. Note that platform control strategy refers to process development and manufacturing control and could in turn incorporate QbD approaches.

10.6.2 PRACTICAL CASES: SELECTING AND DESIGNING OPTIMAL LEAD CANDIDATES

The following sections provide two examples of the application of some of the developability tools described earlier to either select or engineer biopharmaceutical candidates with enhanced properties. We have chosen two different case studies to illustrate their implementation to assess or mitigate immunogenicity or aggregation risks. In both cases, we comment on the application of the respective risk assessment, risk mitigation, and validation steps and how both *in silico* and suitable surrogate or proxy *in vitro* methodologies can be combined in a unified workflow.

10.6.2.1 Case Study 1: Selecting Biopharmaceuticals with Reduced Immunogenicity Risk: Albumin Binding Domain

The Epibase *in silico* immunogenicity prediction platform has been used successfully in a number of biopharmaceuticals. These include examples such as the engineering of Albumod™,[123] the deimmunization of a bispecific IgG antibody,[124] and the performance of a risk assessment regarding the impact of mutations introduced during antibody engineering on immunogenicity.[60]

In this case, we describe the application of *in silico* predictive tools in selecting variants of a biopharmaceutical with reduced immunogenicity potential.[125] A known strategy for half-life extension of therapeutic proteins is to take advantage of the long circulatory half-life of human serum albumin (HSA) in plasma.[126, 127] The Albumod technology developed by Affibody is a proprietary albumin binding technology and is based on a small albumin binding domain (ABD). This domain consists of a 5 kDa protein that has been engineered to bind HSA with high affinity, and is designed to enhance the efficacy of biopharmaceuticals by extending their circulatory half-life in patients. The ABD domain was isolated from a bacterial protein, streptococcal protein G (SpG), which has the capacity to bind serum albumin. The wild-type protein ABD001 (albumin binding region of SpG strain 148, ABD3) had undergone affinity maturation, and one of the resulting engineered mutants, ABD035, demonstrated excellent stability along with an increased affinity for serum albumin of several species, including femtomolar affinity for human serum albumin.[128] Since it has been experimentally confirmed that ABD001 contains a known T-cell epitope,[129] the high-affinity variant ABD035 has been subjected to deimmunization via protein engineering. A number of variants were designed to remove/reduce the number of T- and B-cell epitopes while maintaining thermal stability, solubility, expression yield, and affinity to HSA. The protein engineering stages were guided by B- and T-cell epitope prediction programs and available literature on ABD, and included iterative rounds of protein expression and analytical characterization.[123]

The wild-type ABD001 and a total of 133 different engineered variants were subsequently screened for immunogenicity using the Epibase *in silico* platform, with profiling performed for the Caucasian population using 42 HLA class II allotypes. The variants were ranked based on their immunogenicity score incorporating DRB and DQ allotypes. Three variants, ABD088, ABD094, and ABD095, were then selected from the collection of variants based on their sequence, HSA affinity, thermal stability, solubility, and predicted lower immunogenicity risk. Figure 10.10a shows the predicted immunogenicity scores for these variants and their comparison to the parental ABD001. Deimmunized variants were predicted to have a reduction of approximately 40% in their immunogenicity risk compared to the parental molecule ABD001.

The wild-type ABD001 and three deimmunized variants, ABD088, ABD094, and ABD095-DOTA (DOTA—chelator for divalent metal ions), were further assessed *in vitro* for their ability to activate CD4+ T-cells. During the *in vitro* immunogenicity assessment, proliferation of CD4+ T-cells was used to monitor T-cell activation response induced by the ABD variants. CD4+ T-cell responses were assessed in PBMCs from 52 healthy donors representing the Caucasian population (frequencies based on HLA-DRB1 allotype distribution). Keyhole limpet hemocyanin (KLH) was used as a highly immunogenic benchmark protein and recombinant human albumin (rHSA) as a control reference. Data analysis included identifying the number of individual donors eliciting a significant CD4+ T-cell response to each ABD variant and

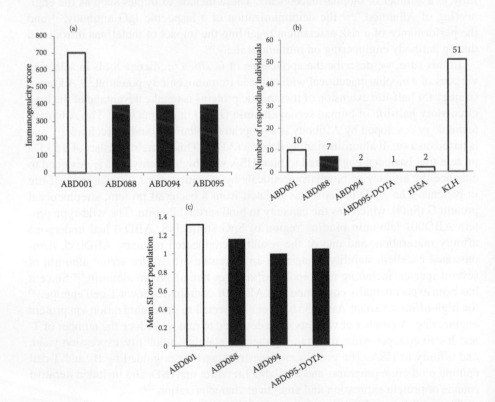

FIGURE 10.10 (a) Predicted immunogenicity scores for three ABD variants and parental sequence ABD001. (b) CD4+ T-cell proliferation responses to ABD variants in a cohort of 52 donors, highlighting the number of donors with proliferative responses to ABD variants compared to KLH and rHSA controls. (c) CD4+ T-cell proliferation responses to ABD variants in a cohort of 52 donors expressed as mean stimulation indices (SIs) over the population. rHSA is used as a reference (SI = 1). (Adapted from Zurdo, J., et al., *Biomed Res Int*, Vol. 2015, Article ID 605 427, 2015. doi:10.1155/2015/605427.)

a measure of the mean CD4+ T-cell response over the whole 52-donor population. Figure 10.10b shows the number of donors with statistically significant proliferative responses using a blank control as reference. When compared to a blank control, only 2 out of 52 individuals responded to ABD094 and ABD095-DOTA, versus 10 donors responding to the wild-type ABD001. On the other hand, a total of 51 donors out of 52 responded to the KLH positive control, and only 2 donors responded to rHSA. Figure 10.10c shows the mean stimulation index (SI) over the population using rHSA as a reference for the four ABD variants.

All three deimmunized variants show a reduction in T-cell proliferation in comparison with the wild-type ABD001. The mean population response for ABD001 and ABD088 is statistically different ($p < .05$) from that for rHSA (mean SIs of 1.31 and 1.15, respectively). The SIs for ABD094 and ABD095-DOTA variants were not significantly different over the test population (mean SIs of 0.99 and 1.04, respectively).

No significant *in vitro* CD4+ T-cell response was detected against the lead candidate ABD094, indicating that the removal of T-cell epitopes via engineering was successful in reducing the immunogenicity of the molecule. As a result of these studies, the candidate ABD094 was selected to progress into development and expected to enter clinical trials.

This project demonstrates the successful use of a combination of *in silico* predictions and *in vitro* immunogenicity assessment tools as suitable platforms to guide protein reengineering to remove T-cell epitopes and to enable lead selection based on the relative immunogenicity risk of different candidates.

10.6.2.2 Case Study 2: Engineering Antibodies with Reduced Aggregation Potential and Improved Manufacturability

Below we describe an example to validate the implementation of computational and protein engineering approaches in a simple workflow to address known manufacturability issues in monoclonal antibodies.[125] This project describes the generation of enhanced antibody variants of an existing parental anti-IFNγ humanized antibody[130] exhibiting significant aggregation problems under native conditions (post-protein A) as determined by gel permeation high-performance liquid chromatography (GP-HPLC).

Three-dimensional structural homology models were built for the Fv regions of the humanized anti-IFNγ antibody. Sequence and structural properties of the molecule were analyzed using Lonza's Aggresolve *in silico* platform to assess relative aggregation propensities and also to identify potential hotspots or weak regions that could justify stability and aggregation issues. Such analysis highlighted several potential aggregation hotspots on the humanized antibody when compared to reference sets of known behavior. Subsequently, a library of different sequence variants was generated targeting aggregation hotspots as well as potential structural liabilities or weak points that could influence the behavior of such hotspots. This library was refined by discarding unsuitable modifications that could have an impact on the stability and structural integrity of the molecule or potentially impact its biological activity, utilizing *in silico* inputs and available information on the parental molecule. Out of this second screening, a reduced number of variants was selected for further *in vitro* analysis.

Productivity and aggregation of the selected variants were evaluated using small-scale transient transfections of CHOK1SV cells in 96-well plates. Attending to their productivity and apparent aggregation, two variants were shortlisted for further evaluation. Variants were expressed in 200 ml shake flasks, and more detailed protein stability (aggregation) and activity studies were conducted.

The reengineered antibodies displayed significantly improved developability properties when compared to the parental humanized anti-IFNγ. GP-HPLC analysis showed an almost complete elimination of aggregation for the two selected candidates (Figure 10.11a). The same type of analysis after an accelerated stability study, in which the antibodies were incubated at 60°C for 2 h, also showed a significant reduction in aggregation, with the levels of monomer recovery increasing for both variants when compared to the parental molecule (Figure 10.11b). Furthermore, the observed expression yield of the reengineered variants increased almost threefold when compared to that of the parental molecule (Figure 10.11c), in line with earlier observations in our group, linking antibody stability and aggregation to productivity. Also, very importantly, the reengineered variants also retained or improved their biological activity when compared to that of the parental humanized sequence (Figure 10.11d).

There is a growing concern about the presence of subvisible particles in biopharmaceutical preparations because of their potential impact on immunogenicity risk.[131]

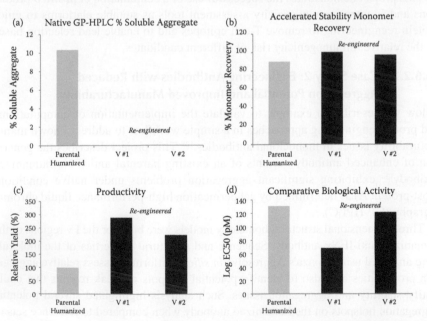

FIGURE 10.11 Reengineered antibodies display improved developability properties. (a) Aggregation is completely eliminated. (b) Monomer recovery after 2 h incubation at 60°C increases to almost 100% in the reengineered variants, suggesting improved stability. (c) Expression yield of reengineered variants increases dramatically compared to that of the parental sequence. (d) Biological activity is either retained or even slightly improved in the variant sequences. (Adapted from Zurdo, J., et al., *Biomed Res Int*, Vol. 2015, Article ID 605 427, 2015. doi:10.1155/2015/605427.)

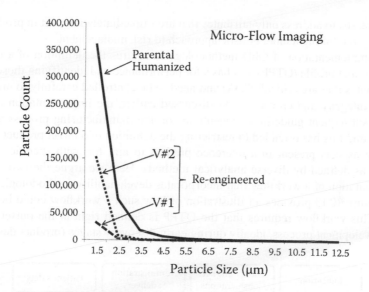

FIGURE 10.12 Subvisible particle distribution in the original humanized and reengineered antibody variants as defined by microflow imaging (MFI). V#1 and V#2 variants display a 5- to 10-fold reduction in the number of particles observed when compared to the humanized anti-IFNγ. (Adapted from Zurdo, J., et al., *Biomed Res Int*, Vol. 2015, Article ID 605 427, 2015. doi:10.1155/2015/605427.)

To address this, all variants were analyzed using microflow imaging (MFI). MFI is able to quantify a distribution of subvisible particles in a given protein solution based on their size. In our tests, the two reengineered antibody variants showed a significant reduction in particles across the spectrum when compared to the parental molecule (Figure 10.12).

These results highlight how applying the right computational and analytical tools during the initial stages of drug development can lead to a significant improvement in developability of a given candidate. It further shows how a short period of time invested in redesign work can significantly increase the productivity and stability of a product, thus decreasing quality and safety issues that could creep in during later stages of preclinical and clinical development.

10.7 IMPLEMENTING A DEVELOPABILITY WORKFLOW

Many battles have been lost by not fully understanding the tactical landscape, and equivalently, more prospective drug therapies have somehow or another failed in their pipeline because of failures to define and control endpoint factors (e.g., What is the disease target? What is the anticipated or desired modality of administration of the therapy? What are the demographics of the patient population?). Without understanding the impact of such characteristics, therapies can proceed all the way through to the end user and then not perform as expected for the patient populations. It is for this reason that we draw attention to the criticality of including these parameters into the intended performance profile (QTPP), and not limiting systems and processes to address only attributes that are immediately relevant in production or testing, but taking a more holistic approach to risk management.

The implementation of QbD methodologies requires the definition of a quality target product profile (QTPP) as a basis for performance and identifying those quality attributes that are critical (CQA) and need to be controlled carefully to maintain product integrity and efficacy.[132] As discussed earlier, the implementation of ICH Q8 and subsequent guidelines concentrates on the manufacturing process aspects of QbD, and this has often led to restricting the definition of a target product profile (TPP) to aspects present in a reference product to aim for, with specific characteristics as defined by diverse analytical methods. Here we argue the case for the implementation of a workflow that incorporates developability methodologies at its core. Figure 10.13 provides an illustration of how such a workflow could be structured. This workflow requires that the QTPP is defined right at the outset of the drug development process, ideally during early discovery stages (product design) in

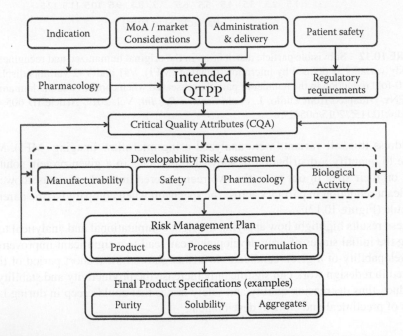

FIGURE 10.13 Developability flowchart. The flowchart illustrates how a development plan could be articulated to effectively integrate developability risk assessment tools. Setting an intended performance profile (QTPP) based on indication, pharmacology, mode of action, market, and delivery, among other considerations, allows the developer to determine the CQAs against which the lead candidate and process should be evaluated. A developability risk assessment would help identify specific risks impacting those CQAs and design and implement a risk mitigation plan. This might involve modification in the selection or design of a lead candidate, potential reengineering (product), designing of specific elements of the manufacturing process aimed to minimize or control risk, or perhaps some specific formulation requirements. All these steps will define the final product specifications in terms of measurable and controllable characteristics. (Adapted from Zurdo, J., et al., *Biomed Res Int*, Vol. 2015, Article ID 605 427, 2015. doi:10.1155/2015/605427.)

order to formulate in the highest possible detail the intended performance, safety, and economic target profile, which will ultimately determine the final product target to aim for during the drug development process. From this starting point, CQAs can be derived and a suitable developability risk assessment implemented to either derive optimal candidates matching the required CQAs or redesign lead candidates to adapt to the target profile. Indeed, during the design stage, these attributes should be mapped out and introduced or selected in the lead candidates, but of course, they should also be part of the design and optimization of the manufacturing process, so that such attributes can be properly controlled in an effective way. One might expect that the de-risking methodologies introduced early on during candidate design and selection will, in turn, increase the robustness of the manufacturing processes, making it easier to control specific CQAs and minimizing the incidence of deviations or out-of-specification (OOS) excursions.

The process of constructing the development plan revolves around a series of distinct phases that in turn address the following four questions:

1. What quality characteristics need to be incorporated into the drug substance and the drug product profile? (Define QTPP)
2. Which characteristics and attributes pose a risk and are considered areas of concern and why? (Define CQAs)
3. Which attributes are critical for the desired performance of the therapeutic molecule? (Developability assessment, criticality)
4. What mitigation plans need to be introduced and when, in discovery and development stages? (Remediation plan)

10.7.1 Starting with the End in Mind: QTPP and CQAs

To illustrate the whole process, we have envisioned a fictional biopharmaceutical product to help put some context around the type of top-level characteristics that should be considered to define a QTPP and the articulation of a suitable workflow. We would like to stress here that the definition of a QTPP is not a simple task. It requires the involvement of technical experts from multiple disciplines and areas of development (discovery, manufacturing, clinical development), supply chain, distribution, and so forth. Most importantly, it should also incorporate key input needs and requirements from end users, or what in QbD nomenclature is known as the voice of the customer. It is important to note that end users, or stakeholders, should be defined in a broad sense to ensure success (in concordance with the definition of "big Q" by Juran discussed above)[6] and should include patients, clinicians, payers, healthcare provision agencies, and so forth. The QTPP will ultimately define what the requirements for a given product are, so it can be considered to be fit for purpose. These can be split into three categories that align with the three pillars of developability presented earlier:

- **Fit for indication.** Suitable for required disease condition, dosing regime, patient population, route of administration, and desirable PK/PD. (Mode of action and pharmacology)

- **Fit for process.** Can be manufactured at the required scale using standard processes. It is sufficiently robust to endure process excursions without significantly impacting CQAs. It is stable enough to endure process and formulation requirements. (Manufacturability)
- **Fit for patient (safety).** Achieves desired therapeutic outcome without compromising patient safety. Does not introduce potentially dangerous side effects. (Safety)

10.7.1.1 An Imaginary Biopharmaceutical

To illustrate the process, we have selected an imaginary biopharmaceutical product and some fictional properties and requirements that would need to be fulfilled, so we can then propose a suitable development workflow incorporating developability methodologies. Let us consider Company A that develops innovative therapeutics against autoimmune conditions. The company has identified two novel drug targets (A and B) with potential for developing new treatments for refractory (i.e., do not respond to existing treatments) rheumatoid arthritis (RA) patients. A collection of potential binders has been selected for each of the two targets using a Fab fragment antibody display library platform.

Based on promising data obtained in animal studies, the company intends to select a suitable binding Fab fragment for each of the targets A and B and combine them into a bispecific antibody format, with a neutralizing/blocking mode of action. The company is exploring different multispecific platforms for the final design. An Fc-containing architecture is favored due to potential for extended half-life and manufacturing readiness using conventional antibody downstream platforms. The program is positioned initially as a treatment for refractory RA, but should also be able to remain a competitive alternative to existing treatments in terms of both affordability and patient compliance, facilitated by a penlike delivery device that could be used at home. The developing company has recently completed a series B funding round with a number of venture capital funds to complete proof-of-concept studies (phase 2 clinical trials) for two of its programs, including the RA program mentioned here, and it intends to partner one of them with a large pharma or biotech company. Investment has been staggered in a number of different payments and is milestone triggered.

In this particular context, a number of potential challenges are likely to impact the development plans proposed by Company A. These challenges can be fully explored by defining an adequate QTPP that will guide a development plan tailored for this product. Such a development plan will incorporate adequate risk mitigation elements to directly address, or at least ameliorate, the risks posed by these challenges highlighted by the QTPP.

10.7.1.2 Defining a Suitable QTPP

As we indicated earlier, the definition of a QTPP for a therapeutic product is not a trivial matter and requires input from multiple stakeholders and disciplines. Here, we just provide a list with a few characteristics that could be relevant to defining a QTPP for our fictional product. Its relevance is highlighted as potential areas of concern or requirements. Note that the requirements are not only related to manufacturing aspects, but primarily have to do with the end-user requirements and characteristics, such as the way the product is going to be utilized and the patient population targeted, among other things. In short, QTPP characteristics will relate to the desired indication, patient population, drug target and dosing regime, route and method of delivery, target indication and market, manufacturing platform, specific molecular format to be used, and inherent properties of the product. Table 10.5 compiles a list of characteristics and potential requirements that could help define a QTPP.

It is important that the QTPP arises from consideration of the whole life cycle of the drug, from design and manufacturing to distribution and patient administration, and even, very importantly, its potential utilization in additional indications in the future (e.g., pediatric Crohn's) that could incorporate very different specific requirements for the product. In this particular case, Company A's business model also

TABLE 10.5
Potential Characteristics That Could Be Used to Define a QTPP for the Product Discussed and Their Respective Potential Areas That Could Be Impacted, as well as Different Requirements to Meet QTPP

	QTPP Characteristics	Potential Areas of Concern/Requirements
1	Intended route of administration (RoA)	Intended patient self-administration. Subcutaneous administration might require special formulation. Important to be able to compete with existing products (cost of treatment needs to be kept as low as possible).
2	Dosage form/presentation to patient	Related to RoA and potential device requirements and patient population to be targeted. Formulation, supply chain, storage, distribution, and administration.
3	Bioavailability/half-life	Reduce dosing requirements. Enhance patient compliance and convenience. Potential reduction of cost of treatment.
4	Safety/immunogenicity	Chronic administration, subcutaneous RoA, highly immunoreactive patients. All contributing factors to high risk of immunogenicity. Potential differentiator with existing products in market.
5	Shelf life/stability	See RoA requirements. Stable product to meet patient self-administration needs. Adequate formulation/delivery device and shelf life important for patient safety and compliance.
6	Productivity/costs	Product needs to be able to compete in price with existing solutions.
7	Mode of action (safety)	Potential side effects due to interference with immune components (primarily agonistic).
8	Mode of action (format)	Potential (residual) effector function in molecule. Also potential influence of multispecific structure in MoA.
9	Delivery system/ container closure system	Potentially linked to RoA and stability, as well as patient requirements (ergonomics).
10	Patient population	Potential secondary pathologies (particularly in refractory patients). Specific care/attention requirements (link with administration). Link with device requirements (patient ergonomics/compliance). Potential pediatric/elderly requirements. Comorbidities.
11	Other	

brings additional constraints to be considered. Particularly in this case, special attention should also be paid to potential requirements from future envisioned partners (i.e., large pharma or biotech) in terms of format, manufacturing platform, quality and regulatory aspects, mode of action, and so forth. Indeed, this could potentially compromise commercialization attempts and future investment rounds.

10.7.1.3 Deriving CQAs

By defining an intended performance profile for our product that covers all the areas defined by the QTPP, relevant CQAs can be derived to design an adequate risk assessment and control management program during the entire drug development process. From a control or risk mitigation perspective, CQAs can be classified in three main categories:

- Controllable by product design (i.e., half-life, T-cell epitopes)
- Controllable by manufacturing process (i.e., glycoform distribution)
- Controllable by a combination of both product design and manufacturing process (i.e., stability, degradation)

Table 10.6 summarizes a list of basic CQAs and impacted QTPP areas, as well as potential control approaches. We used ICH Q6B and ICH Q8(R2) as a basis for defining CQAs, with some additional modifications. For example, note that we include immunogenicity as a CQA. CQAs are often restricted to those aspects or characteristics of a drug substance or drug product that can be assessed through specific analytical methods. Aggregation, stability, impurities, and even biological

TABLE 10.6
Defining Potential CQAs That Are Particularly Important to Meet QTPP Requirements (impacted QTTP areas and potential control approach are also indicated)

	CQAs of Concern	Impacted QTPP Area	Control Approach
1	Biological activity (design)	MoA (format)	Product design
2	Impurities (process)	Productivity, safety	Product design, process
3	Aggregation	Productivity, safety, RoA, shelf life	Product design, process
4	Viscosity	RoA	Product design, process
5	Half-life	MoA, dosage form, delivery system, costs	Product design
6	T-cell epitopes (immunogenicity)	Safety, MoA	Product design, process
7	Solubility	Productivity, safety, RoA, shelf life	Product design, process
8	Stability	Productivity, safety, RoA, shelf life	Product design, process
9	Glycosylation	Productivity, safety, MoA	Product design, process
10	Effector function	Safety, MoA	Product design, process
11	Yield	Productivity, dosage form, delivery system, costs	Product design, process
12	Other?		

activity (through suitable proxy assays for biological activity, such as binding to an immobilized target protein) can be easily measured and characterized. However, as we discussed earlier, safety concerns about a biopharmaceutical product, such as immunogenicity, are a less tangible aspect that traditionally could only be properly addressed in a clinical setting. We believe, however, that the emergence of new technologies and assays (such as those described in this chapter and elsewhere in this book) can now provide suitable platforms to be used in describing other product properties and their relative risks. Also, for the purpose of this early risk assessment, we have restricted the number of CQAs to areas that can be approached in late discovery/early development using the methodologies described earlier in this chapter and elsewhere in the book.

In our particular fictional example, the target indication (an autoimmune disease), dosing regime required (chronic treatment), and selected route of administration (subcutaneous) are all known contributing factors to immunogenicity, which could be further exacerbated by the potential immunomodulatory role of the product. The required pharmacological properties, such as half-life, need to be addressed early on, as they could impact the intended dosing regime, route of administration, formulation, and so forth. The solution of choice by Company A in this case is to use a molecule format that contains an Fc moiety, which would positively impact half-life, as well as facilitate downstream processing. This is a good example of how some CQAs can be easily addressed by designing them directly in the product. Furthermore, the route and format of administration will very likely put strong constraints on formulation, long-term stability, and storage of the product. A number of CQAs, such as solution viscosity, product solubility, stability (thermal, photosensitivity), and aggregation, will need to be properly addressed early on, ideally from the design stages, as they could impact a number of different QTPP areas.

From a manufacturing perspective, purity and yield are also important elements to be considered. These should meet expected market demand for the target indication and fit a reasonable manufacturing capacity. This means that productivity characteristics for the cell line could potentially be important, particularly if the product of choice (bispecific antibody) has signs of low productivity/process yield or incorporates unusually complex downstream process requirements.

10.7.2 Developability Assessment: Defining Areas of Risk and Mitigation Strategies

Following the workflow depicted in Figure 10.13, once CQAs have been defined, then a developability risk assessment comes into place. In Table 10.7, we show a few examples of potential areas of risk that can be approached by implementing the different developability methods described in this chapter. Furthermore, we have shown how relative risk (criticality) can be defined by assessing the probability of occurrence (*in silico* prediction score or *in vitro* assay) and the potential impact of a given risk. Here we have made an assumption of the potential impact that some of the risks described would have on the viability/success of the product, based on the QTPP requirements defined earlier on.

TABLE 10.7
Example of Potential Areas of Risk and CQAs Affected

	Risk	CQA	Assessment	Impact	Mitigation Plan
1	T-cell epitopes	Immunogenicity	*In silico/in vitro*	High	Select/design candidates with low T-cell epitope content
2	Aggregation hotspots	Stability, aggregation, solubility, immunogenicity	*In silico/in vitro*	High	Select/design candidates with low aggregation propensity
Design process with more stringent conditions to control aggregation (screen larger number of clones, additional purification steps, formulation)
3	Nondesired glycosylation	Immunogenicity, impurities	*In silico/in vitro*	Medium/high	Select/design candidates without undesired glycosylation
Perform a more thorough safety risk assessment in relevant *ex vivo* or *in vivo* model
Select alternative expression system
4	Productivity yield*	Purity, impurities	*In silico/in vitro*	Medium/high	Select/design candidates with higher productivity
Screen larger number of clones to favor better producers
Design culture and feed process to improve productivity
5	Viscosity	Viscosity	*in vitro*	High	Select candidates with low viscosity
Design suitable formulation
6	Chemical degradation	Purity, impurities, aggregation, immunogenicity	*In silico/in vitro*	Medium/high	Select candidates with lower incidence of modifications
Adapt manufacturing process or formulation to minimize their occurrence
7	Product isoforms/impurities	Purity, impurities, aggregation, immunogenicity, MoA	*In silico/in vitro*	Medium/high	
8	Other?				

Note: Required tools to assess severity (probability of occurrence) are indicated, as well as expected impact of such risk factors in the CQAs. Some potential mitigation plans to address the risks are also highlighted.

We have introduced productivity/yield as an important risk to be assessed given the lack of experience with this particular structure (multispecific antibodies). This is based on the understanding that multispecific constructs often present manufacturing difficulties of different sorts compared to standard antibody molecules.[133]

Given that the product is likely to be designed by combining two different specificities (from two separate Fab display libraries) into a single molecule, there is potential for addressing possible CQA concerns separately for each of the two constituting Fab fragments before combining them into a single final molecule. For example, the individual monospecific Fabs could be selected/engineered in terms of their behavior in several areas, such as aggregation potential; stability, solubility, and viscosity; potential PTMs (undesired glycosylation); T-cell epitopes; and relative productivity.

From the set of all identified CQAs, it is possible to narrow down the focus to those risks that are particularly relevant for the desired performance of the molecule. Using the strategies outlined in earlier sections of this chapter, it would be possible to address these potential risks by assessing their probability of occurrence with *in silico* predictive methodologies or suitable proxy analytics or *in vitro* tests. Criticality assessment (defined earlier) would combine such a probability of occurrence and its corresponding impact in the product QTPP. If a given property is considered critical, then adequate mitigation plans can be introduced.

For example, an *in silico* analysis could identify a potential occurrence of degradation in the binding region of one of the Fab fragments. This could imply loss of antigen binding as well as product instability, aggregation, and increased risk of immunogenicity; as such, this risk could be considered as having a high impact, and a resulting criticality assessment would recommend the need to implement a mitigation plan. One potential mitigation strategy could be to progress the molecule to a formulability assessment, in order to identify pH ranges, solution conditions, and excipients that could minimize its occurrence. Alternatively, the degradation risk could be potentially eliminated by protein engineering, hopefully without altering biological activity or other essential product characteristics. In some cases, a twin-track approach of protein engineering and formulability assessment could be advisable, particularly in cases where other required CQAs include high solubility or low viscosity (particularly at high concentrations).

As discussed in Section 10.5, a combined assessment could be implemented by defining relative weights for different risks, based on their relative impact in the desired QTPP. This could facilitate the assessment of many different candidates (i.e., Fab fragments identified by display methods) in a very short period of time and focus the selection of best possible candidates based on a combination of characteristics relevant to QTPP.

10.7.3 Designing a Suitable Development Process Workflow

Ultimately, there are a number of constraints that will determine how a developability workflow and development plan can be articulated. For example, elements such as number of potential candidates, material availability, analytical platforms utilized, and so forth, can change dramatically when and how different risks are assessed. In an ideal world, having relevant information on all relevant CQAs from the start would make it possible to find an optimal combination of different properties to select/design a lead candidate to move forward in development. That day has not yet arrived, but in this chapter we propose a number of methodologies that can be utilized to widen the scope of selection criteria for biotherapeutics.

In any case, it is important to define a development plan that incorporates adequate risk mitigation elements. This would include:

- Define clearly all essential quality attributes that will need to be incorporated in the product as part of the QTPP. As discussed earlier, this should be as soon as a product concept is created to maximize probability of success.
- Decide which departments would be involved in the implementation of different risk assessment and mitigation strategies. It is likely this will require a multidisciplinary approach across discovery and development functions that will need to be properly defined.
- Agree on a risk mitigation approach, including developability tools to be implemented and where and when they can be introduced, as well as risk impact assessment and decision-making methods to be utilized.
- Agree on the type of risk mitigation approaches that are acceptable or adequate for the product and its stage of development. For example, engineering strategies could be desirable in some cases, whereas early process design and optimization might be advisable in others.
- Define a road map to establish where in the development process some of the CQAs will need to be addressed and how this should be done, making sufficient allowances in terms of time, resources, investment, and additional development or enabling technologies that could be required.

10.8 CONCLUSIONS AND FUTURE PROSPECTS

There are voices questioning whether the traditional model to develop new drugs is adequate to meet the evolving needs of healthcare provision and patients and more stringent regulatory requirements. In fact, when compared to other industries, pharmaceutical drug development is a particularly long, costly, and highly inefficient process with potentially serious financial and societal consequences.[134] The development of new drugs follows a highly rigid and hierarchical process subdivided according to specific disciplines or development stages. As a result, this structure puts a lot of stress on solving problems ad hoc when they arise (quite often), rather than on preempting them or introducing corrective measures earlier on. This lack of integration between different areas of development is particularly evident during the early stages of drug design and development, where functions usually operate in complete isolation from one another. These structures separated into functional silos are in fact a significant obstacle to the introduction of new practices in the industry, such as the implementation of translational medicine, quality by design, or portfolio life cycle management.[10, 135, 136]

Developability is ultimately a structured approach to introduce a broader, more holistic interpretation of quality into biopharmaceutical development. As such, it emerges as an important tool to expand on the current, still-limited implementation of QbD in pharmaceutical development and biotherapeutic drug development in particular.

The existing ICH guidelines concerning the application of QbD to drug development (Q8–Q11) deal primarily with providing a framework for process understanding and characterization. However, we believe that a true QbD starts with the design of the product itself. In this chapter we try to show how, by defining a meaningful quality target product profile (QTPP) very early on, in the inception of drug development, meaningful CQAs and effective risk management strategies can easily be deployed. Only by having such clear sets of design requirements at the beginning can one embark upon the development of an appropriate manufacturing process more efficiently and with a higher probability of success.

This is particularly important in the case of biopharmaceutical drug development, where many important properties are not properly controlled at the design stages and are left to be determined during the manufacturing process itself. Indeed, the manufacturing process can be used, ideally, to control specific quality attributes in the product, but this considerably increases risks and the potential for failure that sooner or later might need to be addressed, in many cases at a high cost. In this chapter, we suggest a potential workflow that moves away from the classical linear–hierarchical development model into one that is more integrated and where adequate early risk assessment tools can help control CQAs at a very early stage. In this workflow there is a change in emphasis, defining QTPP right at the outset, and with a larger number of criteria that will ultimately define the success of a given product (Mode of Action [MoA], target patient population, delivery requirements, etc.). Second, it involves the introduction of additional de-risking tools that increase the stringency of candidate selection, in order to meet the required QTPP and properly control CQAs in the product from the beginning of development.

Furthermore, we believe that the introduction of such early risk assessment paradigms can not only be financially beneficial and reduce development costs, but also even accelerate the development of new product candidates, for example, by speeding up their transition from preclinical to clinical development, and ultimately provide patients with more effective and affordable medicines.

10.8.1 INDUSTRY TRENDS AND GAPS

Over the last years there has been a steady increase in scientific knowledge about biotherapeutics, their mechanisms of action, as well as a substantial improvement in manufacturing processes, which are experiencing a new resurgence with the application of "omics" (genomics, transcriptomics, metabolomics) and new gene editing technologies to tailor the properties of hosts to specific process and product requirements. These include the generation of auxotrophic knockouts, to favor stringent selection conditions and higher productivities, or the manipulation of the glycosylation patterns presented by the product.[137] Equally, better understanding of metabolic requirements and new media and feeds have multiplied productivities to levels unimaginable 20 years ago.[76]

Given this background, there is a growing need to extend the criteria utilized in the design and selection of biopharmaceutical products to avoid, whenever possible, potential problems that would jeopardize the viability of a given drug discovery program. As we have discussed in this chapter, the increasing use of computational tools to model product performance is potentially a powerful approach to integrate product and process knowledge aspects of QbD into a seamless structure in which discovery and development are seen as integral and interconnected parts to achieve a successful therapeutic product. Still, their implementation is in its infancy. New and better computational tools are needed, particularly around modeling the interaction between product and process, to allow a better understanding and design of suitable manufacturing processes. Furthermore, there is a need to increase validation efforts for these types of computational methods, in order to facilitate their acceptance and implementation by the industry as a whole. New approaches are also needed for early analytical assessment of product quality attributes, allowing higher through-puts, shorter timelines, and requiring substantially lower amounts of product. For example, early formulability assessment is becoming an area of increased interest, but its success will be tightly linked to the development of adapted analytical and computational methods that will facilitate its early application, perhaps even before a final candidate has been selected for manufacturing development.

10.8.2 OBSTACLES TO DEVELOPABILITY

There are a number of obstacles to implement such an early risk management vision and realize the true potential promised by developability methodologies. One is per-haps that, to many, this concept is perhaps too fuzzy to be implemented in the preva-lent drug development structure in the industry. As we have mentioned earlier, the fragmentation of the development process is a key obstacle. Particularly important in this context is the massive gap that still exists between discovery and develop-ment functions for biopharmaceutical drugs. There are historical and technological reasons for this split, but new technical and methodological advances described in this chapter and elsewhere in this book are bringing these two separate universes together in ways that were unimaginable only a few years ago, creating new inter-faces where they could interact and address common problems. Another important obstacle is an endemic reticence to adopt innovation, sometimes justified on the basis of regulatory concerns, costs, or inability to fit to existing processes or infrastruc-ture. This is obvious in biopharmaceutical manufacturing, where small changes can have devastating effects in product quality attributes, with the concomitant risks that could potentially come into play ranging from failed batches to product recalls. Indeed, there is a disproportionate pressure on biopharmaceutical manufacturing to deliver the required quality for a given product, and this is probably due to the fact that during design stages, there is little awareness of the difficulties and complica-tions that could emerge later on in development.

Here, developability emerges as a bridge, raising CQA awareness very early on. But it also can help integrate discovery and development activities, as well as align their objectives toward developing better and safer products (Figure 10.14). Fortunately, as reflected in this book, there are some successful examples in the industry of a growing level of integration between these two functions.

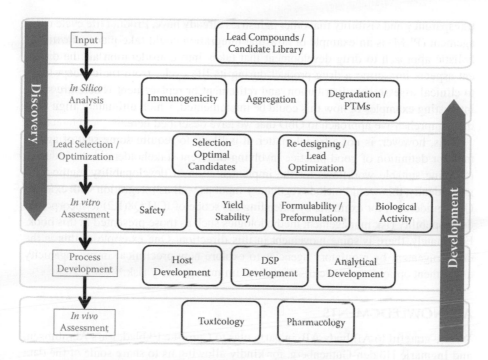

FIGURE 10.14 Developability as a bridging tool. This schematic shows how overlapping layers of assessment integrate in between classical areas of responsibilities, bringing a number of benefits: (1) facilitating the interaction across functions and the transition of candidates between discovery and development stages, (2) facilitating early and inexpensive elimination of problematic candidates, and (3) helping design required characteristics that would reduce failure later on in development, whether manufacturing or clinical. Activities highlighted (boxed) correspond to standard stages in traditional biopharmaceutical development. (Adapted from Zurdo, J., *Pharm Bioprocess* 1: 29–50, 2013.)

10.8.3 Looking at the Future: Developability, QbD, and Product Life Cycle Management

This integration between design and development will require an increase in the utilization of more sophisticated computational models and the expansion of our knowledge about the interplay between product and process and even product and patient. This would enable a better understanding of host biology and how to control it and design it to meet specific manufacturing and product requirements. Also, the ability to model the behavior of a product in its biological context could help identify potential problems that today can only be discovered in the clinic. In other areas, modeling protein–protein intermolecular interactions as well as protein–interface interactions will open the door to more advanced and refined tools to select and design more effective manufacturing processes and formulations.

Most importantly, the development of robust decision-making tools capable of integrating both scientific input and statistical data analysis will streamline the progression of candidates across different development stages, and provide a level of transparency and visibility that other industries already have. Product life cycle management (PLM) is an example of how that integration could take place, allowing a holistic approach to drug development that takes into consideration all the different aspects impacting a drug product during its life cycle, from the concept stage to clinical assessment, distribution, and retirement or replacement. There are some interesting examples of how this could be implemented, which ultimately align with the comprehensive approach to QbD that we have described here.[136]

This, however, is not a trivial matter. It is likely to require some sort of agreement or definition of good practice involving different stakeholders in the industry to define suitable workflows for the implementation of developability methodologies. Ideally, this would include regulatory input as well as integration with existing QbD guidelines, for example, by extending the scope of ICH Q8(R2) to incorporate developability risk management methodologies such as those presented in this book. Fortunately, there is some movement in this direction. One example of this is the encouragement by regulatory agencies to explore new preclinical immunogenicity assessment options as predictors of clinical immunogenicity risk.[87–89]

ACKNOWLEDGMENTS

We are grateful to Affibody AB, in particular to Caroline Ekblad, Lars Abrahmsén, and Ingmarie Höidén-Guthenberg, for kindly allowing us to share some of the data contained in this chapter and providing valuable comments. We would also like to thank Rebecca Michael for her involvement in some of the work reflected in this chapter and Yvette Stallwood, Rajesh Beri, Anne Moschella, and Hilary Metcalfe, as well as many customers and collaborators, for their very valuable input in discussions on the topics included here.

REFERENCES

1. Hay M, Thomas DW, Craighead JL, Economides C, Rosenthal J. (2014). Clinical development success rates for investigational drugs. *Nat Biotechnol* 32: 40–51.
2. Kola I, Landis J. (2004). Can the pharmaceutical industry reduce attrition rates? *Nat Rev Drug Discov* 3: 711–715.
3. Arrowsmith J, Miller P. (2013). Trial watch: Phase II and phase III attrition rates 2011–2012. *Nat Rev Drug Discov* 12: 569.
4. Sacks LV, Shamsuddin HH, Yasinskaya YI, Bouri K, Lanthier ML, Sherman RE. (2014). Scientific and regulatory reasons for delay and denial of FDA approval of initial applications for new drugs, 2000–2012. *JAMA* 311: 378–384.
5. Tversky A, Kahneman D. (1974). Judgment under uncertainty: Heuristics and biases. *Science* 185: 1124–1131.
6. Juran JM, Godfrey AB. (1999). *Quality Handbook*. New York: McGraw-Hill.
7. Schiffauerova A, Thomson V. (2006). A review of research on cost of quality models and best practices. *Int J Qual Reliab Manag* 23: 647–669.
8. Kozlowski S, Swann P. (2006). Current and future issues in the manufacturing and development of monoclonal antibodies. *Adv Drug Deliv Rev* 58: 707–722.
9. Defeo J, Juran JM. (2010). *Quality Handbook: The Complete Guide to Performance Excellence*. 6th ed. New York: McGraw Hill Professional.

10. Zurdo J. (2013). Developability assessment as an early de-risking tool for biopharmaceutical development. *Pharm Bioprocess* 1: 29–50.
11. Zurdo J. (2013). Developability. Surviving the valley of death. *EBR* 195: 50–54.
12. Rathore AS, Mhatre R. (2009). *Quality by Design for Biopharmaceuticals: Principles and Case Studies.* Hoboken, NJ: John Wiley & Sons.
13. Zurdo J, Michael R, Stallwood Y, Hedman K, Aastrup T. (2011). Improving the developability of biopharmaceuticals. *Innov Pharm Technol* 37: 34–40.
14. Mazza CB, Sukumar N, Breneman CM, Cramer SM. (2001). Prediction of protein retention in ion-exchange systems using molecular descriptors obtained from crystal structure. *Anal Chem* 73: 5457–5461.
15. Dismer F, Hubbuch J. (2010). 3D structure-based protein retention prediction for ion-exchange chromatography. *J Chromatogr A* 1217: 1343–1353.
16. Satish H, Angell N, Lowe D, Shah A, Bishop S. (2013). Application of biophysics to early developability assessment of therapeutic candidates and its application to enhance developability properties. In Narhi LO, ed., *Biophysics for Therapeutic Protein Development.* New York: Springer, pp. 127–146.
17. Bermudez O, Forciniti D. (2004). Aggregation and denaturation of antibodies: A capillary electrophoresis, dynamic light scattering, and aqueous two-phase partitioning study. *J Chromatogr B Analyt Technol Biomed Life Sci* 807: 17–24.
18. Mahler HC, Friess W, Grauschopf U, Kiese S. (2009). Protein aggregation: Pathways, induction factors and analysis. *J Pharm Sci* 98: 2909–2934.
19. Johnson DH, Parupudi A, Wilson WW, DeLucas LJ. (2009). High-throughput self-interaction chromatography: Applications in protein formulation prediction. *Pharm Res* 26: 296–305.
20. He F, Becker GW, Litowski JR, Narhi LO, Brems DN, Razinkov VI. (2010). High-throughput dynamic light scattering method for measuring viscosity of concentrated protein solutions. *Anal Biochem* 399: 141–143.
21. Connolly BD, Petry C, Yadav S, Demeule B, Ciaccio N, Moore JM, Shire SJ, Gokarn YR. (2012). Weak interactions govern the viscosity of concentrated antibody solutions: High-throughput analysis using the diffusion interaction parameter. *Biophys J* 103: 69–78.
22. Neergaard MS, Kalonia DS, Parshad H, Nielsen AD, Moller EH, van de Weert M. (2013). Viscosity of high concentration protein formulations of monoclonal antibodies of the IgG1 and IgG4 subclass: Prediction of viscosity through protein-protein interaction measurements. *Eur J Pharm Sci* 49: 400–410.
23. Wagner M, Reiche K, Blume A, Garidel P. (2013). Viscosity measurements of antibody solutions by photon correlation spectroscopy: An indirect approach—limitations and applicability for high-concentration liquid protein solutions. *Pharm Dev Technol* 18: 963–970.
24. Meek-Brown G, Qureshi R, Myers R, Patel B, Berrt P, Rawal S, Aldana S, Barredo M, Higgins L, Alam I. (2013). Development of a high throughput platform to assess product quality in mammalian cells producing custom and biosimilar monoclonal antibodies. Presented at International Bioanalytical Congress, Berlin, September 12.
25. Cromwell ME, Hilario E, Jacobson F. (2006). Protein aggregation and bioprocessing. *AAPS J* 8: E572–E579.
26. Vazquez-Rey M, Lang DA. (2011). Aggregates in monoclonal antibody manufacturing processes. *Biotechnol Bioeng* 108: 1494–1508.
27. Sniegowski M, Mandava N, Kahook MY. (2010). Sustained intraocular pressure elevation after intravitreal injection of bevacizumab and ranibizumab associated with trabeculitis. *Open Ophthalmol J* 4: 28–29.
28. Chirino AJ, Ary ML, Marshall SA. (2004). Minimizing the immunogenicity of protein therapeutics. *Drug Discov Today* 9: 82–90.

29. Rosenberg AS. (2006). Effects of protein aggregates: An immunologic perspective. *AAPS J* 8: E501–E507.

30. Schellekens H, Jiskoot W. (2006). Erythropoietin-associated PRCA: Still an unsolved mystery. *J Immunotoxicol* 3: 123–130.

31. Seidl A, Hainzl O, Richter M, Fischer R, Bohm S, Deutel B, Hartinger M, Windisch J, Casadevall N, London GM, Macdougall I. (2012). Tungsten-induced denaturation and aggregation of epoetin alfa during primary packaging as a cause of immunogenicity. *Pharm Res* 29: 1454–1467.

32. MacDougall IC, Roger SD, de Francisco A, Goldsmith DJ, Schellekens H, Ebbers H, Jelkmann W, London G, Casadevall N, Horl WH, Kemeny DM, Pollock C. (2012). Antibody-mediated pure red cell aplasia in chronic kidney disease patients receiving erythropoiesis-stimulating agents: New insights. *Kidney Int* 81: 727–732.

33. Hermeling S, Crommelin DJ, Schellekens H, Jiskoot W. (2004). Structure-immunogenicity relationships of therapeutic proteins. *Pharm Res* 21: 897–903.

34. Chung CH, Mirakhur B, Chan E, Le QT, Berlin J, Morse M, Murphy BA, Satinover SM, Hosen J, Mauro D, Slebos RJ, Zhou Q, Gold D, Hatley T, Hicklin DJ, Platts-Mills TA. (2008). Cetuximab-induced anaphylaxis and IgE specific for galactose-alpha-1,3-galactose. *N Engl J Med* 358: 1109–1117.

35. Philo JS, Arakawa T. (2009). Mechanisms of protein aggregation. *Curr Pharm Biotechnol* 10: 348–351.

36. Gsponer J, Vendruscolo M. (2006). Theoretical approaches to protein aggregation. *Protein Pept Lett* 13: 287–293.

37. Caflisch A. (2006). Computational models for the prediction of polypeptide aggregation propensity. *Curr Opin Chem Biol* 10: 437–444.

38. Kumar S, Wang X, Singh SK. (2010). Identification and impact of aggregation-prone regions in proteins and therapeutic monoclonal antibodies. In Wang W, Roberts CJ, eds., *Aggregation of Therapeutic Proteins*. Hoboken, NJ: John Wiley & Sons, pp. 103–118.

39. Buck PM, Kumar S, Wang X, Agrawal NJ, Trout BL, Singh SK. (2012). Computational methods to predict therapeutic protein aggregation. *Methods Mol Biol* 899: 425–451.

40. Chiti F, Stefani M, Taddei N, Ramponi G, Dobson CM. (2003). Rationalization of the effects of mutations on peptide and protein aggregation rates. *Nature* 424: 805–808.

41. Dubay KF, Pawar AP, Chiti F, Zurdo J, Dobson CM, Vendruscolo M. (2004). Prediction of the absolute aggregation rates of amyloidogenic polypeptide chains. *J Mol Biol* 341: 1317–1326.

42. Tartaglia GG, Pawar AP, Campioni S, Dobson CM, Chiti F, Vendruscolo M. (2008). Prediction of aggregation-prone regions in structured proteins. *J Mol Biol* 380: 425–436.

43. Fowler SB, Poon S, Muff R, Chiti F, Dobson CM, Zurdo J. (2005). Rational design of aggregation-resistant bioactive peptides: Reengineering human calcitonin. *Proc Natl Acad Sci USA* 102: 10105–10110.

44. Michael R, Stallwood Y, Jimenez JL, Demir M, Arnell A, Zurdo J. (2008). Protein engineering to reduce antibody aggregation. CHI Antibody Europe. Lisbon, Portugal.

45. Chennamsetty N, Voynov V, Kayser V, Helk B, Trout BL. (2009). Design of therapeutic proteins with enhanced stability. *Proc Natl Acad Sci USA* 106: 11937–11942.

46. Lauer TM, Agrawal NJ, Chennamsetty N, Egodage K, Helk B, Trout BL. (2012). Developability index: A rapid *in silico* tool for the screening of antibody aggregation propensity. *J Pharm Sci* 101: 102–115.

47. Perchiacca JM, Ladiwala AR, Bhattacharya M, Tessier PM. (2012). Aggregation-resistant domain antibodies engineered with charged mutations near the edges of the complementarity-determining regions. *Protein Eng Des Sel* 25: 591–601.

48. Chaudhri A, Zarraga IE, Kamerzell TJ, Brandt JP, Patapoff TW, Shire SJ, Voth GA. (2012). Coarse-grained modeling of the self-association of therapeutic monoclonal antibodies. *J Phys Chem B* 116: 8045–8057.

49. Yadav S, Liu J, Shire SJ, Kalonia DS. (2010). Specific interactions in high concentration antibody solutions resulting in high viscosity. *J Pharm Sci* 99: 1152–1168.

50. Wang X, Das TK, Singh SK, Kumar S. (2009). Potential aggregation prone regions in biotherapeutics: A survey of commercial monoclonal antibodies. *MAbs* 1: 254–267.

51. Wang X, Singh SK, Kumar S. (2010). Potential aggregation-prone regions in complementarity-determining regions of antibodies and their contribution towards antigen recognition: A computational analysis. *Pharm Res* 27: 1512–1529.

52. Wu SJ, Luo J, O'Neil KT, Kang J, Lacy ER, Canziani G, Baker A, Huang M, Tang QM, Raju TS, Jacobs SA, Teplyakov A, Gilliland GL, Feng Y. (2010). Structure-based engineering of a monoclonal antibody for improved solubility. *Protein Eng Des Sel* 23: 643–651.

53. Conley GP, Viswanathan M, Hou Y, Rank DL, Lindberg AP, Cramer SM, Ladner RC, Nixon AE, Chen J. (2011). Evaluation of protein engineering and process optimization approaches to enhance antibody drug manufacturability. *Biotechnol Bioeng* 108: 2634–2644.

54. Ewert S, Huber T, Honegger A, Pluckthun A. (2003). Biophysical properties of human antibody variable domains. *J Mol Biol* 325: 531–553.

55. Ewert S, Cambillau C, Conrath K, Pluckthun A. (2002). Biophysical properties of camelid V(HH) domains compared to those of human V(H)3 domains. *Biochemistry* 41: 3628–3636.

56. Wang N, Smith WF, Miller BR, Aivazian D, Lugovskoy AA, Reff ME, Glaser SM, Croner LJ, Demarest SJ. (2009). Conserved amino acid networks involved in antibody variable domain interactions. *Proteins* 76: 99–114.

57. Obrezanova O, Arnell A, Gomez de la Cuesta R, Berthelot ME, Gallagher TRA, Zurdo J, Stallwood Y. (2015). Aggregation risk prediction for antibodies and its application to biotherapeutic development. *MAbs*, 7: 352–363.

58. Liu H, Gaza-Bulseco G, Faldu D, Chumsae C, Sun J. (2008). Heterogeneity of monoclonal antibodies. *J Pharm Sci* 97: 2426–2447.

59. Manning MC, Chou DK, Murphy BM, Payne RW, Katayama DS. (2010). Stability of protein pharmaceuticals: An update. *Pharm Res* 27: 544–575.

60. Buchanan A, Clementel V, Woods R, Harn N, Bowen MA, Mo W, Popovic B, Bishop SM, Dall'acqua W, Minter R, Jermutus L, Bedian V. (2013). Engineering a therapeutic IgG molecule to address cysteinylation, aggregation and enhance thermal stability and expression. *MAbs* 5: 255–262.

61. Robinson NE, Robinson AB. (2001). Molecular clocks. *Proc Natl Acad Sci USA* 98: 944–949.

62. Harris RJ, Kabakoff B, Macchi FD, Shen FJ, Kwong M, Andya JD, Shire SJ, Bjork N, Totpal K, Chen AB. (2001). Identification of multiple sources of charge heterogeneity in a recombinant antibody. *J Chromatogr B Biomed Sci Appl* 752: 233–245.

63. Shire SJ. (2009). Formulation and manufacturability of biologics. *Curr Opin Biotechnol* 20: 708–714.

64. Yang X, Xu W, Dukleska S, Benchaar S, Mengisen S, Antochshuk V, Cheung J, Mann L, Babadjanova Z, Rowand J, Gunawan R, McCampbell A, Beaumont M, Meininger D, Richardson D, Ambrogelly A. (2013). Developability studies before initiation of process development: Improving manufacturability of monoclonal antibodies. *MAbs* 5: 787–794.

65. Haberger M, Bomans K, Diepold K, Hook M, Gassner J, Schlothauer T, Zwick A, Spick C, Kepert JF, Hienz B, Wiedmann M, Beck H, Metzger P, Molhoj M, Knoblich C, Grauschopf U, Reusch D, Bulau P. (2014). Assessment of chemical modifications of sites in the CDRs of recombinant antibodies: Susceptibility vs. functionality of critical quality attributes. *MAbs* 6: 327–339.
66. Li H, d'Anjou M. (2009). Pharmacological significance of glycosylation in therapeutic proteins. *Curr Opin Biotechnol* 20: 678–684.
67. van Beers MM, Bardor M. (2012). Minimizing immunogenicity of biopharmaceuticals by controlling critical quality attributes of proteins. *Biotechnol J* 7: 1484.
68. Walsh G. (2010). Post-translational modifications of protein biopharmaceuticals. *Drug Discov Today* 15: 773–780.
69. Zurdo J. (2005). Polypeptide models to understand misfolding and amyloidogenesis and their relevance in protein design and therapeutics. *Protein Pept Lett* 12: 171–188.
70. Arnell A, Jimenez JL, Michael R, Stallwood Y, Zurdo J, inventors. (2011). Variant immunoglobulins with improved manufacturability. U.S. Patent Application WO2011021009 A1.
71. Pybus LP, James DC, Dean G, Slidel T, Hardman C, Smith A, Daramola O, Field R. (2014). Predicting the expression of recombinant monoclonal antibodies in Chinese hamster ovary cells based on sequence features of the CDR3 domain. *Biotechnol Prog* 30: 188–197.
72. Condino AA, Fidanza S, Hoffenberg EJ. (2005). A home infliximab infusion program. *J Pediatr Gastroenterol Nutr* 40: 67–69.
73. Grainger DW. (2004). Controlled-release and local delivery of therapeutic antibodies. *Expert Opin Biol Ther* 4: 1029–1044.
74. Narasimhan C, Mach H, Shameem M. (2012). High-dose monoclonal antibodies via the subcutaneous route: Challenges and technical solutions, an industry perspective. *Ther Deliv* 3: 889–900.
75. Shire SJ, Shahrokh Z, Liu J. (2004). Challenges in the development of high protein concentration formulations. *J Pharm Sci* 93: 1390–1402.
76. Birch JR, Racher AJ. (2006). Antibody production. *Adv Drug Deliv Rev* 58: 671–685.
77. Banks DD, Latypov RF, Ketchem RR, Woodard J, Scavezze JL, Siska CC, Razinkov VI. (2012). Native-state solubility and transfer free energy as predictive tools for selecting excipients to include in protein formulation development studies. *J Pharm Sci* 101: 2720–2732.
78. Lehermayr C, Mahler HC, Mader K, Fischer S. (2011). Assessment of net charge and protein-protein interactions of different monoclonal antibodies. *J Pharm Sci* 100: 2551–2562.
79. Maddux NR, Joshi SB, Volkin DB, Ralston JP, Middaugh CR. (2011). Multidimensional methods for the formulation of biopharmaceuticals and vaccines. *J Pharm Sci* 100: 4171–4197.
80. Clarke JB. (2010). Mechanisms of adverse drug reactions to biologics. *Handb Exp Pharmacol* 196: 453–474.
81. Scherer K, Spoerl D, Bircher AJ. (2010). Adverse drug reactions to biologics. *J Dtsch Dermatol Ges* 8: 411–426.
82. Rosenberg AS, Worobec A. (2004). A risk-based approach to immunogenicity concerns of therapeutic protein products. Part 2. Considering host-specific and product-specific factors impacting immunogenicity. *Biopharm Int* 17: 34–42.
83. Suntharalingam G, Perry MR, Ward S, Brett SJ, Castello-Cortes A, Brunner MD, Panoskaltsis N. (2006). Cytokine storm in a phase 1 trial of the anti-CD28 monoclonal antibody TGN1412. *N Engl J Med* 355: 1018–1028.
84. Selewski DT, Shah GV, Segal BM, Rajdev PA, Mukherji SK. (2010). Natalizumab (Tysabri). *AJNR Am J Neuroradiol* 31: 1588–1590.

85. Miki H, Okamoto A, Ishigaki K, Sasaki O, Sumitomo S, Fujio K, Yamamoto K. (2011). Cardiopulmonary arrest after severe anaphylactic reaction to second infusion of infliximab in a patient with ankylosing spondylitis. *J Rheumatol* 38: 1220.

86. Schellekens H. (2002). Bioequivalence and the immunogenicity of biopharmaceuticals. *Nat Rev Drug Discov* 1: 457–462.

87. Committee for Medicinal Products for Human Use (CHMP). (2007). Guideline on immunogenicity assessment of biotechnology-derived therapeutic proteins. EMA CHMP. EMEA/CHMP/BMWP/14327/2006.

88. Committee for Medicinal Products for Human Use CHMP. (2009). Concept paper on immunogenicity assessment of monoclonal antibodies intended for *in vivo* clinical use. EMA CHMP. EMEA/CHMP/BMWP/114720/2009.

89. Buttel IC, Chamberlain P, Chowers Y, Ehmann F, Greinacher A, Jefferis R, Kramer D, Kropshofer H, Lloyd P, Lubiniecki A, Krause R, Mire-Sluis A, Platts-Mills T, Ragheb JA, Reipert BM, Schellekens H, Seitz R, Stas P, Subramanyam M, Thorpe R, Trouvin JH, Weise M, Windisch J, Schneider CK. (2011). Taking immunogenicity assessment of therapeutic proteins to the next level. *Biologicals* 39: 100–109.

90. Parenky A, Myler H, Amaravadi L, Bechtold-Peters K, Rosenberg A, Kirshner S, Quarmby V. (2014). New FDA draft guidance on immunogenicity. *AAPS J* 16: 499–503.

91. Van Walle I, Gansemans Y, Parren PW, Stas P, Lasters I. (2007). Immunogenicity screening in protein drug development. *Expert Opin Biol Ther* 7: 405–418.

92. Jawa V, Cousens LP, Awwad M, Wakshull E, Kropshofer H, De Groot AS. (2013). T-cell dependent immunogenicity of protein therapeutics: Preclinical assessment and mitigation. *Clin Immunol* 149: 534–555.

93. Rohn TA, Reitz A, Paschen A, Nguyen XD, Schadendorf D, Vogt AB, Kropshofer H. (2005). A novel strategy for the discovery of MHC class II-restricted tumor antigens: Identification of a melanotransferrin helper T-cell epitope. *Cancer Res* 65: 10068–10078.

94. Brennan FR, Morton LD, Spindeldreher S, Kiessling A, Allenspach R, Hey A, Muller PY, Frings W, Sims J. (2010). Safety and immunotoxicity assessment of immunomodulatory monoclonal antibodies. *MAbs* 2: 233–255.

95. Rombach-Riegraf V, Karle AC, Wolf B, Sorde L, Koepke S, Gottlieb S, Krieg J, Djidja MC, Baban A, Spindeldreher S, Koulov AV, Kiessling A. (2014). Aggregation of human recombinant monoclonal antibodies influences the capacity of dendritic cells to stimulate adaptive T-cell responses *in vitro*. *PLoS One* 9: e86322.

96. Findlay L, Eastwood D, Ball C, Robinson CJ, Bird C, Wadhwa M, Thorpe SJ, Thorpe R, Stebbings R, Poole S. (2011). Comparison of novel methods for predicting the risk of pro-inflammatory clinical infusion reactions during monoclonal antibody therapy. *J Immunol Methods* 371: 134–142.

97. Leng SX, McElhaney JE, Walston JD, Xie D, Fedarko NS, Kuchel GA. (2008). ELISA and multiplex technologies for cytokine measurement in inflammation and aging research. *J Gerontol A Biol Sci Med Sci* 63: 879–884.

98. Cupedo T, Stroock A, Coles M. (2012). Application of tissue engineering to the immune system: Development of artificial lymph nodes. *Front Immunol* 3: 1–6.

99. Holland MC, Wurthner JU, Morley PJ, Birchler MA, Lambert J, Albayaty M, Serone AP, Wilson R, Chen Y, Forrest RM, Cordy JC, Lipson DA, Bayliffe AI. (2013). Autoantibodies to variable heavy (V_H) chain Ig sequences in humans impact the safety and clinical pharmacology of a V_H domain antibody antagonist of TNF-alpha receptor 1. *J Clin Immunol* 33: 1192–1203.

100. Jahnmatz M, Kesa G, Netterlid E, Buisman AM, Thorstensson R, Ahlborg N. (2013). Optimization of a human IgG B-cell ELISpot assay for the analysis of vaccine-induced B-cell responses. *J Immunol Methods* 391: 50–59.

101. Weiss WF, Young TM, Roberts CJ. (2009). Principles, approaches, and challenges for predicting protein aggregation rates and shelf life. *J Pharm Sci* 98: 1246–1277.

102. Hermeling S, Schellekens H, Crommelin DJ, Jiskoot W. (2003). Micelle-associated protein in epoetin formulations: A risk factor for immunogenicity? *Pharm Res* 20: 1903–1907.

103. Creeke PI, Farrell RA. (2013). Clinical testing for neutralizing antibodies to interferon-beta in multiple sclerosis. *Ther Adv Neurol Disord* 6: 3–17.

104. Kumar S, Singh SK, Wang X, Rup B, Gill D. (2011). Coupling of aggregation and immunogenicity in biotherapeutics: T- and B-cell immune epitopes may contain aggregation-prone regions. *Pharm Res* 28: 949–961.

105. Kumar S, Mitchell MA, Rup B, Singh SK. (2012). Relationship between potential aggregation-prone regions and HLA-DR-binding T-cell immune epitopes: Implications for rational design of novel and follow-on therapeutic antibodies. *J Pharm Sci* 101: 2686–2701.

106. Frigault MJ, June CH. (2011). Predicting cytokine storms: It's about density. *Blood* 118: 6724–6726.

107. Bailey L, Moreno L, Manigold T, Krasniqi S, Kropshofer H, Hinton H, Singer T, Suter L, Hansel TT, Mitchell JA. (2013). A simple whole blood bioassay detects cytokine responses to anti-CD28SA and anti-CD52 antibodies. *J Pharmacol Toxicol Methods* 68: 231–239.

108. Locwin B. (2013). Quality risk assessment and management strategies for biopharmaceutical companies. *Bioprocess Int* 11: 52–57.

109. Sober E. (1981). The principle of parsimony. *Brit J Phil Sci* 32: 145–156.

110. Ishikawa K, Oftus JH. (1990). Introduction to quality control. Chapman & Hall, London, UK.

111. Miller GA. (1956). The magical number seven, plus or minus two: Some limits on our capacity for processing information. *Psychol Rev* 63: 81–97.

112. Czitrom V, Spagon PD. (1997). Statistical case studies for industrial process improvement. SIAM, Philadelphia.

113. Hann MM, Keseru GM. (2012). Finding the sweet spot: The role of nature and nurture in medicinal chemistry. *Nat Rev Drug Discov* 11: 355–365.

114. Segall MD. (2012). Multi-parameter optimization: Identifying high quality compounds with a balance of properties. *Curr Pharm Des* 18: 1292–1310.

115. Nicolaou CA, Brown N. (2013). Multi-objective optimization methods in drug design. *Drug Discov Today Technol* 10: e427–e435.

116. Ekins S, Honeycutt JD, Metz JT. (2010). Evolving molecules using multi-objective optimization: Applying to ADME/Tox. *Drug Discov Today* 15: 451–460.

117. Segall MD, Champness E, Obrezanova O, Leeding C. (2009). Beyond profiling: Using ADMET models to guide decisions. *Chem Biodivers* 6: 2144–2151.

118. CMC Biotech Working Group. (2009). A-Mab: A case study in bioprocess development. Version 2.1. http://c.ymcdn.com/sites/www.casss.org/resource/resmgr/imported/A-Mab_Case_Study_Version_2–1.pdf.

119. Ponsel D, Neugebauer J, Ladetzki-Baehs K, Tissot K. (2011). High affinity, developability and functional size: The holy grail of combinatorial antibody library generation. *Molecules* 16: 3675–3700.

120. Tiller T, Schuster I, Deppe D, Siegers K, Stuohnur R, Herrmann T, Berenguer M, Poujol D, Stehle J, Stark Y, Hessling M, Daubert D, Felderer K, Kaden S, Kolln J, Enzelberger M, Urlinger S. (2013). A fully synthetic human Fab antibody library based on fixed V_H/V_L framework pairings with favorable biophysical properties. *MAbs* 5: 445–470.

121. Steinwand M, Droste P, Frenzel A, Hust M, Dubel S, Schirrmann T. (2013). The influence of antibody fragment format on phage display based affinity maturation of IgG. *MAbs* 6: 204–218.

122. Lonberg N. (2005). Human antibodies from transgenic animals. *Nat Biotechnol* 23: 1117–1125.
123. Höidén-Guthenberg I, Jonsson A, Ståhl S, Friedman M, Wållberg H, Bass T, Tolmachev V, Frejd F, Ekblad C. (2012). Combining Affibody® molecules and the Albumod™ technology to create long acting multispecific protein therapeutics. *MAbs* 4: 14–16.
124. Sampei Z, Igawa T, Soeda T, Okuyama-Nishida Y, Moriyama C, Wakabayashi T, Tanaka E, Muto A, Kojima T, Kitazawa T, Yoshihashi K, Harada A, Funaki M, Haraya K, Tachibana T, Suzuki S, Esaki K, Nabuchi Y, Hattori K. (2013). Identification and multidimensional optimization of an asymmetric bispecific IgG antibody mimicking the function of factor VIII cofactor activity. *PLoS One* 8: e57479.
125. Zurdo J, Arnell A, Obrezanova O, Smith N, Gómez de la Cuesta R, Gallagher TRA, Michael R, Stallwood Y, Ekblad C, Höidén-Guthenberg I. (2015). Early implementation of QbD in biopharmaceutical development: A practical example. *Biomed Res Int*, Vol. 2015, Article ID 605 427, 2015. doi:10.1155/2015/605427.
126. Frejd FY. (2012). Half-life extension by binding to albumin through an albumin binding domain. In Kontermann RE, ed., *Therapeutic Proteins: Strategies to Modulate Their Plasma Half-Lives*. Weinheim: Wiley-Blackwell, pp. 269–283.
127. Strohl WR, Strohl LM. (2012). *Therapeutic Antibody Engineering*. Cambridge, UK: Woodhead Publishing.
128. Jonsson A, Dogan J, Herne N, Abrahmsen L, Nygren PA. (2008). Engineering of a femtomolar affinity binding protein to human serum albumin. *Protein Eng Des Sel* 21: 515–527.
129. Goetsch L, Haeuw JF, Champion T, Lacheny C, N'Guyen T, Beck A, Corvaia N. (2003). Identification of B- and T-cell epitopes of BB, a carrier protein derived from the G protein of *Streptococcus* strain G148. *Clin Diagn Lab Immunol* 10: 125–132.
130. Fan ZC, Shan L, Goldsteen BZ, Guddat LW, Thakur A, Landolfi NF, Co MS, Vasquez M, Queen C, Ramsland PA, Edmundson AB. (1999). Comparison of the three-dimensional structures of a humanized and a chimeric Fab of an anti-gamma-interferon antibody. *J Mol Recognit* 12: 19–32.
131. Singh SK, Afonina N, Awwad M, Bechtold-Peters K, Blue JT, Chou D, Cromwell M, Krause HJ, Mahler HC, Meyer BK, Narhi L, Nesta DP, Spitznagel T. (2010). An industry perspective on the monitoring of subvisible particles as a quality attribute for protein therapeutics. *J Pharm Sci* 99: 3302–3321.
132. Rathore AS, Winkle H. (2009). Quality by design for biopharmaceuticals. *Nat Biotechnol* 27: 26–34.
133. Gu J, Ghayur T. (2010). Rationale and development of multispecific antibody drugs. *Expert Rev Clin Pharm* 3: 491–508.
134. Kaitin KI. (2010). Deconstructing the drug development process: The new face of innovation. *Clin Pharmacol Ther* 87: 356–361.
135. Rees H. (2011). *Supply Chain Management in the Drug Industry: Delivering Patient Value for Pharmaceuticals and Biologics*. Hoboken, NJ: John Wiley & Sons.
136. Fraser J, Kerboul G. (2012). A holistic approach to pharmaceutical manufacturing: Product lifecycle management support for high yield processes to make safe and effective drugs. *Pharma Eng* 32: 1–5.
137. Zhu J. (2012). Mammalian cell protein expression for biopharmaceutical production. *Biotechnol Adv* 30: 1158–1170.

Index

Printed and bound by CPI Group (UK) Ltd, Croydon, CR0 4YY

17/10/2024

01775709-0006